Hybrid Energy Systems

NATURAL
GAS
SOLAR
WIND
GEOTHERMAL
BIOMASS
NUCLEAR
FOSSIL
FUELS
ELECTRICITY

Bahman Zohuri

Hybrid Energy Systems

Driving Reliable Renewable Sources of Energy Storage

Springer

Bahman Zohuri
Galaxy Advanced Engineering Inc.,
Department of Electrical and Computer Engineering,
University of New Mexico
Albuquerque, NM, USA

ISBN 978-3-319-70720-4 ISBN 978-3-319-70721-1 (eBook)
https://doi.org/10.1007/978-3-319-70721-1

Library of Congress Control Number: 2017959609

Printed on acid-free paper

This Springer imprint is published by Springer Nature
The registered company is Springer International Publishing AG
The registered company address is: Gewerbestrasse 11, 6330 Cham, Switzerland

This book is dedicated to my son Sasha

Preface

The United States is a rich country when it comes to energy supply. It has an abundance of coal, oil, and natural gas, along with exceptional wind, solar, and hydropower resources. Recently, new technology has driven innovation in nuclear power. We have also seen an increasing focus on local air quality and global climate change.

Today's world is at a turning point. Resources are running low, pollution is increasing, and the climate is changing. Fossil fuels are depleting quickly, and it is necessary to find substitutes that will guarantee wealth and growth. Modern technology is already providing us with alternatives like wind turbines, photovoltaic cells, biomass plants, and more. But these technologies have flaws. Compared to traditional power plants, they produce much smaller amounts of electricity. Even more problematic is the inconsistency of energy production. The global demand for electricity is huge, and growing by approximately 3.6% annually; however, the sun isn't always shining nor is the wind always blowing. For technical reasons, the amount of electricity fed into the power grid must remain on the same level as demanded to prevent blackouts and damage to the grid. This leads to situations where the production is higher than the consumption or vice versa. This is where storage technologies come into play—they are the key element to balance out these flaws.

With the growing importance of renewable energy sources, scientists and engineers are anxious to enhance efficiencies and lower the costs of these technologies. Yet, there seems to be only a handful of technologies available that are efficient and economical. Storing energy isn't an easy task, as most of us know. Our smartphone battery only lasts for about a day, and laptops last only a few hours. The range for electric cars is limited to little more than a hundred kilometers. These are only examples for comparatively small devices. Now imagine the problem of storing energy at the level of hundreds to thousands of wind turbines and photovoltaic cells is much more complex.

Many new products and services that reduce emissions for new and existing power plants have been created. One of our most exciting products is the combined-cycle gas turbine power plant, which uses jet engine technology combined with steam turbine technology to rotate generators to produce electricity. A similar innovative technological approach has been suggested by many scientists and engineers in the field of nuclear energy, using the new generation of nuclear power plant (NPP) that is known as Generation IV. These innovative approaches allow the combination of turbines in single or multiple shaft installation to provide the most cost-effective way to generate electricity from either natural gas or nuclear energy. The turbines used in this manner will provide fuel efficiency of greater than 63% and produce approximately 65% less carbon dioxide than the coal-fired power plants that they replace.

The Conergy hybrid energy storage system (CHESS) provides the key to transitioning large-scale sites to integrated solar energy supply and solar energy storage. CHESS is designed to provide on-demand, stable power supply for on-grid, fringe-of-grid, and remote off-grid sites. CHESS is a practical, renewable energy solution, engineered to lower operating costs and insulate businesses against future volatility in energy and fuel prices.

It is a modular and expandable technology that allows for flexibility and growth demand with full remote control and monitoring functionality. The technology is not only durable but easily transportable, making it ideal for remote locations.

The energy storage system (ESS) in a conventional, stand-alone Renewable Energy Power System (REPS) usually has a short lifespan due to irregular output of renewable energy sources. In certain systems, the ESS is oversized to reduce the stress level and to meet the intermittent peak power demand. A hybrid energy storage system (HESS) is a better solution in terms of durability, practicality, and cost-effectiveness for the overall system implementation. The structure and common issues of stand-alone REPS with ESS are discussed in this paper. This paper presents different structures of stand-alone REPS with HESS such as passive, semi-active, and active HESS. As there are a variety of energy storage technologies available in the market, decision matrixes are introduced in this paper to evaluate the technical and economic characteristics of the energy storage technologies. A detailed review of the state-of-the-art control strategies such as classical control strategies and intelligent control strategies for REPS with HESS is highlighted. The future trends for REPS with HESS combination and control strategies are also discussed.

This book also describes, energy storage at various levels. Energy storage technology has great potential to improve electric power grids, enable growth in renewable electricity generation, and provide alternatives to oil-derived fuels in the nation's transportation sector. In the electric power system, the promise of this technology lies in its potential to increase grid efficiency and reliability—optimizing power flows and supporting variable power supplies from wind and solar generation. In transportation, vehicles powered by batteries or other electric technologies have the potential to displace vehicles burning gasoline and diesel fuel, reducing associated emissions and demand for oil.

Albuquerque, NM, USA B. Zohuri
2016

Acknowledgments

I am indebted to the many people who aided me, encouraged me, and supported me beyond my expectations. Some are not around to see the results of their encouragement in the production of this book, yet I hope they know of my deepest appreciations. I especially want to thank all my friends who have continuously given their support without hesitation, to whom I am deeply indebted. They have always kept me going in the right direction.

Above all, I offer very special thanks to my late mother and father and to my children, in particular, my son Sasha. They have provided constant interest and encouragement, without which this book would not have been written. Their patience with my many absences from home and long hours in front of the computer to prepare the manuscript is especially appreciated.

Contents

1 **Hybrid Renewable Energy Systems** ... 1
1.1 Introduction to Hybrid Energy System ... 1
1.1.1 Hybrid System as Source of Renewable Energy ... 8
1.2 Energy Storage Systems ... 10
1.3 Compressed Air Energy Storage (CAES) ... 11
1.3.1 Compressed Air Energy Storage (CAES) ... 12
1.3.2 Advanced Adiabatic Compressed Air Energy Storage (AA-CAES) ... 16
1.4 Variable Electricity with Base-Load Reactor Operation ... 19
1.5 Why We Need Nuclear Power ... 26
1.5.1 The Merits of Total Transformation ... 27
1.5.2 The Downsides of Monoculture ... 29
1.5.3 The Other Zero-Carbon Energy: Nuclear ... 30
1.5.4 A Diverse Portfolio ... 33
1.6 Security of Energy Supply ... 34
1.7 Environmental Quality ... 35
References ... 37

2 **Cryogenic Technologies** ... 39
2.1 Introduction ... 39
2.2 Low Temperature in Science and Technology ... 41
2.3 Defining Cryogenic Fluids or Liquids ... 46
2.3.1 Defining Cryogenic Fluids or Liquids ... 47
2.3.2 Thermophysical Properties ... 51
2.3.3 Liquid Boil-off ... 51
2.3.4 Cryogen Usage for Equipment Cooldown ... 52
2.3.5 Phase Domains ... 53
2.3.6 Personal Protective Equipment to Be Worn ... 54
2.3.7 Handling Cryogenic Liquids ... 54
2.3.8 Storing Cryogenic Liquids ... 55
2.3.9 Hazards of Cryogenic Liquids ... 55

2.3.10 General Hazards of Cryogenic Liquids 56
2.4 Heat Transfer and Thermal Design 56
2.4.1 Solid Conduction 57
2.4.2 Radiation 58
2.4.3 Convection 59
2.4.4 Gas Conduction 60
2.4.5 Multilayer Insulation 61
2.4.6 Vapor Cooling of Necks and Supports. 62
2.5 Refrigeration and Liquefaction 64
2.5.1 Thermodynamics of Refrigeration. 64
2.5.2 Helium Refrigerators Versus Liquefiers. 66
2.5.3 Real Cycles and Refrigeration Equipment. 67
2.6 Industrial Applications. 70
2.6.1 Cryogenic Processing for Alloy Hardening. 73
2.6.2 Cryogenic Fuels 74
2.6.3 Cryogenic Application in Nuclear Magnetic Resonance Spectroscopy (NMR) 74
2.6.4 Cryogenic Application in Magnetic Resonance Image (MRI) 74
2.6.5 Cryogenic Application in Frozen Food Transport 75
2.6.6 Cryogenic Application in Forward Looking Infrared (FLIR) 76
2.6.7 Cryogenic Application in Space. 78
2.6.8 Cryogenic in Blood Banking, Medicine, and Surgery. 80
2.6.9 Cryogenic in Manufacturing Process. 82
2.6.10 Cryogenic in Recycling of Materials 83
2.7 Cryogenic Application in Research 83
2.7.1 Research Overview 83
2.7.2 Right: Lightweight, High Efficiency Cryocooler. 84
2.7.3 Background 85
2.7.4 Right Liquefier Demo and Cryogenic Insulation Test Facility 85
2.8 Cryogenic Fluid Management 85
2.8.1 Benefits 86
2.9 Conclusion 86
References. 87

3 Reliable Renewables with Cryogenic Energy Storage. 89
3.1 Introduction 89
3.2 Cryogenic Application in Electric Power Transmission within Big Cities 91
3.3 The Basic of Cryogenic Energy Storage (CES). 93
3.4 Cryogenic Energy Storage (CES) 93
3.5 Cryogenic Energy Storage (CES) Characteristics 94
3.5.1 Cryogenic Energy Storage (CES) a Wise Investment 95

3.6 Cryogenic Energy Storage (CES) in Nuclear Power Plants. 97
3.6.1 Proposed Combined Cryogenic Energy Storage (CES) in Nuclear Power Plants 101
References 103

4 Types of Renewable Energy 105
4.1 Introduction 105
4.2 What Are the Different Types of Renewable Energies? 106
4.2.1 Biomass 107
4.2.2 Solar Power 109
4.2.3 Wind Power 113
4.2.4 Tidal Power 114
4.2.5 Geothermal 115
4.3 Top Ten Renewable Energy Sources 116
4.3.1 Nuclear Power 117
4.3.2 Compressed Natural Gas 118
4.3.3 Biomass 119
4.3.4 Geothermal Power 119
4.3.5 Radiant Energy 122
4.3.6 Hydroelectricity Power Source 122
4.3.7 Wind Power 124
4.3.8 Solar Power 125
4.3.9 Wave Power 126
4.3.10 Tidal Power 127
4.4 How to Indirectly Participate in Any or All of These Sustainable Energy Solutions 128
4.5 Renewable Energy Certificates 129
4.5.1 Which Technologies Qualify for Certification? 130
4.5.2 Bottom Line on Renewable Energy Certification 131
4.6 Issues with Adoption Forms of Renewable Source of Energy 133
References 133

5 Hydrogen Energy Technology, Renewable Source of Energy 135
5.1 Introduction 135
5.2 Hydrogen as an Energy Carrier 136
5.3 Hydrogen Fuel Cell 138
5.4 Fuel Cells 142
5.4.1 Different Types of Fuel Cells 143
5.5 The Fuel Cell Technologies 149
5.6 Fuel Cell Backup Power Systems 150
5.7 Fuel Cell Systems for Stationary Combined Heat and Power Applications 151
5.8 Fuel Cell Systems for Portable Power and Auxiliary Power Applications 151
5.9 Hydrogen Storage 152
5.9.1 Why Study Hydrogen Storage 153

5.9.2 How Hydrogen Storage Works 153
5.9.3 Research and Development Goals 154
5.9.4 Hydrogen Storage Challenges 156
5.10 Hydrogen Energy Storage 158
5.10.1 Hydrogen Production 159
5.10.2 Hydrogen Re-electrification 159
5.11 Underground Hydrogen Storage 160
5.12 Materials-Based Hydrogen Storage 161
5.12.1 Technical Targets and Status 162
5.13 Industrial Application of Hydrogen Energy 164
5.14 Electrical Energy Storage 165
5.14.1 Characteristic of Electricity 166
5.14.2 Electricity and the Roles of Electrical Energy Storages . . . 167
5.15 Strategic Asset Management of Power Networks 170
5.16 Orchestrating Infrastructure for Sustainable Smart Cities 171
5.16.1 Smart Technology Solution Create Value 173
5.16.2 New Approach to Smart City Solution 173
5.16.3 Stakeholders Are Key Drivers to Smart City Solution 173
5.16.4 Without Integration Rising to the Level of Systems There Cannot Smart City 174
5.16.5 Horizontal and Vertical Integration a Key to Interoperability 175
5.16.6 Interoperability Is the Key to Open Markets and to Competitive Solutions 175
5.16.7 Guiding Principles and Strategic Orientation 176
References 177

6 Energy Storage for Peak Power and Increased Revenue 181
6.1 Introduction 181
6.2 Variable Electricity and Heat Storage 183
6.3 Implications of Low-Carbon Grid and Renewables on Electricity Markets 184
6.4 Strategies for a Zero-Carbon Electricity Grid 186
6.5 Nuclear Air-Brayton Combined Cycle Strategies for Zero-Carbon Grid 186
6.6 Salt-Cooled Reactors Coupled to NACC Power System 187
6.7 Sodium-Cooled Reactors Coupled to NACC Power System 190
6.8 Power Cycle Comparisons 192
6.9 Summary 193
References 194

7 Fission Nuclear Power Plants for Renewable Energy Source 195
7.1 Introduction 196
7.2 Electricity Markets 198
7.2.1 Global Electricity Consumption Set to Explode 203
7.3 California and Others Are Getting It Wrong 205

7.4 Outlook for Power Generation 206
7.5 Why We Need Nuclear Power Plants 206
7.6 Is Nuclear Energy Renewable Source of Energy 208
7.6.1 Argument for Nuclear as Renewable Energy 209
7.6.2 Argument for Nuclear as Renewable Energy 210
7.6.3 Conclusion 210
References 211

8 Energy Storage Technologies and Their Role in Renewable Integration 213
8.1 Introduction 213
8.2 The Electric Grid 216
8.3 Power Generation 223
8.4 Transmission and Distribution 224
8.5 Load Management 224
8.6 Types of Storage Technology 226
8.6.1 Kinetic Energy Storage or Flywheel Concept 230
8.6.2 Superconducting Magnetic Energy Storage 232
8.6.3 Batteries 237
8.6.4 Other and Future Batteries in Development 243
8.7 A Battery-Inspired Strategy for Carbon Fixation 251
8.8 Saliva-Powered Battery 253
8.9 Summary 254
References 255

Appendix A: Global Energy Interconnection 257

Appendix B: Grid Integration of Large Capacity 263

Appendix C: Energy Storage for Power Grids and Electric Transportation 271

Appendix D: Coping with the Energy Challenge 279

Index 285

About the Author

Bahman Zohuri currently works for Galaxy Advanced Engineering, Inc., a consulting firm that he started in 1991 when he left both the semiconductor and defense industries after many years working as a chief scientist. After graduating from the University of Illinois in the field of physics and applied mathematics, then he went to the University of New Mexico, where he studied nuclear engineering and mechanical engineering. He joined Westinghouse Electric Corporation, where he performed thermal hydraulic analysis and studied natural circulation in an inherent shutdown heat removal system (ISHRS) in the core of a liquid metal fast breeder reactor (LMFBR) as a secondary fully inherent shutdown system for secondary loop heat exchange. All these designs were used in nuclear safety and reliability engineering for a self-actuated shutdown system. He designed a mercury heat pipe and electromagnetic pumps for large pool concepts of a LMFBR for heat rejection purposes for this reactor around 1978, when he received a patent for it. He was subsequently transferred to the defense division of Westinghouse, where he oversaw dynamic analysis and methods of launching and controlling MX missiles from canisters. The results were applied to MX launch seal performance and muzzle blast phenomena analysis (i.e., missile vibration and hydrodynamic shock formation). Dr. Zohuri was also involved in analytical calculations and computations in the study of nonlinear ion waves in rarefying plasma. The results were applied to the propagation of so-called soliton waves and the resulting charge collector traces in the rarefaction characterization of the corona of laser-irradiated target pellets. As part of his graduate research work at Argonne National Laboratory, he performed computations and programming of multi-exchange integrals in surface physics and solid-state physics. He earned various patents in areas such as diffusion processes and diffusion furnace design while working as a senior process engineer at various semiconductor companies, such as Intel Corp., Varian Medical Systems, and National Semiconductor Corporation. He later joined Lockheed Martin Missile and Aerospace Corporation as senior chief scientist and oversaw research and development (R&D) and the study of the vulnerability, survivability, and both radiation and laser hardening of different components of the Strategic Defense Initiative, known as Star Wars.

This included payloads (i.e., IR sensor) for the Defense Support Program, the Boost Surveillance and Tracking System, and Space Surveillance and Tracking Satellite against laser and nuclear threats. While at Lockheed Martin, he also performed analyses of laser beam characteristics and nuclear radiation interactions with materials, transient radiation effects in electronics, electromagnetic pulses, system-generated electromagnetic pulses, single-event upset, blast, thermomechanical, hardness assurance, maintenance, and device technology.

He spent several years as a consultant at Galaxy Advanced Engineering serving Sandia National Laboratories, where he supported the development of operational hazard assessments for the Air Force Safety Center in collaboration with other researchers and third parties. Ultimately, the results were included in Air Force Instructions issued specifically for directed energy weapons operational safety. He completed the first version of a comprehensive library of detailed laser tools for airborne lasers, advanced tactical lasers, tactical high-energy lasers, and mobile/tactical high-energy lasers, for example.

He also oversaw SDI computer programs, in connection with Battle Management C^3I and artificial intelligence, and autonomous systems. He is the author of several publications and holds several patents, such as for a laser-activated radioactive decay and results of a through-bulkhead initiator. He has published the following works: *Heat Pipe Design and Technology: A Practical Approach* (CRC Press); *Dimensional Analysis and Self-Similarity Methods for Engineers and Scientists* (Springer); *High Energy Laser (HEL): Tomorrow's Weapon in Directed Energy Weapons Volume I* (Trafford Publishing Company); and recently the book on the subject directed energy weapons and physics of high-energy laser with Springer. He has other books with Springer Publishing Company: *Thermodynamics in Nuclear Power Plant Systems* and *Thermal-Hydraulic Analysis of Nuclear Reactors*.

Dr. Zohuri presently holds the position of research professor at University of New Mexico, Department of Electrical and Computer Science.

Chapter 1
Hybrid Renewable Energy Systems

This chapter gives an elementary account of hybrid renewable energy systems (HRES). This type of system according to today's demand on providing new source of electricity On-pick and storage of energy as a source of such demandable energy of electricity Off-pick. Hybrid renewable energy systems (HRES) are becoming popular as stand-alone power systems for providing electricity in remote areas due to advances in renewable energy technologies and subsequent rise in prices of petroleum products. A hybrid energy system, or hybrid power, usually consists of two or more renewable energy sources used together to provide increased system efficiency as well as greater balance in energy supply [1]. A renewable energy is energy that is collected from renewable resources, which are naturally replenished on a human timescale, such as sunlight, wind, rain, tides, waves, and geothermal heat. Renewable energy often provides energy in four important areas: electricity generation, air and water heating/cooling, transportation, and rural (off-grid) energy services.

1.1 Introduction to Hybrid Energy System

Hybrid energy systems combine two or more forms of energy generation, storage, or end-use technologies, and they can deliver a boatload of benefits compared with single source systems. The option of having variety in our day-to-day life could be considered as the spice of life; therefore, why limit ourselves to just one energy source or storage option? In these cases, hybrid energy systems are an ideal solution since they can offer substantial improvements in performance and cost reduction and can be tailored to varying end-user requirements.

The energy storage system (ESS) in a conventional stand-alone renewable energy power system (REPS) usually has a short lifespan mainly due to irregular output of renewable energy sources. In certain systems, the ESS is oversized to reduce the stress level and to meet the intermittent peak power demand. A hybrid energy storage system (HESS) is a better solution in terms of durability, practicality, and

B. Zohuri, *Hybrid Energy Systems*,
https://doi.org/10.1007/978-3-319-70721-1_1

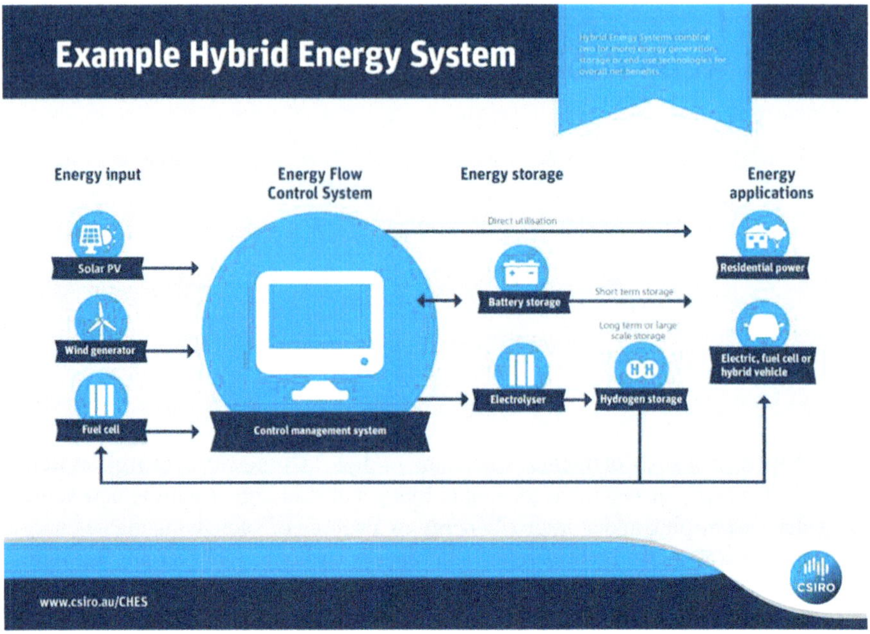

Fig. 1.1 Example of hybrid energy system (Courtesy of CSIROscope Corporation)

cost-effectiveness for the overall system implementation. The structure and the common issues of stand-alone REPS with ESS are discussed in this paper. This paper presents different structures of stand-alone REPS with HESS such as passive, semi-active, and active HESS. As there are a variety of energy storage technologies available in the market, decision matrixes are introduced in this paper to evaluate the technical and economic characteristics of the energy storage technologies based on the requirements of stand-alone REPS. A detailed review of the state-of-the-art control strategies such as classical control strategies and intelligent control strategies for REPS with HESS are highlighted. The future trends for REPS with HESS combination and control strategies are also discussed.

Configurations could include renewable or nonrenewable energy sources, electrical and chemical energy storage, and fuel cells, often connected via a smart grid. They have the potential to dramatically reduce cost and emissions from energy generation and distribution for households but can be held back by the limitations of individual power generation or storage technologies—this may include cost, inconsistent supply (like interrupted solar on a cloudy day), etc. This means there is substantial demand for hybrid energy solutions to lower cost and improve efficiency while still meeting performance requirements. Figure 1.1 is a presentation of an example for hybrid energy system (HES), which is depicted by CSIROscope corporation.

CSIROscope an Australian corporation researcher is claiming that there is now an increased availability of renewable and modular power generation and storage technologies such as batteries, fuel cells, and household solar. "These technologies are becoming cost competitive, but the key to greater use is to combine them in connected hybrid systems," Dr. Badwal a researcher at this company says. He also goes a further step by stating that "By doing this, we can offer substantial improvements in performance and cost."

Consequently, the early player in this game will be ahead of their business and research ball, by keeping their heads together with industry partners, and the collaborative space could be used to share the benefits of emerging hybrid energy systems with industry and government to maximize the value of local energy sources.

Having such foundation under consideration, the first questions that come to mind are what is hybrid system and the word hybrid stands for and what do we really mean by looking at a hybrid energy system as a new source of renewable energy and usage of such source during on-peak demand for electricity. Going toward the next century demand for more electricity is on rise, and consequently the on-peak hours of such demand impose a challenging duty on-grid; thus, an alternative source of energy needs to be found to meet such supply and demand constraints. Hence, looking for a new source of renewable energy is more and more appealing.

The word hybrid can be referred to as some phenomena that are a combination of two different elements that may consist of:

1. Modern science has seen dramatic advances in hybrid technology, giving birth to hybrid cars.
2. Incorporating information and communications technology (ICT) systems that automate smart-houses and eco homes.

Similarly, hybrid energy systems have been designed to generate electricity from different sources, such as solar panels and wind turbines, and now tap into sources such as hydrogen that is stored in a different manner and standing by as a class of renewable energy. Therefore, a demand for its production is most efficient and cost-effective in the scope of every researcher and scientist at university, industry, and national laboratory level who are working in this field.

However, one of the biggest downfalls of renewable energy is that energy supply is not constant; sources like solar and wind power fluctuate in intensity due to the weather and seasonal changes. Therefore, a reliable backup system is necessary for renewable energy-generating stations that are not connected to a national power grid, and they can produce energy during off-peak and store them for utilization during on-peak period, and that is the driving factor behind the idea of producing hydrogen via nuclear power which is indeed a solution to reduce carbon emission. The price to pay includes the cost of nuclear waste storage and other related issues such as proliferation and security of the fissionable weapon grade waste coming out of the reactor core at the end of burnup residue of fuel used in them or for that matter air- or land based on manmade event (i.e., Three Mile Island, Chernobyl accident due to operator errors) or natural disaster (i.e., Fukushima Daiichi in Japan).

Nevertheless, something could probably be done to avoid at least part of this pollution and reduce the public fear of a nuclear disaster by increasing the safety design of these power plants going forward with GEN-IV designs, while we can tap into waste of thermal energy generated by these reactors and put them in use for producing a new source of renewable energy such as hydrogen production plants that are coupled particularly to these very-high-temperature reactors.

As we said, finding a reliable backup system for renewable energy is an inevitable condition, and the systems that consist of a variety of power control methods and storage equipment which include battery bank and diesel generators among others do not have reliable endless life cycle enough to meet the demand on-grid during on-peak or even at small scale looking at these sorts of power storage for usage at residential level or remote areas.

The power systems that are connected to the national grid do not have this problem because, in most cases, there are many different sources of power contributing to the national electricity supply.

Then the question about reducing the demand for energy stepping into future time or meeting that demand is on the table, and somehow as solutions need to be found, thus hybrid technology for the production of electrical energy seems very appealing, and research around these systems to make them cost-effective and efficient has gathered a huge momentum these days.

It is undoubtedly true that big centralized power stations are still needed to generate enough power for big industrial sites. However, if we managed to dramatically reduce the amount of energy that the entire residential and small commercial building stock withdraws every year out of the national energy grid, we may probably need less nuclear power plants. That is arguably the viewpoint of antinuclear folks, but it is something to be remained to see and should not be a showstopper for solutions such as hybrid systems to be in place, and research to make them more productive and efficient must continue.

Hybrid energy systems often consist of a combination of fossil fuels and renewable energy sources and are used in conjunction with energy storage equipment (batteries) or hydrogen storage tanks. This is often done either to reduce the cost of generating electricity from fossil fuels or to provide backup for a renewable energy system, ensuring continuity of power supply when the renewable energy source fluctuates.

There are several types of hybrid energy systems such as wind-solar hybrid, solar-diesel, wind-hydro, and wind-diesel, which are among present in production plants. The design of a system or the choice of energy sources depends on several considerations. The factors affecting the choice of hybrid power technology can also tell us why people use hybrids and some of the advantages. The main factors are cost and resources available. As some of localized advantages as stand-alone operation and off the grid in a self-sustain mode with respect to need for electricity is worth mentioning is a solar system barn in a remote or isolated area, where the framers can take advantages of independency on electricity feed from the grid. Solar energy can be produced on or off the grid.

On-grid means a house remains connected to the state electricity grid. Off-grid has no connection to the electricity grid, so the house, business, or whatever being powered is relying solely on solar or solar-hybrid.

Fig. 1.2 A solar barn in a remote area

Fig. 1.3 An illustration of a solar farm

The ability to produce electricity off the grid is a major advantage of solar energy for people who live in isolated and rural areas. Power prices and the cost of installing power lines are often exorbitantly high in these places, and many have frequent power cuts. Figure 1.2 is an illustration of a solar barn that can go off-grid, and solar power is a huge advantage for people in isolated locations, while Fig. 1.3 is an illustration of a solar farm as part of the electrical grid for providing the electricity power.

The cost of hybrid power technology greatly affects the choices people make, particularly in developing countries.

This also depends on the aim of the project. People who are planning to set up a hybrid energy project for their own use often focus on lowering the total investment and operational costs, while those planning to generate electricity for sale focus on the long-term project revenue.

As such, systems that incorporate hydrogen storage and fuel cells are not very common with small-scale projects. The viability of one hybrid energy system over another is usually pegged on the cost of generating each kilowatt [2, 3].

The availability of the natural resources plays an enormous part when selecting the components of a hybrid energy system—the right power, generation location, and method must be chosen [4].

Often, a hybrid system is opted for because the existing power resource is not enough to generate the amount of power needed—which is often the case when using micro-hydro plants.

In some developing countries, such as parts of Ethiopia, a wind-solar hybrid power system, consisting of wind turbines and solar photovoltaic (PV) panels, was found to be most viable. This was because the wind resource alone was not sufficient to meet the electric load.

Solar PV panel is used primarily for grid-connected electricity to operate residential appliances, commercial equipment, lighting, and air conditioning for all types of buildings. Through stand-alone systems and the use of batteries, it is also well-suited for remote regions where there is no electricity source. Solar PV panels can be ground mounted, installed on building rooftops, or designed into building materials at the point of manufacturing. Solar PV cells were very expensive, so it was not feasible for the project developers to use solar power alone [5].

The efficiency of solar PV increases in colder temperatures and is particularly well-suited for Canada's climate. Many technologies are available which offer different solar conversion efficiencies and pricing.

Solar PV modules can be grouped together as an array of series and parallel-connected modules to provide any level of power requirements, from mere watts (W) to kilowatt (kW) and megawatt (MW) size.

Many city dwellers are also choosing to go off the grid with their alternate energy as part of a self-reliant lifestyle. See Fig. 1.4.

In the next page, you may observe some of the hybrid energy system (HES) sources, where some industry conducting research around that includes the enhancement of these systems by improving them technologically to present better return on investment (ROI) and total cost of ownership (TCO) for energy owners of these resources to meet supply and demand for the electricity.

- *Coal mining and energy production*
  Improving mine safety and developing smarter extraction and carbon capture techniques, which help lower mission.
- *Electricity grid and modeling*
  Improve energy efficiency through intelligence models, systems, and management.
- *Energy storage and battery technologies*
  Cutting-edge energy storage technologies that utilize heat, ceramics, and batteries.

Fig. 1.4 A house with a solar system

- *Solar energy*
  Making solar a reliable, stable power source for future energy—including solar thermal and photovoltaics.
- *Oil and gas*
  Understanding and unlocking resources of such energy both onshore and offshore gas and oil and enabling safe, efficient, and sustainable development of these wealth of resources.
- *Nuclear energy*
  The new development based on research on new generation of nuclear power plant known as GEN-IV has built up a new momentum to increase the thermal efficiencies of these power plants higher than their previous generation of GEN-III, while they are more cost-effective to be manufactured [6].
- *Cryogenic for renewable energy*
  The cryogenic energy facility stores power from renewables or off-peak generation by chilling air into liquid form.
  When the liquid air warms up, it expands and can drive a turbine to make electricity. The company behind the scheme, Highview Power Storage, believes that the technology has a great potential to be scaled up for long-term use with green energy sources.
- *Low emissions technologies*
  New technologies that facilitate the development of low emissions energy sources and improve emissions from existing sources.

Hybrid systems are most suitable for small grids and isolated or stand-alone and self-reliable systems as hybrid power generation is, by definition, a solution for getting around problems where one energy source is not sufficient.

Fig. 1.5 Enercon E-Ship 1

The popularity of hybrid energy systems has grown so much that it is now a niche industry in itself—with custom systems being engineered for specific functions.

For instance, Enercon, a German wind power company, has come up with a unique factory-designed hybrid power technology, including the world's first hybrid wind-diesel-powered ship, the E-Ship 1 [7].

The German wind turbine manufacturer Enercon launched and christened its new rotor ship E-Ship 1 on August 2, 2008. The vessel has now been in service for 5 years to transport wind turbines and other equipment to locations around the world and is shown in Fig. 1.5.

1.1.1 Hybrid System as Source of Renewable Energy

As we have mentioned in the previous section, hybrid energy system is a combination of energy sources of different characteristics and an energy storage medium. When it comes to stand-alone (Fig. 1.6) applications, depending on hybrid energy system is a challenging process due to number of reasons such as determining the best combination, which reduces the initial capital investment, maintaining power supply reliability, reducing the maintenance of system components, etc. [8]

A combination of energy sources having different characteristics reduces the impact of time-varying energy potential of renewable energy sources. Simply solar PV (SPV) energy is available in the daytime, but when it comes to night, you need to fine some other alternatives or store some SPV energy during daytime. When it comes to wind energy, it is also having similar qualities but

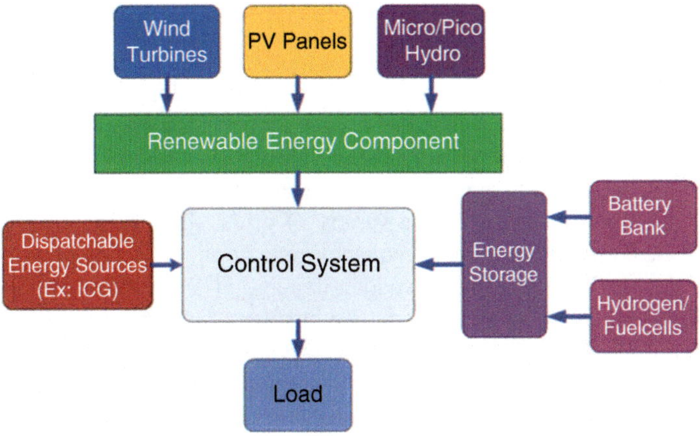

Fig. 1.6 Schematic diagram of a stand-alone hybrid energy system

generally with a much chaotic variation. The time-varying nature of the renewable energy potential makes it essential to incorporate energy storage and dispatchable energy sources.

Energy systems are playing a major role in day-to-day life. It may be your refrigerator, air conditioner, power generator that you use to get electricity, etc. that we discuss. It is prudent that though we use power and energy, very few of us are concerned on energy conservation. Even though we always try to match it with financial aspects, I feel that there is something more on it especially when considering the social responsibility. Fossil fuel resources are depleting at a rapid speed, and at the same time, we are facing lots of problems created by the emission of fossil fuel combustion. Therefore, we are in a period where special attention should be given to conservation of energy.

Optimal designs of energy systems become vital in such circumstances, which is always a challenging process where a number of techno-economic and environmental aspects need to be considered. Most of the time, modeling related with such energy systems is a difficult task. Meanwhile the number of design parameters is to be considered. This makes the optimization work hard, and it is essential to move away from classical methods.

Current commercial, utility-scale hybrid energy systems include:

- Geothermal + solar PV
- Biomass + solar CSP
- Solar PV + fuel cells
- Wind + solar PV
- Biodiesel + wind
- Gas + solar CSP
- Coal + solar CSP

More information of any of these commercial plants can be found here [9].

1.2 Energy Storage Systems

The benefits of energy storage are significant and have long been recognized as necessary for the coordinated and reliable operation of utility grids. Energy storage is especially important to the integration of distributed renewable generation technologies. Storage protects against errors in forecasting, removes barriers to connecting renewable energy resources to a variety of grids, shifts demand on-peaks by storing off-peak energy, provides frequency regulation, and can delay expensive grid upgrades or downtime due to sudden demand or any trip-off of any sources attached to the nationwide grid system. See chapter 17 of reference by Zohuri and McDaniel [10].

It is important to know that there is no "national power grid" in the United States. In fact, the continental United States is divided into three main power grids (Fig. 1.7):

1. The Eastern Interconnected System or the Eastern Interconnect
2. The Western Interconnected System or the Western Interconnect
3. The Texas Interconnected System or the Texas Interconnect

Current commercial, utility-scale energy storage technologies include:

- Pumped hydropower storage
- Compressed air energy storage (CAES)
- Adiabatic compressed air energy storage for electricity (ADELE)
- Molten salt energy storage (MSES)
- Batteries
- Flywheels

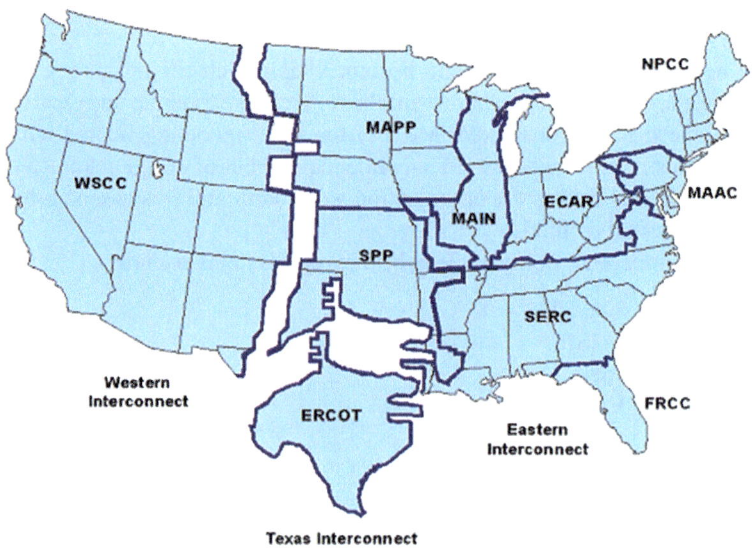

Fig. 1.7 Electrical grid distribution in the US Department of Energy graphics (Courtesy of Department of Energy)

Fig. 1.8 Herdecke pumped-storage power plant (Courtesy of RWE of German Power Company)

Note that: *"Adiabatic" here means additional use of the compression heat to increase efficiency.*

The technology of choice today is the pumped-storage power plant. In any excess power supply, water is electrically pumped into a reservoir on a hill, so that it can be discharged when power demand is high to drive a turbine in the valley downstream. Germany has a pumped-storage power plant producing a total of about 7000 MW with an efficiency which they are claiming is "between" 75% and 86%. The expansion potential is severely limited, especially in northern Germany where the balancing need is greatest.

Figure 1.8 is an illustration of a compressed air energy storage (CAES) in Herdecke, Germany, and its conceptual design is similar in principle to pumped storage: during the phases of excess availability, electrically driven compressors compress air in a cavern to some 70 bars. For discharge of the stored energy, the air is conducted via an air turbine, which drives a generator.

Just as in pumped storage, its power can be released very quickly.

One merit over pumped storage, however, is that the visible impact on the landscape is low. What is more is the facilities can be built near the centuries of wind power production, especially in central and northern Germany. See Fig. 1.9.

1.3 Compressed Air Energy Storage (CAES)

Compressed air energy storage (CAES) is the term given to the technique of storing energy as the potential energy of a compressed gas. Usually it refers to air pumped into large storage tanks or naturally occurring underground formations. While the technique has historically been used to provide the grid with a variety of ancillary

Fig. 1.9 Turbine hall of the Vianden pumped-storage power plant (Courtesy of RWE of German Power Company)

services, it is gaining attention recently as a means of addressing the intermittency problems associated with wind turbine electrical generators. See Fig. 1.10, which is an artistic schematic of the CAES approach.

When energy is available, it is used to run air compressors which pump air into the storage cavern. When electricity is needed, it is expanded through conventional gas turbine expanders. Note that some additional energy (typically natural gas) is used during the expansion process to ensure that maximum energy is obtained from the compressed air (albeit as much as 67% less gas than would be used for an equivalent amount of electricity using gas turbine generators without CAES).

Today, there exist two compressed air energy storage (CAES) plants:

1. *Compressed air energy storage* (*CAES*)
2. *Advanced adiabatic compressed air energy storage* (*AA-CAES*)

CAES plants store energy in form of compressed air. Only two plants of this type exist worldwide, the first one built over 30 years ago in Huntorf, Germany, with a power output of 320 MW and a storage capacity of 580 MWh. The second one is located in McIntosh, Alabama, USA, and began operation in 1991 with a 110 MW output and 2860 MWh of storage capacity. Both are still in operation.

1.3.1 Compressed Air Energy Storage (CAES)

One is in Huntorf (Lower Saxony) since 1978 and another in McIntosh (Alabama, USA) since 1991. The efficiency of the 320 MW plant in Huntorf is about 42% and that of McIntosh around 54%. This means that they are more than 20 percentage points below the efficiency of pumped-storage plants [11].

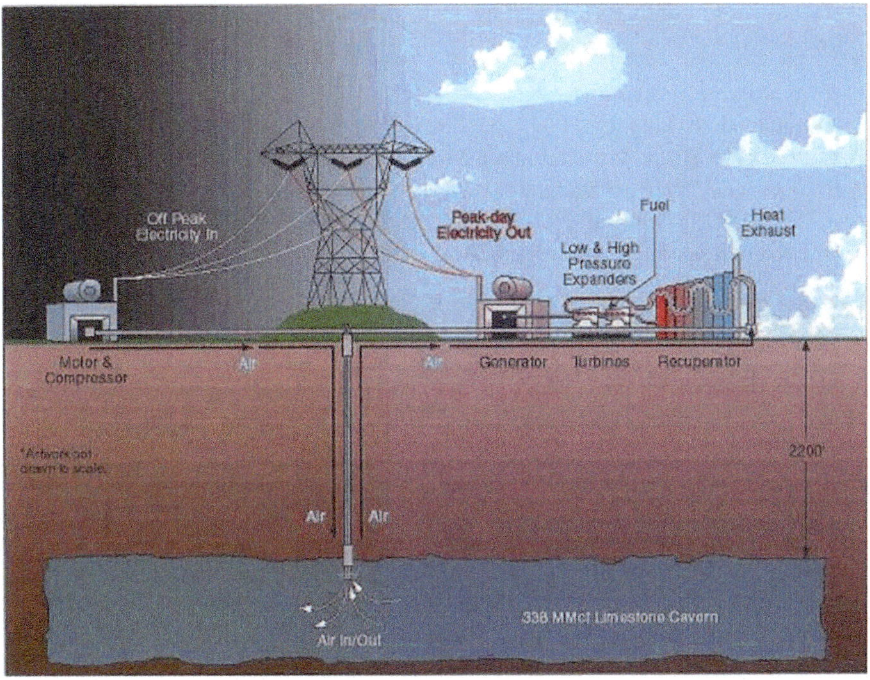

Fig. 1.10 Conceptual representation of CAES

1. *Huntorf Plant*
   The world's first compressed air storage power station, the Huntorf Plant, has been operational since 1978. The 290 MW plant, located in Bremen, Germany, is used to provide peak shaving, spinning reserves, and VAR support. A total volume of 11 million cubic feet is stored at pressures up to 1000 psi in two underground salt caverns, situated 2100–2600 ft below the surface. It requires 12 h of off-peak power to fully recharge and then is capable of delivering full output (290 MW) for up to 4 h. This system operates a conventional cycle and combusts natural gas prior to expansion [12].

2. *McIntosh*
   Alabama's electric cooperative (AEC) has been running the world's second CAES facility since 1991. Called the McIntosh project, it is a 110 MW unit. This commercial venture is used to store off-peak power, generate peak power, and provide spinning reserve. Nineteen million cubic feet is stored at pressures up to 1080 psi in a salt cavern up to 2500 ft deep and can provide full power output for 26 h. This system recovers waste heat which reduces fuel consumption by ~25% compared to the Huntorf Plant [12].

There are more companies investing in CAES approach and they are listed as follow:

3. *Iowa Stored Energy Park*
   Announced in January 2007, the Iowa Stored Energy Park is a partnership between the Iowa Association of Municipal Utilities and the Department of Energy. They plan to integrate a 75–150 MW wind farm with underground CAES, 3000 ft below the surface. The ISEP is currently in design phase with anticipated generation starting in 2011.
4. *General Compression*
   A start-up company in the Boston area has teamed up with a compressor company (Mechanology) to produce the world's first wind turbine-air compressor. These new wind turbines will have the capacity of approximately 1.5 MW, but instead of generating electricity, each wind turbine will pump air into CAES. This approach has the potential for saving money and improving overall efficiency by eliminating the intermediate and unnecessary electrical generation between the turbine and the air compressor.

Conceptually, the basic idea is to use an electric compressor to compress air to a pressure of about 60 bars and store it in giant underground spaces like old salt caverns, aquifers, or pore storage sites and to power a turbine to generate electricity again when demanded. These cavern storages are sealed airtight as proved by the existing two plants and have also been used to store natural gas for years now.

There are few advantages associated with CAES, and the primary benefits of implementing a CAES system are ancillary services provided to the grid. Applications include peak shaving, spinning reserve, VAR support, and arbitrage [12]. By utilizing CAES, the energy from a variety of sources (including wind, solar, and the grid itself) can be temporarily stored to be recovered at a later time, presumably when it is more needed and, perhaps, more valuable. The advantages of CAES are particularly compelling when coupled with an intermittent source such as wind energy. The proposed wind park in Iowa will result in a wind farm which could conceivably be used by utilities to supplement base loads or in meeting hourly load variations and peaks.

Although CAES systems which use underground storage are inherently site specific, it is estimated that more than 80% of the US territory, including most of Idaho, has geology suitable for such underground storage. CAES utilizes proven technology that can be optimized for specific site conditions and competitively delivered by various suppliers.

However, the concept has two major problems when it comes to pressuring air. First, compressing the air leads to a very significant amount of heat generation and subsequent power loss if unused. In addition, the air will freeze the power turbine when decompressed. Therefore, both the existing plants in Huntorf and McIntosh use a hybrid concept with gas combustion as gas turbine power stations require compressed air to work efficiently anyway. Instead of using the combustion of the gas to compress the air like in a conventional gas turbine [13], the stored air in the caverns can be used, meaning that, technically, these CAES plants both store and produce electricity.

As is the case with any energy conversion, certain losses are inevitable. Less energy eventually makes it to the grid if it passes through the CAES system than in a similar system without storage. Some of these losses are mitigated in the approach used by General Compression (using the wind turbine to compress the air directly). In any event, the requirement for additional heating in the expansion process is the most significant disadvantage. By some estimates, 1 kWh worth of natural gas will be needed for every 3 kWh generated from a CAES system. This is particularly problematic if fossil fuels are used for the heat addition. As natural gas prices increase, the economics of CAES, marginal at present, could fail. Again using the wind energy example, one might view a wind farm using CAES as a gas turbine plant with a threefold increase in yield over a conventional gas turbine generator. While this is an impressive improvement, it takes some of the "renewable" luster off the wind farm. It is not clear how policies like the production tax credit or renewable portfolio standards will view this technology.

Now the question is, what is the disadvantages that worth to consider and as part of this kind of storage as it mentioned in above is

What Lowers the Efficiency?
We can seek the answer as follows:

1. First, the air that heats up during compression must be cooled down again to the ambient temperature before it can be stored in the cavern.
2. Second, the cold air must be reheated for discharge of the storage facility since it cools strongly when expanding in a turbine for power generation. Today's plants use natural gas for this. Valuable efficiency percentages are lost.

Rheinisch-Westfälisches Elektrizitätswerk (RWE) Power Corporation, the biggest power producer company in Germany, along with General Electric (GE) is the leading player in the extraction of energy raw materials, and they have teamed up to work on the production of an adiabatic compressed air energy storage (CAES) facility and the project for electricity supply known as *ADELE*. The concept and principle of process steps behind ADELE are as follows, and Fig. 1.11 shows a conceptual layout of such facility.

When the air is compressed, the heat is not released into the surroundings: most of it is captured in a heat storage facility. During discharge, the heat storage device rereleases its energy into the compressed air, so that no gas co-combustion to heat the compressed air is needed. The object is to make efficiencies of around 70% possible. What is more, the input of fossil fuels is avoided.

Hence, this technology permits the CO_2-neutral provision of peak-load electricity from renewable energy. That this technology is doable has been shown by the EU project.

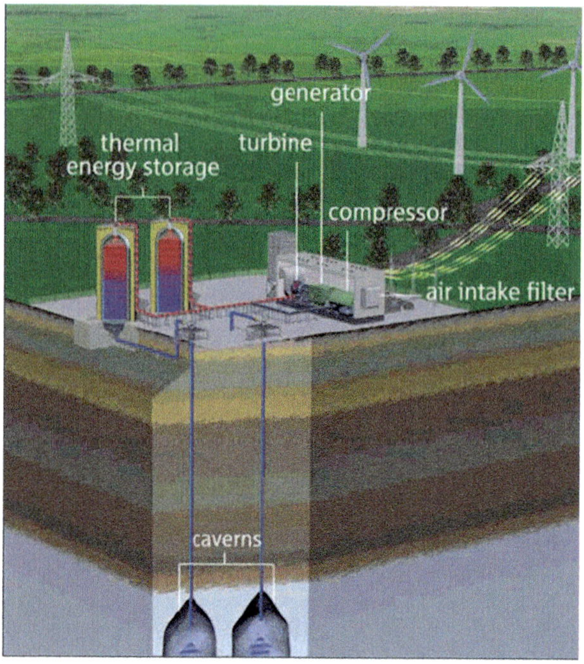

Fig. 1.11 Illustration of ADELE facility (Courtesy of RWE of German Power Company)

1.3.2 Advanced Adiabatic Compressed Air Energy Storage (AA-CAES)

Currently in the development phase is the first ever AA-CAES plant called ADELE [14] in Germany under the direction of the Rheinisch-Westfälisches Elektrizitätswerk (RWE) AG and in cooperation with General Electric (GE), Ed. Züblin AG, and the German Aerospace Center (DLR) [15].

The Advanced Adiabatic Compressed Air Energy Storage (AA-CAES) was under study by General Electric and it was presented in 2008.

The aim of the new joint project set by the German Aerospace Center (DLR), Ed. Züblin AG, Erdgasspeicher Kalle GmbH, GE Global Research, Ooms-Ittner-Hof GmbH, and RWE Power AG—the project being officially sealed in January 2010—is to develop an adiabatic CAES power station up to bidding maturity for a first demonstration plant. The federal ministry for economics has held out a prospect of funding for the ADELE project.

The notable difference to existing CAES plants is that the heat produced by the compressing process, which reaches up to 600 °C (873 K), was dissipated into the environment. Now it is transferred by heat exchangers and stored in heat storage sites. During the discharge, the heat storage releases its energy into the compressed air so that no gas co-combustion to heat the compressed air is needed in order to prevent the turbines from freezing, making it a real energy storage with a theoretical

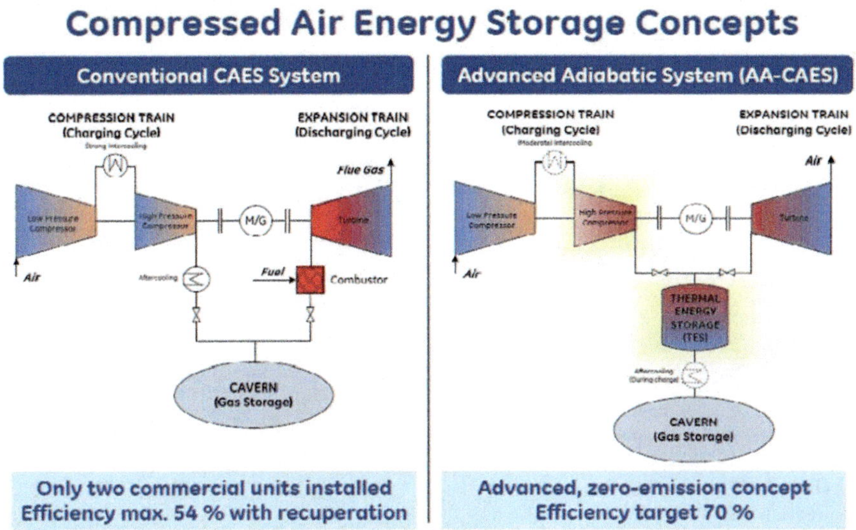

Fig. 1.12 General electric commercially available CAES units

efficiency of approximately 70% and vastly carbon dioxide (CO_2) neutral. Figure 1.12 is an illustration of two commercial units available by General Electric (GE).

In conclusion, if implemented in Idaho, CAES can be used to delay or offset upgrades to the electric transmission grid that would otherwise be necessary. Additionally, it can be used to offset the adverse effects of intermittent renewable energy sources such as wind and solar. The energy community, particularly wind developers and grid operators with significant wind capacity, is watching the Iowa project closely. The economics of the concept appear to work out, but significant research and development efforts could address and mitigate some of the disadvantages. Until the improvements discussed above become commercially available, a biomass source of combustor gas for the expander would bring the approach to a carbon-neutral status. In the light of looming carbon regulations and rising natural gas costs, that would alleviate most of the economic uncertainty of CAES.

As part of business case argument, we can state that:

- The Electric Power Research Institute (EPRI) calls CAES the only energy storage option, apart from pumped hydro, which is available now and can store large amounts of energy and release it over long periods of time—both of which are necessary if you are looking at energy storage for the electrical grid. 150 MW salt-based project is under development in upstate New York.
- Economics of large CAES (100–300 MW underground storage).

    1. Capital \$590–730 per kW
    2. Variable \$1–\$2 per KWh
    3. Hours 10
    4. Total cost \$600–750 per kW \$kW + (hours × \$/kWh)

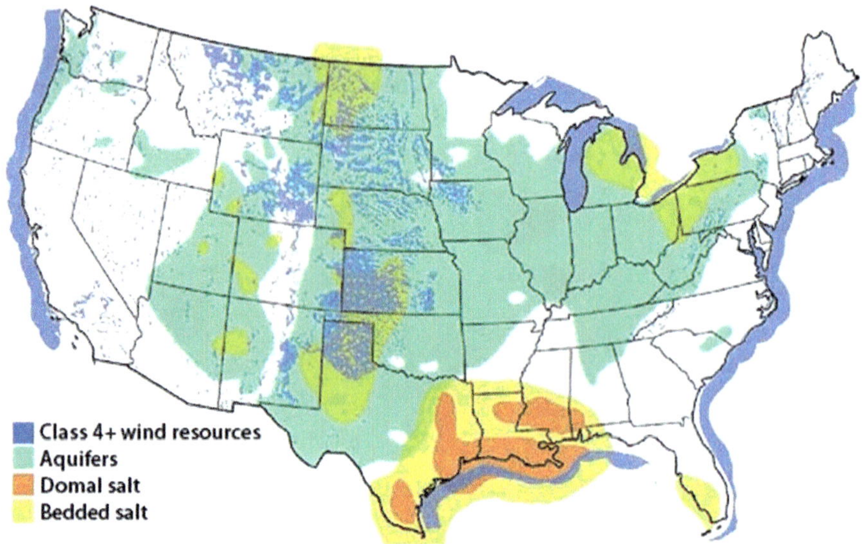

Fig. 1.13 Chart of regions with geology favorable for CAES and class 4+ winds

Figure 1.13 is a presentation of chart of regions with geology favorable for CAES, and class 4+ winds are superimposed to indicate promising CAES plant locations. Source: "Compressed Air Energy Storage: Theory, Resources, and Applications for Wind Power," Samir Succar and Robert H. Williams, Princeton University (published April, 2008).

Pros and cons of the CAES are listed below:

Benefits

- *Efficiency*—CAES plants consume about 35% of the amount of premium fuel utilized by a conventional combustion turbine (CT) and thus produce about 35% of the pollutants per kWh generated from a CT.
- *Availability*—CAES is the only technology available today, other than pumped hydro which can store large amounts of energy and release them over long periods of time. According to a recent study by EPRI, 80% of the US land has geology suitable for underground storage. Pumped hydro is still the most common option for large-scale energy storage, but few new sites are available and they are linked to weather.
- *Potential large scale*—Like pumped hydro, there are no technical limits to the implementation of large projects.
- *Energy price variation*—Playing the spread between on-peak and off-peak prices. The differential between the two prices is the time value of energy storage. This is basically "buy low, sell high." But according to Smith BV, this does not necessarily get you there. And "there" is the ability for a CAES project to generate revenue as a stand-alone project.

- *Capacity.*
- *Ancillary services* such as spinning reserves, upregulation, downregulation, black start, and VAR support.
- *Integrating renewable energy sources.*

Risks and Issues

- *Limited geologic formations*—Unfortunately, the geologic formations necessary for compressed air storage are relatively rare, meaning that it likely will never be a major contributor to the national energy system. At large scale open to similar siting constraints as pumped hydro.
- *Safety*—Mainly concerns with the catastrophic rupture of the tank. Highly conservative safety codes make this a rare occurrence at the trade-off of higher weight. Codes may limit the legal working pressure to less than 40% of the rupture pressure for steel bottles and less than 20% for fiber-wound bottles. Design rules are according to the ISO 11439 standard. High-pressure bottles are fairly strong so that they generally do not rupture in crashes.
- *Cost*—Also subject to financing difficulties due to the nature of underground construction.
- *Proof of concept*—The effectiveness and economy of CAES have not yet been fully proved, especially adiabatic storage.
- *Reheat requirement*—Upon removal from storage, the air must be reheated prior to expansion in the turbine to power a generator. The technology is not truly "clean" because it consumes about 35% of the amount of premium fuel consumed by a conventional combustion turbine and thus produces about 35% of the pollutants on a per kWh basis when compared to it.

1.4 Variable Electricity with Base-Load Reactor Operation

Another way of storing energy to meet a variable electricity with base-load reactor operation is suggested by Charles Forsberg [16] of MIT based on recent technology and research suggested by Forsberg et al. [17] on Nuclear Air-Brayton Combined Cycle which is a continuous collaboration.

The goal of their collaboration is not only to deal with a low-carbon world and use the energy sources such as nuclear but also to look at other means of renewable energy source such as wind, solar, and hydrogen produced by means of very-high-temperature reactor (VHTR) of next-generation nuclear plant coupled with hydrogen production plant (HPP) in a coexisting circumstance.

The defining characteristics of these technologies are:

1. High capital and low operating costs requiring full capacity operation for economic energy production.
2. Output does not match the variable energy needs by men.

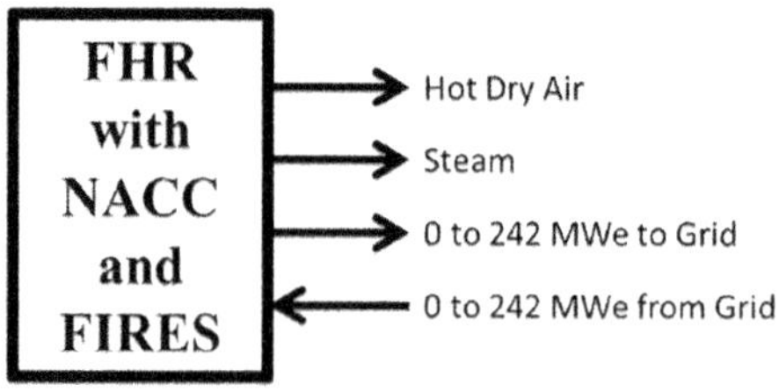

Fig. 1.14 Capability of modular FHR with NACC and FIRES with base-load FHR operation (See references by Forsberg et al. [16, 17])

This challenge suggests a need to develop new nuclear technologies to meet the variable energy needs for low-carbon world while improving economics. Hence, to the above challenge, we have been developing a fluoride-salt-cooled high-temperature reactor (FHR) with a Nuclear Air-Brayton Combined Cycle (NACC) [7, 8] and Firebrick Resistance-Heated Energy Storage (FIRES). The goals are to:

1. Improve nuclear power plant economics by 50–100% relative to a base-load nuclear power plant.
2. Develop the enabling technology for a zero-carbon nuclear renewables electricity grid by providing dispatchable power.
3. Eliminate major fuel failures and hence eliminate the potential for major offsite radionuclide releases in a beyond design basis accident.

Figure 1.14 shows the capabilities of a modular FHR when coupled to the electricity grid. FHR produces base-load electricity with peak electricity produced by a topping cycle using auxiliary natural gas or stored heat or further into the future using hydrogen. The FIRES heat storage capability enables the FHR to replace energy storage technologies such as batteries and pumped storage—a storage requirement for a grid with significant non-dispatchable solar or wind generating systems.

The FHR is a new class of reactors (Fig. 1.15) with characteristics different from light-water reactor (LWR). The fuel is the graphite-matrix coated-particle fuel used by high-temperature gas-cooled reactor (HTGR) resulting in similar reactor core and fuel cycle designs—except the power density is greater because liquids are better coolants than gases. The coolant is a clean fluoride-salt mixture. The coolant salts were originally developed for the molten salt reactor (MSR) where the fuel is dissolved in the coolant. Current coolant-boundary material limitations imply maximum coolant temperatures of about 700 °C. New materials are being developed that may allow exit coolant temperatures of 800 °C or more. The power cycle is like that used in natural gas-fired plants.

The fluoride-salt coolants were originally developed for the US Aircraft Nuclear Propulsion in the late 1950s. The goal was to develop a nuclear-powered jet bomber. These fluoride salts have low nuclear cross sections with melting points of 350–500 °C and boiling points more than 1200 °C—properties for efficient transfer of heat from a reactor to a jet engine. Since then there have been two developments. The first development was high-temperature graphite-matrix coated-particle fuels

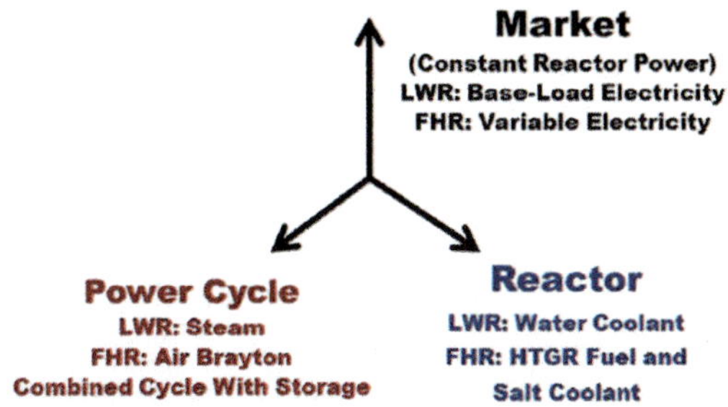

Fig. 1.15 Comparison of the LWR and FHR [7]

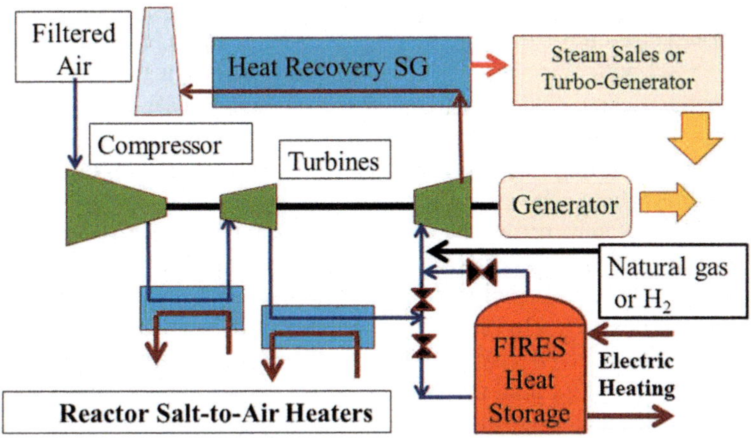

Fig. 1.16 Nuclear air-Brayton combined cycle (NACC) with firebrick resistance-heated energy storage (FIRES) [17]

for HTGRs that are compatible with liquid salt coolants. The second has been a half-century of improvements in utility gas turbines that now make it feasible to couple a nuclear reactor (the FHR) to NACC.

The FHR is coupled to a NACC with FIRES (Fig. 1.16). In the power cycle, external air is filtered, compressed, heated by hot salt from the FHR while going through a coiled tube air heat exchanger (CTAH), sent through a turbine producing electricity, reheated in a second CTAH to the same gas temperature, and sent through a second turbine producing added electricity. Warm low-pressure airflow from the gas turbine system exhaust drives a heat recovery steam generator (HRSG), which provides steam to either an industrial steam distribution system for process heat sales or a Rankine cycle for additional electricity production. The air from the HRSG is exhausted up the stack to the atmosphere. Added electricity can be produced by injecting fuel (natural gas, hydrogen, etc.) or adding stored heat after nuclear heating by the second CTAH. These boost temperatures in the compressed gas stream going to the second turbine and to the HRSG [17].

Since a NACC system looks quite good for a salt cooled reactor, it is worth considering what it might do for a sodium-cooled reactor. With some modifications, it appears that it could be competitive with systems that have been built. A computer model was built based on standard techniques for analyzing Brayton and Rankine systems. System performance was optimized by varying the turbine outlet temperatures for a fixed turbine inlet temperature. A second parameter that can usually be varied to obtain optimum performance is the peak pressure in the steam cycle. For most of the cases considered here, this was held constant at 12.4 MPa (1800 psi) [8].

Fairly detailed design was attempted for the heat exchangers involved in the system as they tend to dominate system size; more details are provided in Chaps. 5, 6, and 7 of this book. The techniques and data were extracted from the text by Kays and London [18].

The stored heat option involves heating firebrick inside a press-tress concrete pressure vessel with electricity to high temperatures at times of low electricity prices; that is, below the price of natural gas. When peak power is needed, compressed air after nuclear heating and before entering the second turbine would be routed through the firebrick, heated to higher temperatures, and sent to the second turbine. The efficiency of converting electricity to heat is 100%. The efficiency of converting auxiliary heat (natural gas or stored heat) to electricity in our current design is 66%. This implies a round-trip efficiency of electricity to heat to electricity of ~66%. Improvements in gas turbines in the next decade are expected to raise that efficiency to 70%. FIRES would only be added to NACC in electricity grids where there are significant quantities of electricity at prices less than the price of natural gas. As discussed later, these conditions are expected in any power grid with significant installed wind or solar capacity.

As we said, much of the FIRES heat storage technology is being developed by General Electric® and its partners for adiabatic compressed air energy storage (CAES) system called ADELE (German abbreviation). The first prototype storage system is expected to be operational by 2018 with 90 MWe peak power and storing 360 MWh. When the price of electricity is low, the air is (1) adiabatically compressed to 70 bars with an exit temperature of 600 °C, (2) cooled to 40 °C by flowing the hot compressed air through firebrick in a prestress concrete pressure vessel, and (3) stored as cool compressed air in underground salt caverns. At times of high electricity prices, the compressed air from the underground cavern goes through the firebrick, is reheated, and sent through a turbine to produce electricity with the air exhausted to the atmosphere. The expected round-trip storage efficiency is 70%. The ADELE project is integrating firebrick heat storage into a gas turbine system. For NACC using FIRES, the differences are (1) the peak pressure would be about a third of the ADELE project, (2) the firebrick is heated to higher temperatures, and (3) electricity is used to heat the firebrick at times of low electricity prices to higher temperatures. The technology for heat storage integration into NACC is partly under development.

To show that utilization of a NACC system is very efficient and cost-effective for such an innovative approach to store energy in the form of FIRES process, the

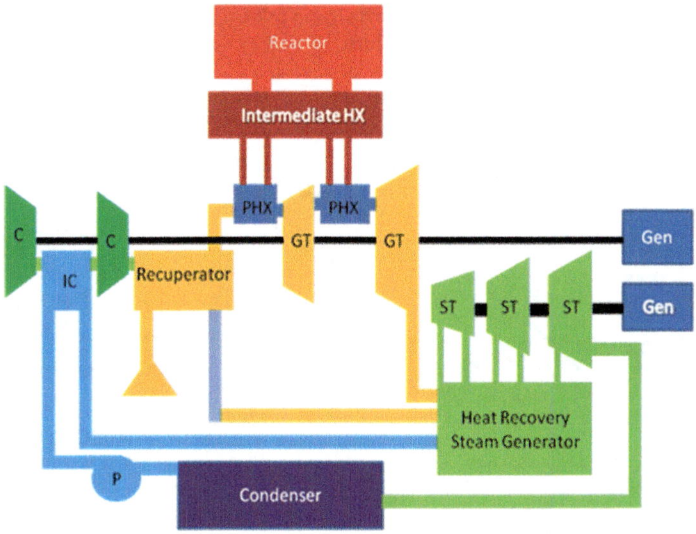

Fig. 1.17 System layout with recuperator and intercooler (*C* compressor, *GT* gas turbine, *ST* steam turbine, *PHX* primary heat exchanger, *IC* intercooler, *P* pump) [11]

following reasoning is presented here, and for further information, the reader should explore the textbook by Zohuri [6].

Since the high-pressure water in the bottoming cycle must be heated and the heating of the air in the air compressor increases the work required, it is possible to split the compressor and add an intercooler that heats the high-pressure water in the bottoming cycle and cools the output from the first part of the compressor. If this is done, the efficiency goes to 40.3%, and the overall compressor pressure ratio goes to 2.0. A system diagram is provided in Fig. 1.17 [6].

The efficiency of NACC power systems continues to increase with increased turbine inlet temperatures. For the foreseeable future, there does not appear to be a limitation to using off-the-shelf materials as it is not likely that a reactor heated system will exceed 1300 K turbine inlet temperature. A comparison of the cycle efficiencies for several cycles that have been proposed for the next-generation nuclear plant (Zohuri) [6] is presented in Fig. 1.18. The calculations for the NACC systems are based on the system described in Fig. 1.18 with a peak steam pressure of 12.4 MPa.

NACC systems can be applied to most of the proposed next-generation systems. Their strongest competitor in terms of cycle efficiency is the supercritical CO_2 system. NACC systems will match or better the efficiency of these systems at or above 700 °C. But NACC systems have the competitive advantage of a large customer base for system hardware, significantly reduced circulating water requirement for rejecting waste heat, and much greater efforts to improve the technology relative to other power cycles [17].

On January 21, 2010, the California Public Utilities Commission (CPUC) approved Pacific Gas and Electric's (PG&E's) request for matching funds of

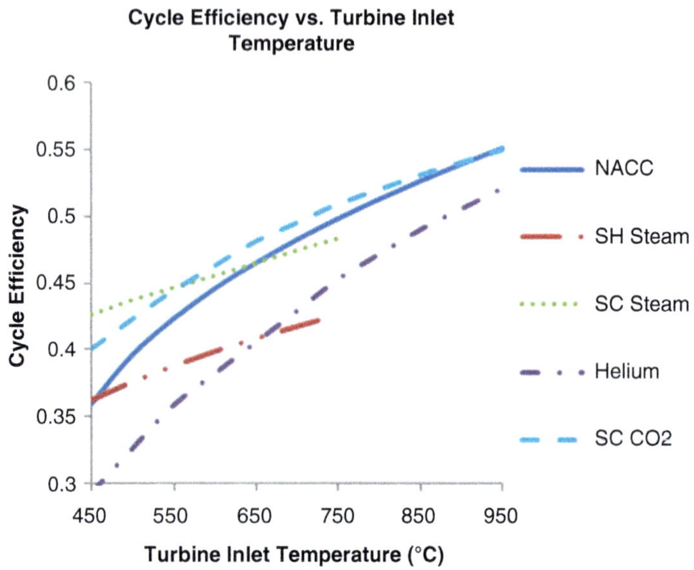

Fig. 1.18 Cycle efficiencies for various advanced cycles [8]

$25 million for the project. The CPUC found that the CAES demonstration project will provide PG&E with a better understanding of a promising energy storage technology, which has the potential to lower costs for customers and reduce greenhouse gas emissions through greater integration of renewable energy sources. The California Energy Commission (CEC) has also shown support for the project with conditional approval of a $1 million grant.

The commercial-scale project has a nominal output capacity of 300 MW—like a mid-sized power plant—for up to 10 h. It is estimated that a commercial plant could come online in the 2020–2021 time frame.

The time frame of this project is laid out here, and Fig. 1.19 is conceptual of such commercial facility.

PG&E is exploring this project in three primary phases:

Phase 1: Reservoir feasibility including site control, reservoir performance, economic viability, and environmental impacts.

Phase 2: Commercial plant engineering, procurement and construction, and commissioning.

Phase 3: Operations monitoring and technology transfer.

However, construction of a prototype brings new obstacles and other challenges. The engineering of heat storage sites capable of holding the energy over longer periods without significant losses, compressors that can handle both high pressures and high temperatures, and turbines with the ability to maintain on a constant output under changing conditions (changing temperatures, decreasing air pressure) are some of the challenges. However, with the current state of the art, it is very doable.

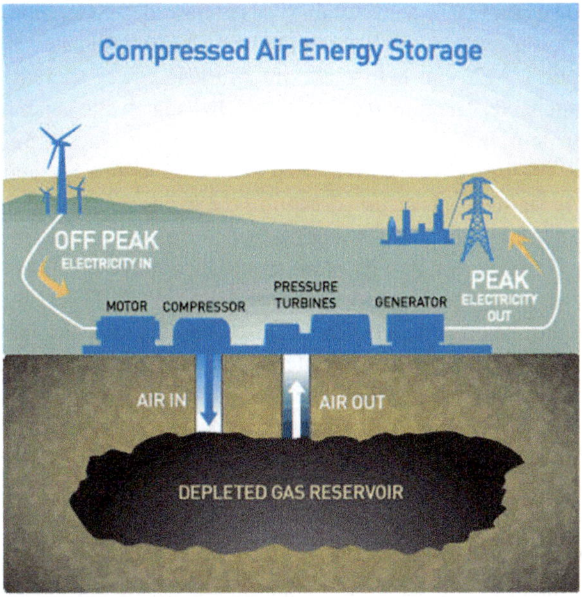

Fig. 1.19 Conceptual illustration for PG&E approach

Before we finish off this section, a few acronyms and definitions should reflect here, which are used in this technology as renewable energy, and they are:

(a) *Adiabatic Storage*—The heat that appears during compression is also stored and then returned to the air when the air is expanded. This is a subject of ongoing study, but no utility-scale plants of this type have been built. The theoretical efficiency of adiabatic energy storage approaches 100% for large and/or rapidly cycled devices, but in practice round-trip efficiency is expected to be 70%. Heat can be stored in a solid such as concrete or stone or more likely in a fluid such as hot oil (up to 300 °C) or a molten salt (600 °C).

(b) *Diabatic Storage*—The extra heat is removed from the air with intercoolers following compression (thus approaching isothermal compression) and is dissipated into the atmosphere as waste. Upon removal from storage, the air must be reheated prior to expansion in the turbine to power a generator. The heat discarded in the intercoolers degrades efficiency, but the system is simpler than the adiabatic one and thus far is the only system which has been implemented commercially.

   The McIntosh CAES plant requires 0.69 kWh of electricity and 1.17 kWh of gas for each 1.0 kWh of electrical output (a non-CAES natural gas plant can be up to 60% efficient therefore uses 1.67 kWh of gas per kWh generated).

(c) *Dispatchable Generation*—Sources of electricity that can be dispatched at the request of power grid operators; that is, it can be turned on or off upon demand.

This should be contrasted with certain types of baseload generation capacity, such as nuclear power, which may have limited capability to maneuver or adjust their power output. CAES can help make intermittent power sources such as wind power dispatchable. The time periods in which dispatchable generation plant may be turned on or off may vary and be considered in time frames of minutes or hours.

(d) *Intercooler* (original UK term, sometimes after cooler in US practice), or charge air cooler, is an air-to-air or air-to-liquid heat exchange device which removes the heat of compression (i.e., the temperature rise) that occurs in any gas when its pressure is raised or its unit mass per unit volume (density) is increased. Compressing air heats it and expanding it cools it. Therefore practical air engines require heat exchangers in order to avoid excessively high or low temperatures and even so do not reach ideal constant temperature conditions.

(e) *Isothermal Compression and Expansion* approaches (which attempt to maintain operating temperature by constant heat exchange to the environment) are only practical for rather low power levels, unless very effective heat exchangers can be incorporated. The theoretical efficiency of isothermal energy storage approaches 100% for small and/or slowly cycled devices and/or perfect heat transfer to the environment.

(f) *Turbo-expander* (also referred to as an expansion turbine) is a centrifugal or axial flow turbine through which a high-pressure gas is expanded to produce work that is often used to drive a compressor. Because work is extracted from the expanding high-pressure gas, the expansion is an isentropic process (i.e., a constant entropy process) and the low-pressure exhaust gas from the turbine is at a very low temperature, sometimes as low as −90 °C or less.

1.5 Why We Need Nuclear Power

Some scientists and engineers in the nuclear engineering field are calling for 100% renewable energy. That is totally the wrong approach. However, the new generation of nuclear power plants (NPP) that are known as GEN-IV is taking a different approach from the design point of view to be more efficient and cost-effective from ownership perspective. It has been suggested by this author [6] and others as an innovative approach to consider combined cycle as way of improving the thermal output of these rectors as a means of small modular reactor (SMR) design configuration to the level if not higher but at least to be close to 60% efficiency, where today's fossil- and gas fuel-type power plants are to produce electricity. GEN-IV NPPs are designed with smaller real estate and footprint in respect to their predecessor of past generation such as GEN-III types. See Fig. 1.20, which illustrates a typical generation three (GEN-III) nuclear power plant, in the southern part of Detroit, Michigan.

As part of the Climate Desk collaboration, the story of renewable energy was published by Julian Spector on July 20, 2015, in Citlab.com site under the title of "The Environmentalist Case Against 100% Renewable Energy Plants" [19].

In his article, he claimed that "It might be technically feasible, but that doesn't mean it's the best plan to pursue," and he continued to state that:

Fig. 1.20 A nuclear power plant south of Detroit, Michigan

> Renewable energy has had a busy year. California and New York have adopted ambitious plans calling for 50% renewable energy by 2030. A group of Stanford and Berkeley scientists has put forth an even bolder vision—encouraging all 50 states to run on wind, water, and solar by 2050, without any nuclear energy or biofuels in the picture. New York City Mayor Bill de Blasio has announced his intention to go fully renewable with the city government's power, too.

A world without any fossil fuel energy would be a much cleaner place for both people and the environment. Right now, renewable energy accounts for just 13% of all U.S. electricity. A significant increase in that share would lead to a major reduction in air pollution and its attendant diseases, not to mention the costs of climate change-induced flooding or wildfires. The lives, time, and property saved could be put to work tackling other social problems.

But it is not entirely clear that a US energy grid based on 100% renewables is the best way to achieve a zero-carbon future. On the contrary, there is a strong environmentalist case for approaching that goal with caution. Limiting a zero-carbon future to wind, water, and solar means greater costs of storing this energy, discarding other existing zero-carbon sources like nuclear, and generally blanketing the Earth with panels and turbines as a mean to save it.

1.5.1 The Merits of Total Transformation

For their renewable energy roadmap study, Stanford professor Mark Jacobson and his team used US Energy Information Administration data to project "business-as-usual" energy consumption in 2050. They then compiled state-by-state energy portfolios needed to meet that projected demand through expanded wind, water, and solar energy generation.

The endpoint is a future in which every driver in America rides an electric car, every stove in every house and restaurant cooks with electricity instead of gas, every plane flies by cryogenic hydrogen (that is what rockets use, and the Soviet Union built an experimental airplane that flew on it, too) [20]. The authors point out that electric energy is more efficient than fuel combustion for heating and motors. When that efficiency is scaled up to an entirely electrified society by 2050, they project a 39.3% reduction in America's electricity load compared to business as usual. See Fig. 1.21.

Mark Jacobson plans a plan chart to convert US energy to 100% renewables by 2050 which involves gradually reducing reliance on fossil fuels and nuclear energy and increasing supply of wind, solar, and water energy.

Heavy reliance on wind (50% of supply) and solar (45.25% of supply) poses the challenge, though, of meeting peak consumer energy demand when the sun is not shining or the winds die down. Jacobson and company propose to do this without any new battery technology by assembling a host of creative energy storage devices, such as piping surplus energy as heat into the ground and pulling it up later for use or using cheap off-peak electricity to make ice which then goes to work cooling buildings during high-demand periods. The roadmap says that by 2050 it can "match load without loss every 30 [seconds] for six years"; the authors have an auxiliary study to support this claim, although it has not been released yet.

This all might sound overwhelmingly expensive, but the researchers counter that with macrolevel accounting of the societal costs avoided by a fully renewable grid. They estimate the price tag for lives lost to fossil fuel-induced air pollution; eliminating that pollution, they find, would save up to $600 billion per year in 2050. They also estimate savings to the United States from avoiding climate change-related damage, such as droughts, wildfires, floods, and severe weather. The shift to

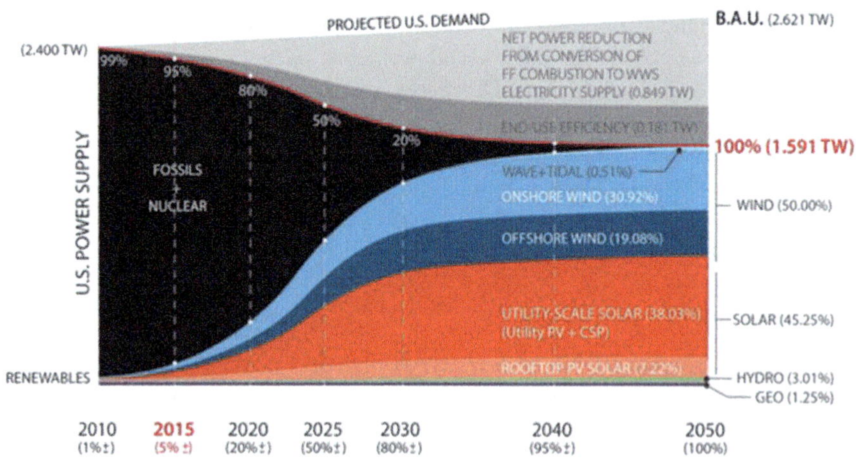

Fig. 1.21 A plan from stanford's Mark Jacobson et al. (Courtesy of Energy & Environmental Science)

renewables will eliminate around 3.9 million jobs associated with the old energy industry but will result in a net gain of around 2 million 40-year jobs.

Jacobson sees this transition as a way to recognize the negative externalities the country has already been paying for. "The people who are running these coal mines have not paid for the health and climate costs they have been causing," he says. "They have been freeloading on society for a long time."

1.5.2 The Downsides of Monoculture

The goal of 100% renewable plans is to achieve a host of social benefits by cutting carbon emissions out of energy production. Rejecting some zero-carbon energy sources, such as nuclear, from the outset makes the problem harder, says MIT doctoral candidate Jesse Jenkins, who researches the electric power sector's transition to a zero-carbon system.

"Why would we want to constrain ourselves to a narrow set of options to confront climate change and air pollution and other energy sector challenges when those challenges are already quite difficult?" he says.

An entirely renewable portfolio creates its own special obstacles. For instance, Jenkins notes, the marginal value of renewables decreases as they penetrate the market. The free energy inputs of wind and solar initially displace the more expensive energy inputs, like natural gas. But assuming renewables successfully displace all coal and natural gas, then the plan would require building more wind and solar to displace nuclear, which provided 19% of US electricity production in 2014. That requires spending more money to achieve the same goal of a clean grid. Figure 1.22 is an illustration of a solar farm with its array of panels to collect sun energy.

The other problem for planning an all-renewable grid is variability: solar produces when there is enough sun and wind produces when there is enough wind. Luckily the sun tends to shine during the day, when there is higher demand for energy. But to ensure power when the renewables are not producing much requires energy storage. That storage could be done through batteries or the heat, ice, and other methods Jacobson mentions.

"What people really miss about storage is it's not just a daily storage problem," says Armond Cohen, executive director of the Clean Air Task Force, a group that researches low-carbon energy technologies. "Wind and solar availability around the world, from week to week and month to month, can vary up to a factor of five or six."

Storage must account for when the wind cuts out for weeks due to seasonal weather variation. It is easy enough to make ice one night to cool your building the next day, Cohen notes, but to save energy for 3 weeks of low wind, you would need to store up enough ice to cool the building for that whole time.

Accounting for sufficient storage, then, increases the costs and scope of the energy transition. Jacobson calls for 605,400 MW of new storage capacity. US grid storage as of August 2013 totaled 24,600 MW, meaning a nearly 25-fold increase would be required to meet the roadmap.

That is not impossible, but it is an effort that would not be necessary with continuous energy sources.

Fig. 1.22 Solar panels soak in rays at a Southern California Edison electricity station in Carson, California (Courtesy of REUTERS/Lucy Nicholson)

1.5.3 The Other Zero-Carbon Energy: Nuclear

Plans calling for 100% renewable energy eliminate nuclear energy from the mix. The new state roadmap casts out nuclear without much discussion, but Jacobson tells CityLab this is because when you factor in the mining and refining of uranium, nuclear energy emits more carbon than wind power. He also cites the difficulty and expense of creating new nuclear plants and other risks like proliferation and meltdown.

The decision to entirely abandon nuclear was particularly galling to Michael Shellenberger, president and cofounder of the Breakthrough Institute, which researches ways that modernism and technology can improve human prosperity and safeguard the environment, and an author of *An Ecomodernist Manifesto* [21]. He argues that nuclear's efficiency, small land-use footprint, and limited resultant pollution make it a vital part of any low-carbon future. This is debatable by nuclear scientists and engineers. Even some environmentalists believe any plan for cleaner energy in the United States should involve nuclear. Figure 1.23 is a picture of Three Mile Island nuclear power plant at night.

"If you care about the environment, you want food and energy production to become more efficient and centralized," he says. "You want to put fewer inputs in and get more outputs out and get less waste."

As primary energy sources advanced from firewood to coal to natural gas to nuclear, Shellenberger says, humans have managed to get consecutively more energy out compared to what they put in. "Neither solar nor wind are substitutes for coal or natural gas or oil," he says. "The new product has to be equal to or

Fig. 1.23 Three Mile Island nuclear power plant at night in 2011 (Courtesy of REUTERS/ Jonathan Ernst)

superior to the predecessor and solar and wind are totally different than those fuels and inferior in that they're intermittent."

Stanford economics professor Frank Wolak, director of the Program on Energy and Sustainable Development [22], agrees that nuclear should play a role in a zero-carbon grid. He notes that American nuclear generators are safer than ever and have an extremely high capacity factor, meaning they produce almost all their potential energy. American nuclear set a record high capacity factor of 91.8% for 2014. Wind and solar have capacity factors less than half as large.

Note that the Program on Energy and Sustainable Development (PESD) is an international, interdisciplinary program that draws on the fields of economics, political science, law, and management to investigate how real energy markets work. This means understanding not only what is technologically possible and economically efficient but also how actual political and regulatory processes lead to outcomes that are costlier and less effective than they could be.

Nuclear energy has a much higher capacity factor than renewable energy does—meaning it produces far more of its potential energy, see Fig. 1.24. "Nuclear energy is an extremely reliable source of zero-carbon energy," Wolak writes via email. "It makes very little economic sense to phase it out, particularly given how successful the U.S. nuclear industry has been over the past 30 years."

The irony of environmentalists cutting out nuclear in favor of primarily wind and solar is that these sources require much more transformation of the landscape to produce the same amount of energy. That footprint draws opposition from other environmental groups and people who just do not want to live near wind turbines.

The Jacobson plan, for instance, envisions 156,200 new 5 MW offshore wind turbines. Cohen from the Clean Air Task Force compares that to the Cape Wind project, which would have installed 468 MW of wind turbines off Cape Cod. That project collapsed following legal and political opposition from millionaire

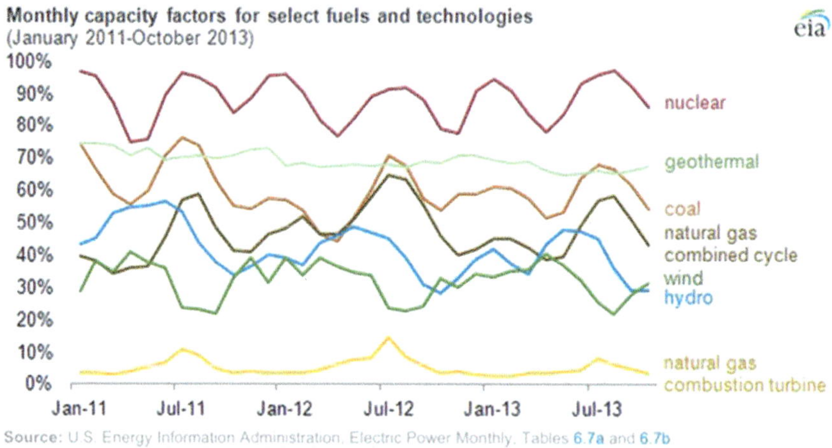

Fig. 1.24 Illustration of monthly capacity factors for select fuels and technologies (Courtesy of U.S. Energy Information Administration)

Fig. 1.25 A proposed rendering of the failed Cape Wind project to install 468 MW of wind turbines off Cape Cod

landowners, but also local townspeople and fishermen. Jacobson's proposal amounts to building nearly 1700 times the offshore capacity of Cape Wind. See Fig. 1.25.

1.5.4 A Diverse Portfolio

The Jacobson roadmap shows that a 100% renewable grid is feasible from an engineering standpoint. The politics of implementing such a plan are much trickier, though. The study itself offers only broad recommendations for easing the transition (e.g., "Incentivize more use of efficient lighting in buildings and on city streets.") and points to the ramp-up of aircraft production during World War II as evidence of America's manufacturing ability under pressure.

Whether or not US urgency about the environment will ever reach wartime heights is another question. Jacobson says it will be up to policymakers to fill in the details and notes that the recent renewable visions outlined by California and New York show it can be done. "We're trying to provide an endpoint and each state is going to have to figure out how to get to that endpoint," he says.

Technical and political feasibility aside, it is also unclear why a fully renewable grid would be more desirable than any other combination of zero-carbon energy sources. "[The 100% renewable roadmap] is not an optimization study," says Jenkins. "It's not saying this is the best pathway forward in terms of any metric, particularly in terms of cost. They say, 'How much can we push renewables and only renewables? And what will be necessary to try to decarbonize with that pathway alone?'"

In other words, if the goal is to cut out carbon emissions, there are other ways to do it. Jenkins is working on models that optimize the electric grid with constraints for cost, technical feasibility, and low CO_2 emissions. For an area with Texas-like wind and solar resources and energy demand, around 67% would come from nuclear (plus hydropower or fossil fuels with carbon capture and storage if available). Wind and solar make up about 19%, and the remaining 13% would be gas utilities that fire up quickly to meet peak demand when the other sources cannot. Those numbers change for different places, and in a scenario with better storage capacity, renewables can take on more of the load from nuclear.

> If you try to push any one of these pieces too far it ends up being more costly and difficult to manage than the optimal system.

"All these pieces work together," Jenkins says. "If you try to push any one of these pieces too far it ends up being more costly and difficult to manage than the optimal system."

As solar and wind technology improves and gets cheaper, other paths to cleaner power are evolving, too. New molten salt nuclear reactors, still in development, promise less uranium-intensive power generation that does not need water for cooling. They would play a significant role in bringing costs down for nuclear plants. Technology to retrofit fossil fuel plants for carbon capture and storage is still scaling up and lowering costs, Jenkins says. That would make it possible to clean up coal and gas plants that still have decades of operation left in their lifetimes, rather than shutting them down and building new capacity in their place.

There is a good environmental argument for replacing dirty fossil fuel systems with renewables, but the reasons for replacing zero-carbon systems with other zero-carbon systems are less clear. Recognizing cost constraints while planning for a zero-carbon grid would force us to do more with less, which is actually a pretty good approach to sustainability.

1.6 Security of Energy Supply

Coordinating Energy Security in Supply Activities (CESSA) is of utmost importance for the United States of America (USA) and European Union (EU) and its member states, in economic, technical, and political terms. Securing energy supply is a cornerstone of the "magic triangle" of energy policy, the two others being competition and sustainability. And in times of rising geopolitical conflicts, supply security has also increased in importance in the external relations of the United States and EU.

The project of CESSA was originally funded by DG Research in EU within Sixth Framework Programme and was also supported by DG TREN through information and access to decision-makers. CESSA was coordinated by the Université Paris-Sud and the École des Mines de Paris/Paris Tech, with work packages attributed to the University of Cambridge, the Universidad Pontificia Comillas in Madrid, and the German Institute for Economic Research (DIW Berlin) in cooperation with the Chair of Energy Economics and Public Sector Management at the University of Technology (TU) Dresden.

The Florence School of Regulation provided input to the project coordination and the conclusions. In addition, scholars from Stanford University and the Massachusetts Institute of Technology, among others, contributed to the work.

The salience of electricity security differs greatly across the member states of the EU. In western member states, history has provided robust and flexible electricity system, and market liberalization is generally well advanced. These countries enjoy a diverse range of energy sources, and much investment is underway to expand this range of supply options. Investment in nuclear energy represents one such option; however, several European Union countries, including Ireland and Austria, remain resolutely opposed to nuclear power. The growth of EU from 12 states to 27 has reduced the proportion of member countries with an antinuclear stance.

In EU, nuclear power policy is shaped by two regulatory pressures: the regulation of electricity markets and the safety regulation of a hazardous and politically contentious technology. While the benefits of a single European electricity market are widely recognized, progress on the question of pan-European safety regulation is much less developed. An international project collaboration is emerging, particularly in Eastern EU member states. CESSA would support moves toward the regionalization and the eventual Europeanization of safety regulation.

There is also some movement in the United states toward nuclear power industry by looking at the new generation of these plants, namely, GEN-IV, and in particular in smaller footprint in design known as small modular reactor (SMR), and companies like Westinghouse (W), General Electric (GE), Babcock & Wilcox (B&W), and in particular a newly established company, namely, NuScale are in lead of these types of reactors.

Economics is central to the future of nuclear power. We stress that nuclear power plants can be developed in a liberalized electricity market with no direct subsidy. This possibility is favored by stable long-term carbon prices; sustained high oil and gas prices and regulatory approval for grid reinforcement by monopoly transmission companies similar to that put in place to assist new renewables projects. During the CESSA project, the relative economic attractiveness of nuclear energy investment has improved significantly, such that economic risks now appear less daunting, although important issues of economic risk do remain, notably arising from the recent rapid escalation in construction costs and remaining uncertainties about the time before commissioning.

The need to expand the supply of domestically produced energy is significant. America's transportation sector relies almost exclusively on refined petroleum products. Approximately 52% of the petroleum consumed for transportation in the United States is imported, [23] and that percentage is expected to rise steadily for the foreseeable future (Fig. 1.25). On a global scale, petroleum supplies will be in higher demand as highly populated; developing countries expand their economies and become more energy-intensive. Hydrogen-powered fuel cell vehicles would virtually eliminate imports of foreign oil, because the hydrogen fuel can be produced almost entirely from the diverse domestic energy sources of renewable resources, fossil fuels, and nuclear power. Hydrogen's role as a major energy carrier would also provide the United States with a more efficient and diversified energy infrastructure that includes a variety of options for fueling central and distributed electric power generation systems.

> America's reliance on imported oil is the key challenge to our energy security. While oil is used in all sectors and for a wide variety of uses, the large majority is used for transportation—and a majority of that is used in light-duty passenger vehicles (cars and light trucks). [24]

1.7 Environmental Quality

The combustion of fossil fuels accounts for the majority of anthropogenic greenhouse gas emissions (chiefly carbon dioxide, CO_2) released into the atmosphere. The largest sources of CO_2 emissions are the electric utility and transportation sectors. Should strong constraints on carbon emissions be required, hydrogen will play an important role in a low-carbon global economy. Distributed hydrogen

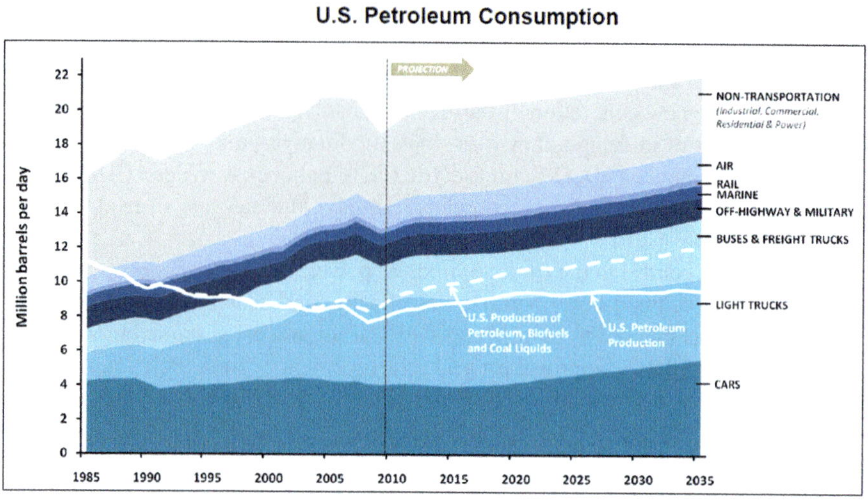

Fig. 1.26 America's widening "oil gap" (Courtesy of Department of Energy)

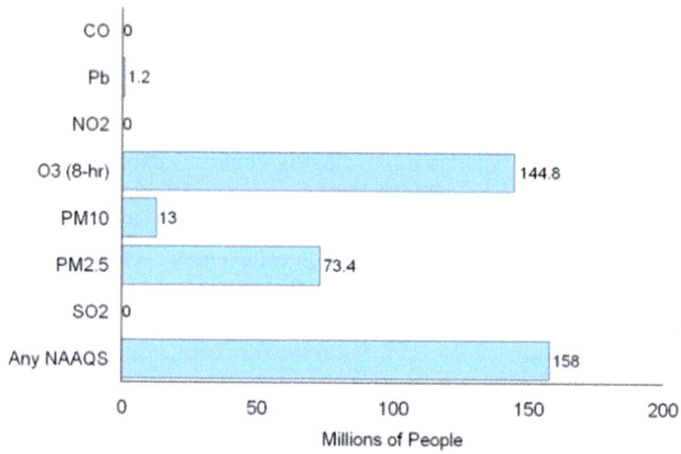

Fig. 1.27 Number of people living in countries with air quality concentrations above the level of the NAAQS in 2007 (Courtesy of Department of Energy)

production from natural gas and central hydrogen production from natural gas (with the potential for capture and sequestration of carbon) and coal (with the capture and sequestration of carbon) can provide the means for domestic fossil fuels to remain viable energy resources. In addition, fuel cells operating on hydrogen produced from renewable resources or nuclear energy result in near-zero-carbon emissions.

Air quality is a major national concern. It has been estimated that about 50% of Americans live in areas where levels of one or more air pollutants are high enough to affect public health and/or the environment. (See Fig. 1.26) [25]. Personal vehicles and electric power plants are significant contributors to the nation's air quality problems. Most states are now developing strategies for achieving national ambient air.

Despite great progress in air quality improvement, approximately 150 million people nationwide lived in counties with pollution levels above the National Ambient Air Quality Standards (NAAQS) in 2007. See references (a) and (b) below. See Fig. 1.27.

(a) U.S. Environmental Protection Agency, "Air Trends: Basic Information" (n.d.), retrieved November 18, 2008, from http://www.epa.gov/airtrends/sixpoll.html
(b) U.S. Censure Bureau, 2007 Population Estimate, retrieved November 18, 2008, from http://www.census.gov

References

1. https://en.wikipedia.org/wiki/Hybrid_renewable_energy_system
2. U.S. Department of Energy – Fuel Cells Technologies Program. https://www1.eere.energy.gov/hydrogenandfuelcells/pdfs/doe_h2_fuelcell_factsheet.pdf
3. M. Khan, M. Iqbal, Pre-feasibility study of stand-alone hybrid energy systems for applications in Newfoundland – Memorial University of Newfoundland St. John's Canada June 2004: Elsevier Renewable Energy, Issue 30 (835–854)
4. http://exploringgreentechnology.com/solar-energy/hybrid-energy-systems/
5. Feasibility for a standalone Solar-Wind-Based hybrid energy system for application in Ethiopia by Getachew Bekele and Bjorn Palm Department of Energy, KTH, Stockholm, Sweden; Elsevier – Applied Energy, 87 (487–495), 2010
6. B. Zohuri, *Combined Cycle Driven Efficiency for Next Generation Nuclear Power Plants: An Innovative Design Approach* (Springer, Cham, 2015)
7. Got Powered – Enercon E-Ship 1: Wind-powered ship. http://gotpowered.com/2011/enercon-e-ship-1-wind-powered-ship/
8. http://hybrid-renewable.blogspot.com/2011/03/importance-of-hybrid-energy-systems.html
9. http://www.cleanenergyactionproject.com/CleanEnergyActionProject/Hybrid_Renewable_Energy_Systems_Case_Studies.html
10. B. Zohuri, P. McDaniel, *Thermodynamics in Nuclear Power Plant* (Springer, Cham, 2015)
11. http://www.rwe.com/web/cms/mediablob/en/391748/data/364260/1/rwe-power-ag/innovations/Brochure-ADELE.pdf
12. *EPRI-DOE Handbook of Energy Storage for Transmission and Distribution Applications* (2003)
13. http://www.youtube.com/watch?v=grPzZ39ZyUI
14. ADELE stands for the German acronym for adiabatic compressed air energy storage for electricity supply
15. http://www.rwe.com/web/cms/de/365478/rwe/innovationen/stromerzeugung/energiespeicherung/druckluftspeicher/projekt-adele/
16. C. Forsberg, *Variable Electricity with Base-load Reactor Operations Fluoride-salt-cooled High-temperature Reactor (FHR) with Nuclear Air-Brayton Combined Cycle (NACC) and Firebrick Resistance-Heated Energy Storage* (MIT, 25 Jan 2014)
17. C. Forsberg, P. McDaniel, B. Zohuri, Variable Electricity and Steam from Salt, Helium, and Sodium Cooled Base-Load Reactors with Gas Turbines and Heat Storage, in *Proceedings of ICAPP 2015,* 03–06 May 2015 – Nice (France) Paper 15115
18. W.M. Kays, A.L. London, *Compact Heat Exchangers* (McGraw Hill, New York, 1964)

19. https://www.citylab.com/environment/2015/07/the-environmentalist-case-against-100-renewable-energy-plans/398906/
20. B. Zohuri, *Physics of Cryogenic an Ultra-Low Temperature Phenomena* (Elsevier Publishing Company, 2017)
21. http://www.ecomodernism.org/
22. http://pesd.fsi.stanford.edu/
23. Sources: Oak Ridge National Laboratory, *Transportation Energy Data Book: Edition 29, ORNL-6985,* July 2010, http://info.ornl.gov/sites/publications/files/Pub24318.pdf; Energy Information Administration, *Petroleum Supply Annual 2009*, July 2010, http://205.254.135.24/petroleum/supply/annual/volume1/archive/2009/pdf/volume1_all.pdf
24. Sources: Oak Ridge National Laboratory, *Transportation Energy Data Book: Edition 29, ORNL-6985,* July 2010, http://info.ornl.gov/sites/publications/files/Pub24318.pdf; Energy Information Administration, *Annual Energy Outlook*, Apr 2010, www.eia.doe.gov/oiaf/aeo/pdf/0383(2010).pdf
25. DOE Hydrogen Program Record 8013, available at: http://www.hydrogen.energy.gov/pdfs/8013_air_quality_population.pdf

Chapter 2
Cryogenic Technologies

Cryogenics is the science that addresses the production and effects of very low temperatures. The word originates from the Greek words "kryos" meaning "frost" and "genic" meaning "to produce." Under such a definition, it could be used to include all temperatures below the freezing point of water (0 °C). However, Prof. Kamerlingh Onnes of the University of Leiden in the Netherlands first used the word in 1894 to describe the art and science of producing much lower temperatures. He used the word in reference to the liquefaction of permanent gases such as oxygen, nitrogen, hydrogen, and helium. Oxygen had been liquefied at −183 °C a few years earlier in 1887, and a race was in progress to liquefy the remaining permanent gases at even lower temperatures. The techniques employed in producing such low temperatures were quite different from those used somewhat earlier in the production of artificial ice. In particular, efficient heat exchangers are required to reach very low temperatures. Over the years the term cryogenics has generally been used to refer to temperatures below approximately −150 °C (123.15 K, −238.00 °F).

2.1 Introduction

According to the laws of thermodynamics, there exists a limit to the lowest temperature that can be, achieved, which is known as absolute zero. Molecules are in their lowest, but finite, energy state at absolute zero. Such a temperature is impossible to reach because the input power required approaches infinity. However, temperatures within a few billionths of a degree above absolute zero have been achieved. Absolute zero is the zero of the absolute or thermodynamic temperature scale. It is equal to −273.15 °C or −459.67 F. The metric or SI (International System) absolute scale is known as the Kelvin scale whose unit is the kelvin (not Kelvin) which has the same magnitude as the degree Celsius. The symbol for the Kelvin scale is K, as adopted by the 13th General Council on Weights and Measures (CGPM) in 1968, and not K. Thus, 0 °C equals 273.15 K. The English absolute scale, known

B. Zohuri, *Hybrid Energy Systems*,
https://doi.org/10.1007/978-3-319-70721-1_2

Table 2.1 Normal boiling and triple and critical points

Cryogen	(K)	(°C)	(°R)	(°F)	Triple point	Critical point
Methane	111.7	−161.5	201.1	−258.6	90.7	190.5
Oxygen	90.2	−183.0	162.4	−297.3	54.4	154.6
Nitrogen	77.4	−195.8	139.3	−320.4	63.1	126.2
Hydrogen	20.3	−252.9	36.5	−423.2	13.8	33.2
Helium	4.2	−269.0	7.6	−452.1	2.2	5.2
Absolute zero	0.0	−273.15	0.0	−459.67	–	–

as the Rankine scale, uses the symbol R and has an increment the same as that of the Fahrenheit scale. In terms of the Kelvin scale, the cryogenic region is often considered to be that below approximately 120 K (−153 °C). The common permanent gases referred to earlier change from gas to liquid at atmospheric pressure at the temperatures shown in Table 2.1, called the normal boiling point (NBP). In this table, we have included the triple point and critical point, which we will explain them in the next chapter. Such liquids are known as cryogenic liquids or cryogens. When liquid helium is cooled further to 2.17 K or below, it becomes a superfluid with very unusual properties associated with being in the quantum mechanical ground state. For example, it has zero viscosity and produces a film that can creep up and over the walls of an open container, such as a beaker, and drip off the bottom as long as the temperature of the container remains below 2.17 K.

The measurement of cryogenic temperatures requires methods that may not be so familiar to the public in general. Normal mercury or alcohol thermometers freeze at such low temperatures and become useless. One of the metal elements that have a well-defined behavior of electrical resistance versus temperature is platinum resistance thermometer. It is commonly used to measure accurately, including cryogenic temperatures down to about 20 K. Certain semiconducting materials, such as doped germanium, are also useful as electrical resistance thermometers for temperatures down to 1 K and below, as long as they are calibrated over the range they are to be used. Such secondary thermometers are calibrated against primary thermometers that utilize fundamental laws of physics in which a physical variable changes in a well-known theoretical way with temperature.

The production of cryogenic temperatures usually utilizes the compression and expansion of gases. In typical air liquefaction process, the air is compressed, causing it to heat, and allowed to cool back to room temperature while still pressurized. The compressed air is further cooled in a heat exchanger before it is allowed to expand back to atmospheric pressure. The expansion causes the air to cool and a portion of it to liquefy. The remaining cooled gaseous portion is returned through the other side of the heat exchanger where it precools the incoming high-pressure air before returning to the compressor. The liquid portion is usually distilled to produce liquid oxygen, liquid nitrogen, and liquid argon. Other gases, such as helium, are used in a similar process to produce even lower temperatures, but several stages of expansion are necessary.

Cryogenics has many applications. Cryogenic liquids, such as oxygen, nitrogen, and argon, are often used in industrial and medical applications. The electrical resistance of most metals decreases as temperature decreases. Certain metals lose all electrical resistance below some transition temperature and become superconductors. An electromagnet wound with a wire of such a metal can produce extremely high magnetic fields with no generation of heat and no consumption of electric power once the field is established and the metal remains cold. These metals, typically niobium alloys cooled to 4.2 K, are used for the magnets of magnetic resonance imaging (MRI) systems in most hospitals. Superconductivity in some metals was first discovered in 1911 by Onnes, but since 1986, another class of materials, known as high-temperature superconductors, has been found to be superconducting at much higher temperatures, currently up to about 145 K. They are a type of ceramic, and because of their brittle nature, they are more difficult to fabricate into wires for magnets.

Other applications of cryogenics include fast freezing of some foods and the preservation of some biological materials such as livestock semen as well as human blood, tissue, and embryos. The practice of freezing an entire human body after death in the hope of later restoring life is known as cryonics, but it is not an accepted scientific application of cryogenics. The freezing of portions of the body to destroy unwanted or malfunctioning tissue is known as cryosurgery. It is used to treat cancers and abnormalities of the skin, cervix, uterus, prostate gland, and liver.

2.2 Low Temperature in Science and Technology

Cryogenics as it was described in the previous section is defined as *that branch of physics, which deals with the production of very low temperatures and their effect on matter* [1], a formulation which addresses both aspects of attaining low temperatures which do not naturally occur on Earth and of using them for the study of nature or the human industry. In a more operational way [2], it is also defined as *the science and technology of temperatures below* 120 K. The reason for this latter definition can be understood by examining characteristic temperatures of cryogenic fluids as it is shown in Table 2.1.

The limit temperature of 120 K comprehensively includes the normal boiling points of the main atmospheric gases, as well as of methane, which constitutes the principal component of natural gas. Today, liquefied natural gas (LNG) represents one of the largest—and fast-growing—industrial domains of application of cryogenics (see Fig. 2.1), together with the liquefaction and separation of air gases (see Fig. 2.2). The densification by condensation and separation by distillation of gases were historically—and remain today—the main driving force for the cryogenic industry. Exemplified not only by liquid oxygen and by nitrogen used in chemical as well as metallurgical processes but also by the cryogenic liquid propellants of rocket engines (see Fig. 2.3) where the proposed use of hydrogen as a "clean" energy vector in transportation (see Fig. 2.4).

As we have stated, the cryogenic technology has the need for smaller cryocoolers because of the advances in the miniaturization of electrical and optical devices and the need for cooling and conducting efficiency. Cryogenic technology deals with materials at low temperatures and the physics of their behavior at these temperatures.

Fig. 2.1 130,000 m^3 LNG carrier with integrated Invar tank

Fig. 2.2 Cryogenic air separation plant with heat exchanger and distillation column towers

In this book, we try to demonstrate the ongoing new applications are being discovered for cryocooled electrical and optical sensors and devices, with particular emphasis on high-end commercial applications in medical and scientific fields as well as in the aerospace and military industries.

Refrigerators, cryocoolers, and micro-coolers are needed by various commercial, industrial, space, and military systems. Cryogenic cooling plays an important role in unmanned aerial vehicle systems, infrared search and track sensors, missile warning receivers, satellite tracking systems, and a host of other commercial and military systems.

(a) Ariane 5
(25 t Liquid Hydrogen, 130 t Liquid Oxygen)

(b) Space Shuttle
(100 t Liquid Hydrogen, 600 t Liquid Oxygen)

Fig. 2.3 Rockets using cryogenic liquid propellants

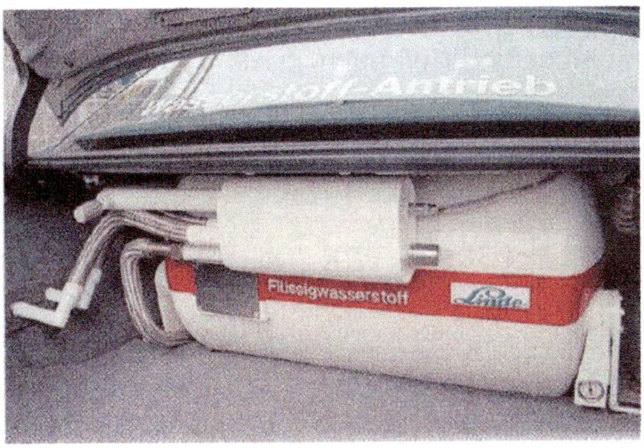

Fig. 2.4 Automotive liquid hydrogen fuel tank

Now with new generation of nuclear power plants that are known as GEN-IV, a lot of attention is focused toward making them more efficient and cost-effective [3] as well as using cryogenic techniques to implement energy storage in nuclear plants [4]. Energy storage in nuclear power plants resides on a novel method of integration of nuclear power generation with cryogenic energy storage (CES) to achieve an effective time shift of the electrical power output. CES stores excess electricity in the form of cryogen (liquid air/nitrogen) through an air liquefaction process at off-peak hours and recover the stored power by expanding the cryogen at peak hours [5].

The quest for low temperatures however finds its origin in early thermodynamics, with Amontons's gas pressure thermometer (1703) opening the way for the concept of absolute zero inferred a century later by Charles and Gay-Lussac and eventually formulated by Kelvin. However, with the advent of Boltzmann's statistical

Fig. 2.5 Ludwig Boltzmann's grave in the Zentralfriedhof Vienna, bearing the entropy formula

thermodynamics in the late nineteenth century, temperature—a phenomenological quantity—could be explained in terms of microscopic structure and dynamics. Consider a thermodynamic system in a macrostate, which can be obtained by a multiplicity W of microstates. The entropy S of the system was postulated by Boltzmann as

$$S = k_B \log W \tag{2.1}$$

with $k_B \simeq 1.38 \times 10^{-23}$J/K. This formula, which founded statistical thermodynamics, is displayed on Boltzmann's grave in Vienna (see Fig. 2.5).

Adding reversibly heat dQ to the system produces a change of its entropy dS, with a proportionality factor T which is precisely temperature

$$T = \frac{dQ}{dS} \tag{2.2}$$

Thus, a low-temperature system can be defined as one to which a minute addition of heat produces a large change in entropy, i.e., a large change in its range of possible microscopic configurations. Boltzmann also found that the average thermal energy of a particle in a system in equilibrium at temperature T is

$$E \sim k_B T \tag{2.3}$$

Consequently, a temperature of 1 K is equivalent to a thermal energy of 10^{-4} eV or 10^{-23} J per particle.

A temperature is therefore low for a given physical process when $k_B T$ is small compared to the characteristic energy of the process that is considered.

Table 2.2 Characteristic temperature of low-energy phenomena

Phenomenon	Temperature (K)
Debye temperature of metals	Few 100
High-temperature superconductors	~100
Low-temperature superconductors	~10
Intrinsic transport properties of metals	<10
Cryopumping	Few
Cosmic microwave background	2.7
Superfluid helium-4	2.2
Bolometers for cosmic radiation	<1
Low-density atomic Bose-Einstein condensates	$\sim 10^{-6}$

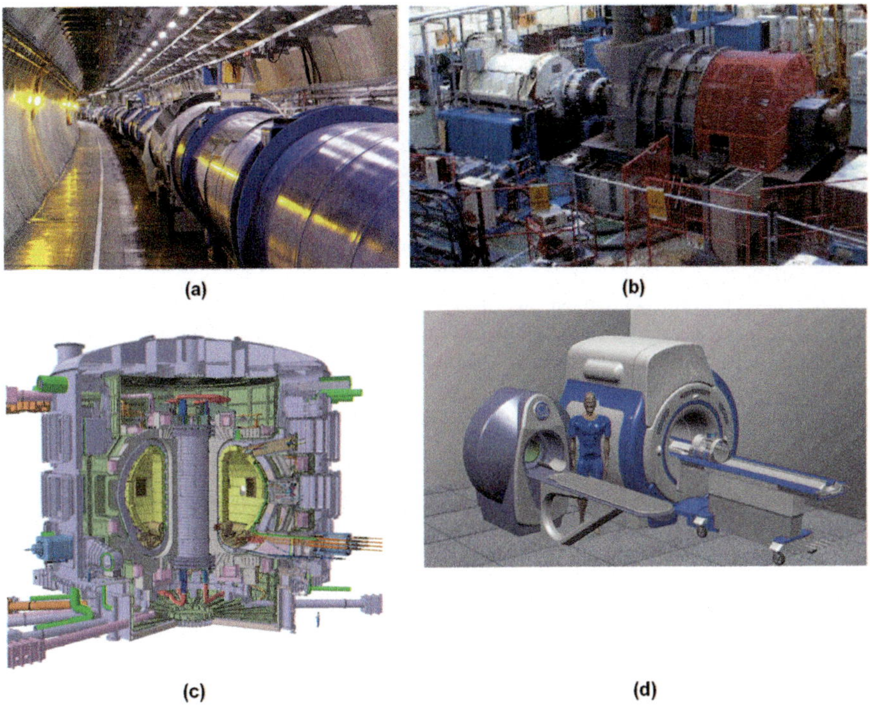

Fig. 2.6 Helium-cooled superconducting devices. (**a**) Large Hadron Collider at CERN. (**b**) 5 MW HTS ship propulsion motor (AMS). (**c**) ITER experimental fusion reactor. (**d**) Whole-body MRI system (Bruker)

Cryogenic temperatures thus reveal phenomena with low characteristic energy (Table 2.2) and enable their application when significantly lower than the characteristic energy of the phenomenon of interest. From Tables 2.1 and 2.2, it is clear that "low-temperature" superconductivity requires helium cryogenics: several examples of helium-cooled superconducting devices are shown in Fig. 2.6. Considering vapor pressures of gases at low temperature (see Fig. 2.7), it is also clear that helium must be the working cryogen for achieving "clean" vacuum with cryopumps.

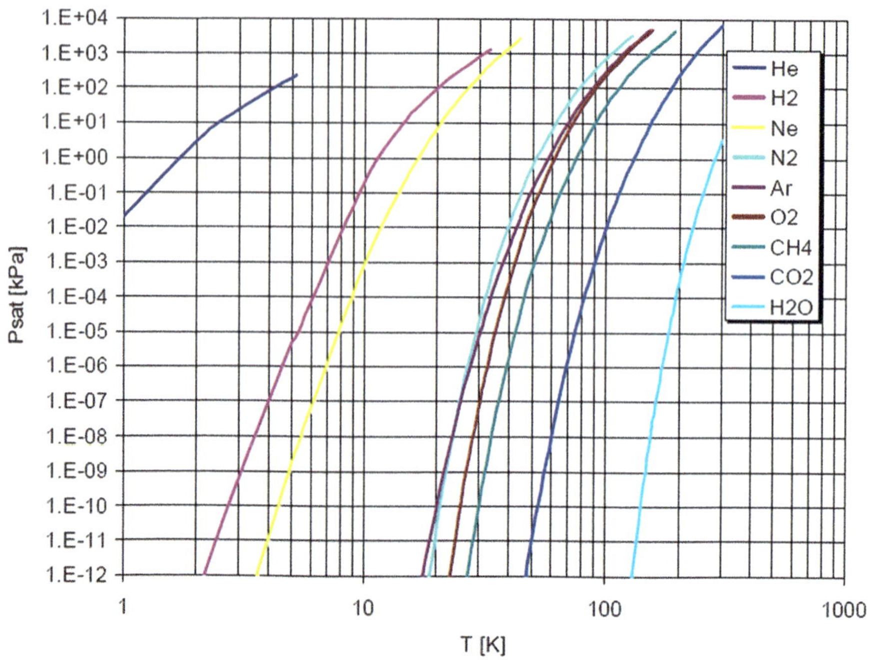

Fig. 2.7 Vapor pressure of common gases at cryogenic temperature

2.3 Defining Cryogenic Fluids or Liquids

Cryogenic liquids, also known as cryogens, are gases at normal temperatures and pressures. However, at low temperatures, they are in their liquid state. These liquids are extremely cold and have boiling points less than −150 °C (−238 °F). Even the vapors and gases released from cryogenic liquids are very cold. They often condense the moisture in air, creating a highly visible fog. Different cryogens become liquids under different conditions of temperature and pressure, but all have two properties in common; extremely cold and small amounts of liquid can expand into very large volumes of gas. Everyone who works with cryogenic liquids must be aware of their hazards and know how to work safely with them. Figure 2.8 is a presentation of liquid nitrogen (LN).

The discovery of superconducting materials with critical temperatures significantly above the boiling point of liquid nitrogen has provided new interest in reliable, low-cost methods of producing high-temperature cryogenic refrigeration.

Fig. 2.8 Liquid nitrogen

The term "high-temperature cryogenic" describes temperatures ranging from above the boiling point of liquid nitrogen, −195.79 °C (77.36 K; −320.42 °F), up to −50 °C (223.15 K; −58.00 °F), the generally defined upper limit of study referred to as cryogenics [6]. Cryogenicists use the Kelvin or Rankine temperature scales present in nature.

2.3.1 Defining Cryogenic Fluids or Liquids

Each cryogenic liquid has its own specific properties, but most cryogenic liquids can be placed into one of the three groups:

- *Inert gases*: Inert gases largely to any extent do not react chemically. They do not burn or support combustion. Examples of this group are nitrogen, helium, neon, argon, and krypton.
- *Flammable gases*: Some cryogenic liquids produce a gas that can burn in air. The most common examples are hydrogen, methane, carbon monoxide, and liquefied natural gas.
- *Oxygen*: Many materials considered as noncombustible can burn in the presence of liquid oxygen. Organic materials can react explosively with liquid oxygen. The hazards and handling precautions of liquid oxygen must therefore be considered separately from other cryogenic liquids.

It is generally agreed that cryogenic fluids are those whose boiling points (bp) at atmospheric pressure are about 120 K or lower, although liquid ethylene with its boiling point of 170 K is often included. A list of the cryogenic fluids, together with some selected properties, is given in Table 2.3. Detailed properties are available commercially on computer disk.

Perhaps the most important and widely used fluids are liquefied natural gas or LNG (bp = boiling point about 120 K), *liquid oxygen* (bp 90.2 K), and liquid nitrogen (bp 77.3 K).

The availability of cryogenic fluids forms an essential part of the infrastructure of a modem industrialized and civilized society. One of the major reasons for using liquid cryogens is to allow transport and storage as liquid at atmospheric pressure, rather than as high-pressure gas in thick-walled vessels, although there is an energy penalty involved in *refrigeration*. However, the distillation of liquid air (air separation) enables the production of very high-purity oxygen and nitrogen. Plants producing up to several hundred tons per day and more of oxygen are commonplace, sometimes connected permanently to a chemical plant or steel works. Liquid nitrogen—formerly a by-product of the process—is now a product in its own right, being used principally as a convenient source of refrigeration, especially in the frozen food industry.

The other important by-product of air separation is liquid argon, which again can be produced at a very high purity. For welding, it is increasingly being stored as liquid at the factory rather than being delivered in high-pressure cylinders.

All cryogenic fluids except *helium* and *hydrogen* behave as "normal" fluids, their common distinguishing features in general being a low specific heat and enthalpy of vaporization. All gaseous cryogens are odorless, and all liquid cryogens are colorless apart from *oxygen*, which is pale blue, and fluorine, which is pale yellow. They are all diamagnetic except oxygen, which is quite strongly paramagnetic.

With the exception of oxygen, all the gases are asphyxiants, and even oxygen will not support human life in concentrations greater than about 60%. Fluorine and oxygen are powerful oxidizers even in liquid form. Some cryogens are flammable; hydrogen is especially delicate to handle.

Hydrogen is an unusual fluid in that the molecule exists in two forms known as ortho and para, with somewhat different properties. The ratio of ortho to para is determined by conventional thermodynamics and is dependent on temperature. There are also different forms of isotopes (deuterium and tritium), and these two isotopes are used in driving fusion energy production via either magnetic confinement fusion (MCF) [7] or inertial confinement fusion (ICF) [8].

An explanation of the behavior of the hydrogen molecule requires knowledge of quantum mechanics and will not be discussed here. At low temperatures, equilibrium hydrogen (e-H^2) is entirely para. At room temperature, the ortho-para ratio is 3. The equilibrium state at room temperature is often known as normal hydrogen or n-hydrogen. The transition from the ortho to the para state involves a heat of conversion—which can be greater than the enthalpy of vaporization—so that the vaporization rates of hydrogen are often much larger than expected. It is for this reason that a catalyst is often included in a hydrogen liquefier to ensure that only para hydrogen is present in the liquid [9].

Table 2.3 Some properties of cryogens at their normal boiling points

	He^4	n-H_2	d_2	Ne	N_2	CO	F_2	Ar	O_2	CH_4	Kr	Xe	C_2H_4
Normal boiling point (K)	4.22	20.4	23.7	21.1	77.3	81.7	85.0	87.3	90.2	111.6	120.0	165.0	169.4
Liquid density (kg/m^3)	125	71.0	163	1205	809	792	1502	1393	1141	423	2400	3040	568
Liquid density-vapour density	7.4	53	71	126	175	181	267	241	255	236	270	297	272
Enthalpy of vaporisation (kJ/kg)	20.42	446	301	86	199	216	175	161	213	512	108	96	482
Enthalpy of vaporisation (kJ/kg-mole)	80.6	899	1211	2333	5565	6040	6659	6441	6798	8206	9042	12,604	13,534
Volume of liquid vaporised by energy input of 1 W-hr. (cm^3)	1410	114	74	35	22	21	14	16	15	17	14	13	13
Dynamic viacosity of liquid (μNsec/m^2)	3.3	13.3	28.3	124	152	–	240	260	195	119	404	506	170
Surface tension (mN/m)	0.10	1.9	~3	4.8	8.9	9.6	14.8	12.5	13.2	13.2	5.5	18.3	16.5
Thermal conductivity of liquid (mW m^{-1} K^{-1})	18.7	100	~100	113	135	–	–	128	152	187	94	74	192
Volume of gas at 15 °C released from 1 volume of liquid	739	830	830	1412	681	806	905	824	842	613	689	520	475

Source: Hands [9]

[a]Pressure of 1.01325 bar

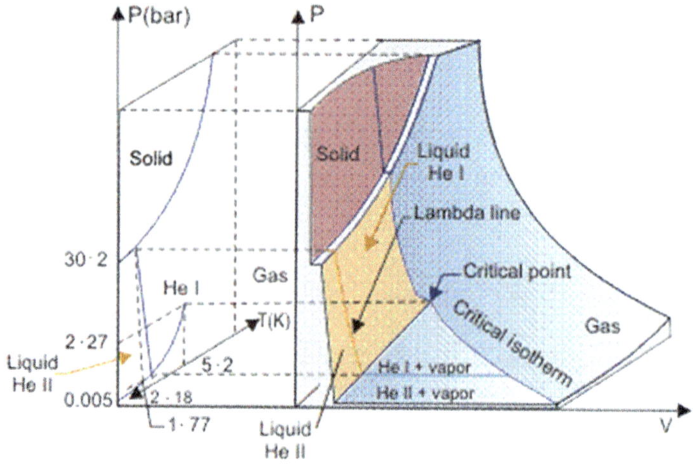

Fig. 2.9 From cryogenic engineering by hand [9]

Helium is the one cryogenic fluid, which can be claimed to be unique. Because of its low molecular weight and chemical inertness, quantum mechanical effects are important. There are two isotopic forms: the natural form He-4, which has a nucleus consisting of two protons and two neutrons, and the comparatively rare manufactured form He-3, with only one neutron. The two isotopes have markedly different properties due to their different nuclear spins. He-3 is not considered here.

Below 2.2 K, He-4 becomes "superfluid" and is often known as He-2, the "normal" liquid being known as "He-1." The locus of the He-1/He-2 transition is known as the *Lambda line or λ line* from the shape of the curve of specific heat as a function of temperature. The phase diagram of He-4 is shown in Fig. 2.9, in which features of particular interest are the absence of a triple point and the fact that the liquid can only be solidified under pressure (greater than about 26 bars).

The temperature of the normal-superfluid transition depends somewhat on pressure. One end of this boundary forms with solid He-1 and He-2 as the "upper lambda point" (at 1.77 K and 30.2 bars). The other end of the line (at 2.18 K, 0.005 bar) where vapor, He-1, and He-2 coexist is known as the "lower lambda point."

He-1 behaves as a conventional liquid (except when near the λ line) but requires much more care in handling than other cryogenic fluids, principally because of its extremely low latent heat of vaporization. He-2 is quite different, having a variety of properties quite different from those of any other liquid. It will, for instance, climb up over the edge of a container and drip off the bottom; it has a small or zero viscosity and a very large thermal conductivity. Flow velocity through fine capillaries is independent of the pressure head and is greater in tubes of smaller diameter. Flow may be induced by a temperature gradient in the absence of any pressure gradient. A consequence of the very high thermal conductivity is that below the λ point, boiling ceases and the liquid becomes "quiescent," although the rate of heat transfer remains very high. Vinen has published a brief but useful review of the properties of superfluid helium.

Table 2.4 Properties of helium and nitrogen compared to water

Property	Helium	Nitrogen	Water
Normal boiling point (K)	4.2	77	373
Critical temperature (K)	5.2	126	647
Critical pressure (bar)	2.3	34	221
Liquid density (kg/m^3)	125	808	960
Liquid density ratio[a]	7.4	175	1600
Heat of vaporization[a] (kJ/kg)	20.4	199	2260
Liquid viscosity[a] (μPl)	3.3	152	278

[a]At normal boiling point

2.3.2 *Thermophysical Properties*

The simplest way of cooling equipment with a cryogenic fluid is to make use of its latent heat of vaporization, e.g., by immersion in a bath of boiling liquid. Consequently, the useful temperature range of cryogenic fluids is that in which there exists latent heat of vaporization, i.e., between the triple point and the critical point, with a particular interest in the normal boiling point, i.e., the saturation temperature at atmospheric pressure. This data is given in Table 2.1. In this introduction to cryogenics, we will concentrate on two cryogens: helium, which is the only liquid at very low temperature, and nitrogen for its wide availability and ease of use for pre-cooling equipment and for thermal shielding.

To develop a feeling about properties of these cryogenic fluids, it is instructive to compare them with those of water (see Table 2.4). In both cases, but particularly with helium, applications operate much closer to the critical point, i.e., in a domain where the difference between the liquid and vapor phases is much less marked: the ratio of liquid to vapor densities and the latent heat associated with the change of phase are much smaller. Due to the low values of its critical pressure and temperature, helium can be used as a cryogenic coolant beyond the critical point, in the supercritical state. It is also interesting to note that, while liquid nitrogen resembles water as concerns density and viscosity, liquid helium is much lighter and less viscous.

This latter property makes it a medium of choice for permeating small channels inside superconducting magnet windings and thus stabilizing the superconductor.

2.3.3 *Liquid Boil-off*

The factor of ten in latent heat of vaporization between helium and nitrogen, combined with the lower density of the former, induces a large difference in vaporization rates under the same applied heat load (Table 2.5). This illustrates the need for implementing much better insulation techniques in liquid helium vessels to achieve comparable holding times.

Table 2.5 Vaporization of liquid helium and liquid nitrogen at normal boiling point under 1 W applied heat load

Cryogen	mg/s	l/h Liquid	l/min gas NTP
Helium	48	1.38	16.4
Nitrogen	5	0.02	0.24

Boil-off measurements constitute a practical method for measuring the heat load of a cryostat holding a saturated cryogen bath. In steady conditions, i.e., provided the liquid level in the bath is maintained constant, the boil-off $\dot{m}_{\text{vap}}$ precisely equals the vapor flow $\dot{m}_{\text{out}}$ escaping the cryostat, which can be warmed up to room temperature and measured in a conventional gas flow meter. At decreasing liquid level though, part of the vapor will take the volume in the cryostat previously occupied by the liquid, which has vaporized, and the escaping flow will be lower than the boil-off. More precisely, if the boil-off vapor is taken at saturation in equilibrium with the liquid

$$\dot{m}_{\text{out}} = m_{\text{vap}}\left(1-\frac{\rho_v}{\rho_\ell}\right) < \dot{m}_{\text{vap}} \tag{2.4}$$

The escaping gas flow measured must therefore be corrected upward to obtain the true boil-off. From values of saturated liquid to vapor density ratios in Table 2.4, this correction factor is only 1.006 for nitrogen and can therefore be neglected. For helium though, it amounts to 1.16 and must clearly be taken into account.

2.3.4 Cryogen Usage for Equipment Cooldown

For both fluids, the sensible heat of the vapor over the temperature range from liquid saturation to ambient is comparable to or larger than the latent heat of vaporization. This provides a valuable cooling potential at intermediate temperature, which can be used for thermal shielding or for precooling of equipment from room temperature. The heat balance equation for cooling a mass of, say, iron m_{Fe} of specific heat $C_{Fe}(T)$ at temperature T by vaporizing a mass dm of cryogenic liquid at saturation temperature T_v, latent heat of vaporization L_v, and vapor specific heat C taken as constant, assumes perfect heat exchange with the liquid and the vapor.

$$m_{Fe}C_{Fe}(T)dT = \left[L_v + C(T-T_v)\right]dm \tag{2.5}$$

Hence, the specific liquid cryogen requirement for cooldown from temperature T_0

$$\frac{m}{m_{Fe}} = \int_{T_0}^{T} \frac{C_{Fe}(T)dT}{L_v + C(T-T_v)} \tag{2.6}$$

Table 2.6 Volume of liquid cryogens required to cool down 1 kg of iron [1]

Using	Latent heat only	Latent heat and enthalpy of vapor
Liquid helium from 290 to 4.2 K	29.5	0.75
Liquid helium from 77 to 4.2 K	1.46	0.12
Liquid nitrogen from 290 to 77 K	0.45	0.29

The term $C(T - T_v)$ adding to L_v in the denominator brings a strong attenuation to the specific liquid requirement, provided there is good heat exchange between the solid and the escaping vapor. Calculated values of specific liquid cryogen requirements for iron are given in Table 2.6, clearly demonstrating the interest of recovering the sensible heat of helium vapor, as well as that of precooling equipment with liquid nitrogen.

2.3.5 Phase Domains

Typical operating domains with cryogenic helium are, as shown in Fig. 2.10, superimposed on the peculiar phase diagram of the substance: the solid phase only exists under pressure, and the normal liquid, He-1, transitions to another liquid phase, He-2, below 2.2 K instead of solidifying. There is no latent heat associated with this phase transition, but a peak in the specific heat, the shape of which gave the name "λ-line" to the phase boundary. He-2 exhibits superfluidity, a macroscopic quantum behavior entailing very high thermal conductivity and very low viscosity. While operating in saturated He I provides fixed (saturation) temperature and high boiling heat transfer at moderate heat flux, it may develop instabilities in two-phase flow and is prone to boiling crisis above the peak nucleate boiling flux (about 1 W/cm2). The use of monophase supercritical helium in forced-flow systems avoids the problems of two-phase flow. However, the strongly varying properties of the fluid near the critical point may create other issues, such as density wave oscillations. More fundamentally, supercritical helium exhibits no latent heat, so that applied heat loads result in temperature increases, which must be contained by high flow rate or periodic recooling in extended systems. At lower temperature, He-2 demonstrates excellent transport properties, which make it a coolant of choice for advanced superconducting devices [10]. Besides the thermodynamic penalty of lower temperature, the use of He-2 imposes that at least part of the cryogenic circuits operate at subatmospheric pressure, thus requiring efficient compression of low-pressure vapor and creating risks of dielectric breakdown and contamination by air in-leaks. See Fig. 2.10.

Thermophysical properties of cryogenic fluids are available from tables, graphs, and software running on personal computers, a selection of which is listed in the bibliography.

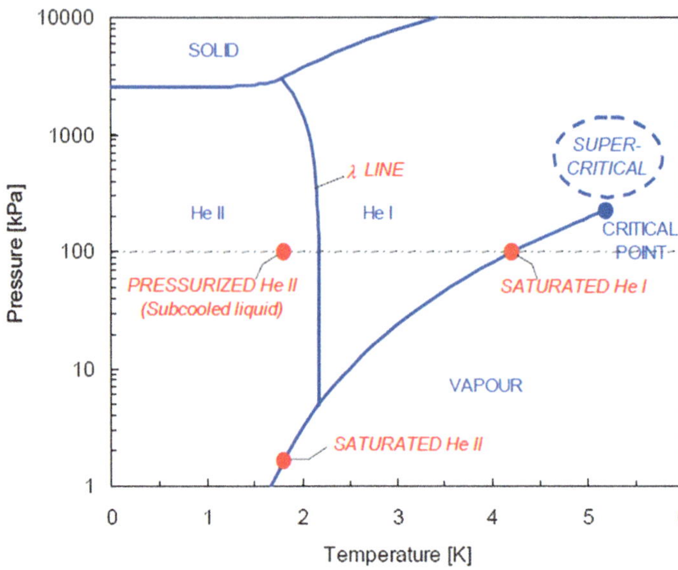

Fig. 2.10 Phase diagram of helium, showing typical operating domains

2.3.6 *Personal Protective Equipment to Be Worn*

As part of safety requirement in order to handle any cryogenic liquids, we need to consider the following personal protective equipments (PPEs) and wear them during handling of such liquids, in order to protect our skin and eyes:

- Be sure to work in a well-ventilated area to prevent oxygen-deficient atmospheres under 19.5% oxygen.
- Wear safety shoes when handling containers along with long sleeve shirts and trousers without cuffs.
- *Always* wear a full-face shield and splash resistant safety goggles. Contact lenses should not be worn.
- Wear a lab coat and an apron when dispensing liquid nitrogen.
- Wear insulated or leather gloves when handling liquid nitrogen or large, cold objects.

2.3.7 *Handling Cryogenic Liquids*

Handling the cryogenic liquids requires the following precautions as part of standard operating procedures, and they are listed here as:

- Never allow any unprotected part of the body to touch noninsulated pipes or vessels, which contain cryogenic fluids. Tissue damage that results is similar to frostbite or thermal burns.

- The extremely cold metal will cause flesh to stick fast and tear when one attempts to withdraw from it.
- Use a suitable hand truck for container movement.
- Do not drop, tip, or roll containers on their sides. Do not remove or interchange connections. If user experiences any difficulty operating container valve or with container connections, discontinue use and contact supplier. Use the proper connection. *Do not use adapters.*
- Many substances become brittle and may shatter when cold, sending pieces of the material flying. Avoid common glass and large, solid plastics.

2.3.8 Storing Cryogenic Liquids

In order to store the cryogenic liquids, we should consider the following steps as pointed out here:

- Store and use with adequate ventilation.
- Do not store in a confined space.
- Cryogenic containers are equipped with pressure relief devices to control internal pressure. Under normal conditions, these containers will periodically vent product. Do not plug, remove, or tamper with pressure relief device for this could cause an explosion.
- Containers shall be handled and stored in an upright position.
- Small quantities of liquid nitrogen can be stored in Dewar bottles. Dewar bottles are hollow-walled glass-lined containers, which provide excellent insulation.

2.3.9 Hazards of Cryogenic Liquids

Hazards of cryogenic liquids are also listed here as follows:

- *Extreme Cold Hazard*: Cryogenic liquids and their associated cold vapors and gases can produce effects on the skin similar to a thermal burn. Brief exposures that would not affect skin on the face or hands can damage delicate tissues such as the eyes. Prolonged exposure of the skin or contact with cold surfaces can cause frostbite. The skin appears waxy yellow. There is no initial pain, but there is intense pain when frozen tissue thaws. Unprotected skin can stick to metal that is cooled by cryogenic liquids. The skin can then tear when pulled away. Even nonmetallic materials are dangerous to touch at low temperatures. Prolonged breathing of extremely cold air may damage the lungs.
- *Asphyxiation Hazard*: When cryogenic liquids form a gas, the gas is very cold and usually heavier than air. This cold, heavy gas does not disperse very well and can accumulate near the floor. Even if the gas is nontoxic, it displaces air. When there is not enough air or oxygen, asphyxiation and death can occur. Oxygen deficiency

is a serious hazard in enclosed or confined spaces. Small amounts of liquid can evaporate into very large volumes of gas.

- *Toxic Hazards*: Each gas can cause specific health effects. Refer to the MSDS for information about the toxic hazards of a particular cryogen.

2.3.10 General Hazards of Cryogenic Liquids

The following points are important to bear in mind, when it comes to general hazards of cryogenic liquids, and they are listed as:

- *Fire Hazard*: Flammable gases such as hydrogen, methane, carbon monoxide, and liquefied natural gas can burn or explode. Hydrogen is particularly hazardous. It forms flammable mixtures with air over a wide range of concentration. It is also very easily ignited.
- *Oxygen-Enriched Air*: When transferring liquid nitrogen through uninsulated metal pipes, the air surrounding a cryogen containment system can condense. Nitrogen, which has a lower boiling point than oxygen, will evaporate first. This evaporation can leave an oxygen-enriched condensate on the surface that can increase the flammability or combustibility of materials near the system, creating potentially explosive conditions. Equipment containing cryogenic fluids must be kept clear of combustible materials in order to minimize the fire hazard potential.
- *Liquid Oxygen Hazard*: Liquid oxygen contains 4000 times more oxygen by volume than normal air. Materials that are usually considered noncombustible (carbon and stainless steels, cast iron, aluminum, zinc, Teflon (PTFE), etc.) may burn in the presence of liquid oxygen. Many organic materials can react explosively, especially if a flammable mixture is produced. Clothing splashed or soaked with liquid oxygen can remain highly flammable for hours.
- *Embrittlement*: Rubber, plastic, and carbon steel are some examples of materials that can be can become brittle and break with very little stress applied to them. Try to avoid using these materials when working with cryogenic. If these materials are used, perform an inspection before use.

2.4 Heat Transfer and Thermal Design

With the exception of superfluid helium, the heat transfer processes at work in cryogenics are basically the same as for any engineering temperature range. The strong variation of thermal properties of materials and fluids at low temperature however has two consequences: the relative and absolute magnitudes of the processes may be very different from those at room temperature, and the equations become nonlinear, thus requiring numerical integration. Cryogenic thermal

Table 2.7 Thermal conductivity integrals of selected materials (W/m)

From vanishing low temperature up to	20 K	30 K	290 K
OFHC copper	11,000	60,600	152,000
DHP copper	395	5890	46,100
Aluminum 1100	2740	23,300	72,100
2024 aluminum alloy	160	2420	22,900
AISI 304 stainless steel	16.3	349	3060
G-10 glass-epoxy composite	2	18	153

design is the art of using these processes adequately, either for achieving thermal insulation (cryostats, transfer lines) or for improving thermal coupling between equipment and coolant (cooldown and warm-up, thermal stabilization, thermometry) [11].

2.4.1 *Solid Conduction*

Heat conduction in solids is represented by Fourier's law as it can be seen in Eq. 2.7 and expressing proportionality of heat flux with thermal gradient. In one dimension, this reads

$$Q = k(T) A \frac{dT}{dx} \tag{2.7}$$

This equation also defines the thermal conductivity $k(T)$ of the material, which varies with temperature. Conduction along a solid rod of length L, cross section A spanning a temperature range $[T_1, T_2]$, e.g., the support strut of a cryogenic vessel, is then given by the integral form as

$$Q = \frac{A}{L} \int_{T_1}^{T_2} k(T)\, dT \tag{2.8}$$

Thermal conductivity integrals $\int_{T_1}^{T_2} k(T)\, dT$ of standard materials are tabulated in the literature. A few examples are given in Table 2.7, showing the large differences between good and bad thermal conducting materials; the strong decrease of conductivity at low temperatures, particularly for pure metals; and the interest of thermal interception to reduce conductive heat in-leak in supports. As an example, the thermal conductivity integral of austenitic stainless steel from 80 K to vanishingly low temperature is nine times smaller than from 290 K, hence the benefit of providing a liquid nitrogen-cooled heat sink on the supports of a liquid helium vessel. The lower thermal conductivity values of nonmetallic composites, combined with their good mechanical properties, make them materials of choice for low heat in-leak structural supports (Fig. 2.11).

Fig. 2.11 Nonmetallic composite support post with heat intercepts for LHC superconducting magnets at CERN

2.4.2 Radiation

Blackbody radiation strongly and only depends on the temperature of the emitting body, with the maximum of the power spectrum given by Wien's law

$$\lambda_{max} T = 2898[\mu m \mathrm{K}] \tag{2.9}$$

and the total power radiated given by Stefan-Boltzmann's law as

$$Q = \sigma A T^4 \tag{2.10}$$

with Stefan-Boltzmann's constant $\sigma \simeq 5.67 \times 10^{-8} \mathrm{Wm^{-2}K^{-4}}$. The dependence of the radiative heat flux on the fourth power of temperature makes a strong plea for radiation shielding of low-temperature vessels with one or several shields cooled by liquid nitrogen or cold helium vapor. Conversely, it makes it very difficult to cool down equipment to low temperature by radiation only: in spite of the 2.7 K background temperature of outer space and notwithstanding the sun's radiation and the Earth's albedo, which can be avoided by proper attitude control, satellites or interplanetary probes can use passive radiators to release heat down to about 100 K and embarked active refrigerators are required to reach lower temperatures.

Technical radiating surfaces are usually described as "gray" bodies and characterized by an emissivity $\varepsilon < 1$

$$Q = \varepsilon \sigma A T^4 \tag{2.11}$$

The emissivity ε strictly depends on the material, surface finish, radiation wavelength, and angle of incidence. For materials of technical interest, measured average values are found in the literatures, a subset of which is given in Table 2.8. As a general rule, emissivity decreases at low temperature, for good electrical conductors and for polished surfaces. As Table 2.7 shows, a simple way to obtain this combination of properties is to wrap cold equipment with aluminum foil. Conversely, radiative thermal coupling requires emissivity as close as possible to that of a blackbody, which can be achieved in practice by special paint or adequate surface treatment, e.g., anodizing of aluminum.

Table 2.8 Emissivity of some technical materials at low temperature

	Radiation from 290 K surface at 77 K	Radiation from 77 K surface at 4.2 K
Stainless steel, as found	0.34	0.12
Stainless steel, mesh polished	01.2	0.07
Stainless, electropolished	0.10	0.07
Stainless steel + aluminum foil	0.05	0.01
Aluminum, black anodized	0.95	0.75
Aluminum, as found	0.12	0.07
Aluminum, mesh polished	0.10	0.06
Aluminum, electropolished	0.08	0.04
Copper, as found	0.12	0.06
Copper, mesh polished	0.06	0.02

The net heat flux between two "gray" surfaces at temperature T_1 and T_2 is similarly given by

$$Q = E\sigma A\left(T_1^4 - T_2^4\right) \tag{2.12}$$

with the emissivity factor E being a function of the emissivities ε_1 and ε_2 of the surfaces, of the geometrical configuration, and of the type of reflection (specular or diffuse) between the surfaces. Its precise determination can be quite tedious, apart from the few simple geometrical cases of flat plates, nested cylinders, and nested spheres.

If an uncooled shield with the same emissivity factor E is inserted between the two surfaces, it will "float" at temperature T_s given by the energy balance equation.

$$Q = E\sigma A\left(T_1^4 - T_s^4\right) = E\sigma A\left(T_s^4 - T_2^4\right) \tag{2.13}$$

Solving for T_s yields the value of $Q_s = Q/2$: the heat flux is halved in presence of the floating shield. More generally, if n floating shields of equal emissivity factor are inserted between the two surfaces, the radiative heat flux is divided by $n+1$.

2.4.3 Convection

The diversity and complexity of convection processes cannot be treated here. Fortunately, in the majority of cases, the correlations established for fluids at higher temperature are fully applicable to the cryogenic domain [12], and reference is made to the abundant technical literature on the subject.

In the case of forced convection, one should keep in mind that the high density and low viscosity of cryogenic fluids often result in flows with high Reynolds number Re and hence strong convection. The Nüsselt number *Ni* which characterizes the efficiency of convective heat transfer relative to conduction in the fluid is an increasing function of the Prandtl Pr and Reynolds numbers, respectively, representing the ratio of mass to heat transport and the ratio of inertial to viscous forces.

$$Ni = f(\mathrm{Pr}, \mathrm{Re}) \tag{2.14}$$

The case of natural convection at low temperature however deserves particular mention, as this mechanism, usually weak at room temperature except on very large scales, becomes dominant in cryogenic equipment. In this case, the Nüsselt number is an increasing function of the Prandtl and Grashof number *Gr*, with the latter representing the ratio of buoyancy to viscous forces.

$$Ni = f(\mathrm{Pr}, Gr) \tag{2.15}$$

For gases, while Pr is about constant and independent of temperature, *Gr* is proportional to the heated volume, temperature difference, and coefficient of volume thermal expansion which scales as $1/T$ in the ideal case. Consequently, there may exist in helium cryostats strong natural convection processes with Grashof numbers up to the 10^{12} range, i.e., higher than those encountered in the general circulation of the Earth's atmosphere. This has been used by hydrodynamics specialists to study turbulent convection in extreme conditions. The cryogenic engineer sees it as a powerful mechanism for cooling equipment and homogenizing its temperature.

2.4.4 Gas Conduction

Since J. Dewar's invention (1898) of the cryogenic vessel, which bears his name, evacuated envelopes provide the best insulation against heat transport in gaseous media. At low pressure, convection becomes negligible and only residual gas conduction is at work. This process operates in two distinct regimes, depending upon the value of the mean free path of gas molecules ℓ relative to the typical distance *d* between the cold and warm surfaces.

When $\ell \ll d$ corresponding to higher residual pressure, the probability of interaction of a given molecule with others before it travels distance *d* is a highly viscous regime, and heat diffuses as in any continuous medium, then we can write

$$Q = k(T) A \frac{dT}{dx} \tag{2.16}$$

Note that the thermal conductivity $k(T)$ of the gas is independent of pressure.

When $\ell \gg d$ at low residual pressure, the molecular regime prevails and the heat transfer between two surfaces at temperatures T_1 and T_2 is given by Kennard's law

Table 2.9 Typical values of heat flux to vanishingly low temperature between flat plates (W/m^2)

Blackbody radiation from 290 K	401
Blackbody radiation from 80 K	2.3
Gas conduction (100 mPa helium) from 290 K	19
Gas conduction (1 mPa helium) from 290 K	0.19
Gas conduction (100 mPa helium) from 80 K	6.8
Gas conduction (1 mPa helium) from 80 K	0.07
MLI (30 layers) from 290 K, pressure < 1 mPa	1–15
MLI (10 layers) from 80 K, pressure < 1 mPa	0.05
MLI (10 layers) from 80 K, pressure 100 mPa	1–2

$$Q = A\alpha(T)\Omega P(T_2 - T_1) \qquad (2.17)$$

where Ω is a parameter depending upon the gas species and α is the "accommodation coefficient" representing the thermalization of molecules on the surfaces; its value depends on T_1, T_2, the gas species, and the geometry of the facing surfaces. Note that the conductive heat flux in molecular regime is proportional to pressure P and independent of the spacing between the surfaces (and therefore not amenable to the concept of thermal conductivity). Typical values of heat flux by gas conduction at cryogenic temperature are given in Table 2.9.

2.4.5 Multilayer Insulation

Multilayer insulation (MLI) is based on multiple reflecting shields wrapped around the cryogenic piece of equipment to be insulated, with the aim of benefiting from the $n+1$ reduction factor in radiative heat in-leak. In practice, this is implemented in the form of aluminum or aluminized polymer films, with low packing density achieved by crinkling or by insertion of a net-type spacer between layers. The wrapping can be made by winding the layers and spacer in situ or by prefabricated blankets installed and held in place by insulating fasteners (Fig. 2.12). In all cases, MLI is a complex thermal system, involving the combination of radiation, solid contact conduction, and residual gas conduction between layers. As a result, increasing the number of layers, while beneficial for cutting radiation, usually results in increased packing with more contacts and

Fig. 2.12 Prefabricated MLI blankets are being installed around an accelerator superconducting magnet

trapped residual gas between layers, two effects which increase heat transfer. In view of the nonlinearity of these elementary processes, thermal optimization requires layer-to-layer modeling and efficient control of the critical parameters. In practice, performance is measured on test samples, and measured data is available from an abundant literature. Typical values for some practical MLI systems are given in Table 2.9.

Of particular interest is the case of operation in degraded vacuum, where the heat in-leak by molecular conduction is directly proportional to the residual pressure. The presence of a multilayer system, which segments the insulation space into many cells thermally in series significantly, contains the increase in heat in-leak to the low-temperature surface (Table 2.8). In this respect, the multilayer system is no longer used for its radiative properties but for the reduction of molecular gas conduction.

In the extreme case of complete loss of vacuum in a liquid helium vessel, MLI also efficiently limits the heat flux, which would otherwise be very high due to condensation of air on the cold wall, thus alleviating the requirements for emergency discharge systems.

2.4.6 Vapor Cooling of Necks and Supports

The enthalpy of cryogen vapor escaping from a liquid bath can be used to continuously intercept conduction heat along solid supports and necks connecting the cryogenic bath with the room temperature environment (Fig. 2.13).

Assuming steady state and perfect heat exchange between the escaping vapor and the solid, the energy balance equation reads

$$k(T)A\frac{dT}{dx} = Q_v + \dot{m}C(T - T_v) \tag{2.18}$$

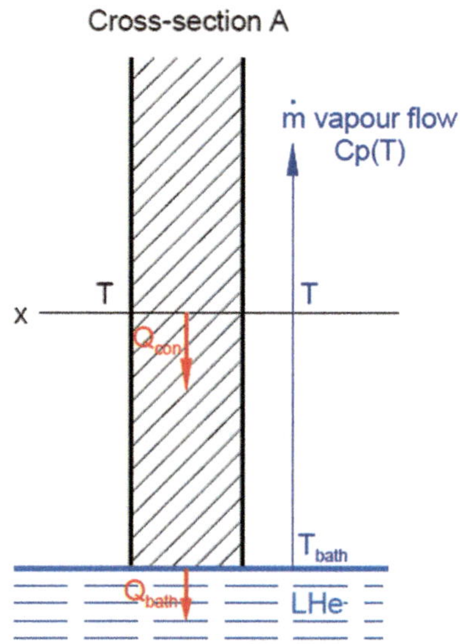

Fig. 2.13 Vapor cooling of necks and supports with perfect heat exchange

Table 2.10 Attenuation of heat conduction between 290 K and 4 K by self-sustained helium vapor cooling (W/cm)

Material	Purely conductive regime	Self-sustained vapor cooling
ETP copper	1620	128
OFHC copper	1520	110
Aluminum 1100	728	39.9
Nickel 99% pure	213	8.65
Constantan	51.6	1.94
AISI 300 stainless steel	30.6	0.92

where Q_v is the heat reaching the liquid bath and $\dot{m}$ is the vapor mass flow rate. In the particular case of self-sustained vapor cooling, i.e., when the vapor mass flow rate $\dot{m}$ precisely equals the boil-off from the liquid bath

$$Q_v = L_v \dot{m} \tag{2.19}$$

Combining Eqs. 2.18 and 2.19 and integrating yields the value of Q_v

$$Q_v = \frac{A}{L} \int_{T_v}^{T_0} \frac{k(T)}{1 + (T - T_v)\dfrac{C}{L_v}} dT \tag{2.20}$$

The denominator of the integrand clearly acts as an attenuation term for the conduction integral. Numerical results for helium and a few materials of technical interest appear in Table 2.10. If properly used, the cooling power of the vapor brings an attenuation of one to two orders of magnitude in the conductive heat in-leak.

Vapor cooling can also be used for continuous interception of other heat loads than solid conduction. In cryogenic storage and transport vessels with vapor-cooled shields, it lowers shield temperature and thus reduces radiative heat in-leak to the liquid bath. In vapor-cooled current leads, a large fraction of the resistive power dissipation by Joule heating is taken by the vapor flow, in order to minimize the residual heat reaching the liquid bath [13].

A worked-out example of how these diverse thermal insulation techniques are implemented in a real design is given in reference [14].

2.5 Refrigeration and Liquefaction

Refrigeration and liquefaction of gases are historically at the root of cryogenics, as they constitute the enabling technology, which gave access to the low-temperature domain. They have developed over the years along several lines to become a specialized subject, which would deserve a thorough presentation. In the following, we shall briefly describe the basic thermodynamics, the cooling processes at work, and the corresponding equipment in the case of helium refrigerators/liquefiers based on the Claude cycle. For a more complete review, see reference [15].

2.5.1 *Thermodynamics of Refrigeration*

A refrigerator is a machine raising heat Q_i from a low-temperature source T_i to a higher-temperature sink (usually room temperature) T_0, by absorbing mechanical work W_i; doing so, it rejects heat Q_0 (see Fig. 2.14). These quantities are related through the application of the first (Joule) and second (Clausius) principles of thermodynamics.

$$Q_0 = Q_i + W_i \tag{2.21}$$

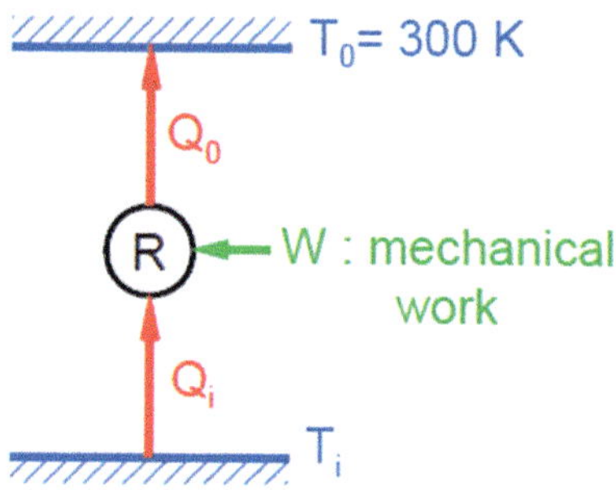

Fig. 2.14 Thermodynamic scheme of a refrigerator

$$\frac{Q_0}{T_0} \geq \frac{Q_i}{T_i} \tag{2.22}$$

In Eq. 2.22, the equality applies to the case of reversible process. From the above

$$W_i \geq T_0 \frac{Q_i}{T_i} - Q_i \tag{2.23}$$

This expression can be written in three different ways. Introducing the reversible entropy variation as $\Delta S_i = Q_i / T_i$, then we have

$$W_i \geq T_0 \Delta S_i - Q_i \tag{2.24a}$$

Another form isolates the group $[(T_0/T_i) - 1]$ as the proportionality factor between Q_i and W_i, i.e., the minimum specific refrigeration work as

$$W_i \geq Q_i \left(\frac{T_0}{T_i} - 1 \right) \tag{2.24b}$$

As Carnot has shown in 1824, this minimum work can only be achieved through a cycle constituted of two isothermal and two adiabatic transforms (Carnot cycle). All other thermodynamic cycles entail higher refrigeration work for the same refrigeration duty.

A third form of Eq. 2.23 is

$$W_i \geq \Delta E_i \tag{2.24c}$$

This introduces the variation of exergy $\Delta E_i = (T_0/T_i) - 1$, a thermodynamic function representing the maximum mechanical work content (Gouy's "energy utilizable") of a heat quantity Q_i at temperature T_i, given an environment at temperature T_0.

Equation 2.24b enables to calculate the minimum mechanical power needed to extract 1 W at 4.5 K (saturated liquid helium temperature at 1.3 bar pressure, i.e., slightly above atmospheric) and reject it at 300 K (room temperature), yielding a value of 65.7 W. This is the power that would be absorbed by a refrigerator operating on a Carnot cycle between 4.5 K and 300 K. In practice, the best practical cryogenic helium refrigerators have an efficiency of about 30% with respect to a Carnot refrigerator, hence a specific refrigeration work of about 220.

Cryogenic refrigerators are often required to provide cooling duties at several temperatures or in several temperature ranges, e.g., for thermal shields or continuous heat interception (see paragraph 3.6 above). Eq. 2.24b can then be applied to the cooling duty at every temperature and every elementary mechanical power W_i summed or integrated in the case of continuous cooling. This also allows comparison of different cooling duties in terms of required mechanical work.

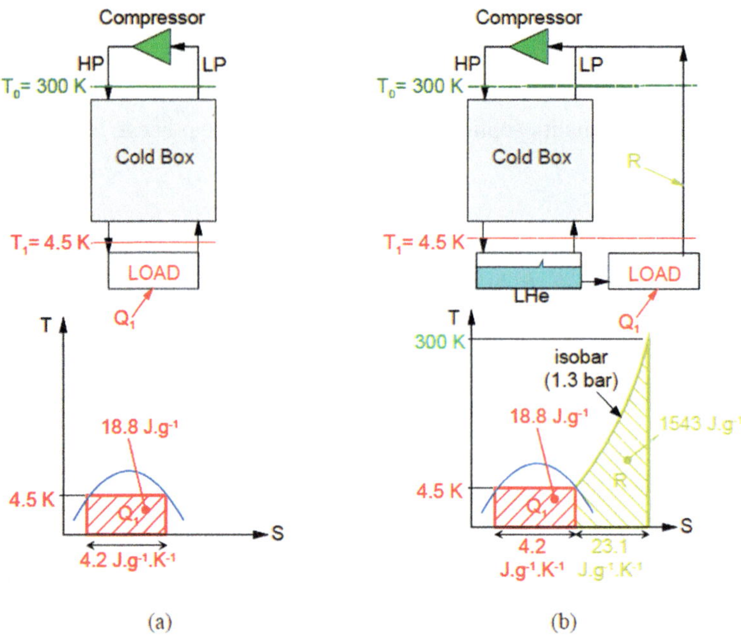

Fig. 2.15 Helium refrigerator (**a**) vs. liquefier (**b**)

2.5.2 Helium Refrigerators Versus Liquefiers

A 4.5 K helium refrigerator absorbs heat isothermally at this temperature, usually by recondensing cold helium vapor at saturation (the saturation pressure is 1.3 bar). By contrast, a liquefier also eventually condenses cold helium vapor at saturation, but starting from gaseous helium at 300 K which it must first precool to 4.5 K (Fig. 2.15). From Eq. 2.24a, the minimum mechanical power for helium liquefaction is

$$W_{\text{liq}} = W_{\text{condens}} + W_{\text{precool}} \tag{2.25}$$

$$W_{\text{liq}} = T_0 \Delta S_{\text{condens}} - Q_{\text{condens}} + T_0 \Delta S_{\text{precool}} - Q_{\text{precool}} \tag{2.26}$$

The heat quantities Q_{condens} and Q_{precool} exchanged at constant pressure are given by the enthalpy variations $\Delta H_{\text{condens}}$ and $\Delta H_{\text{precool}}$. With $T_0 = 300$ K and the entropy and enthalpy differences taken from thermodynamic tables, one finds $W_{\text{liq}} = 6628$ W per g/s of helium liquefied. Given the minimum specific mechanical work of 65.7 at 4.5 K, this yields an approximate equivalence of about 100 W at 4.5 K for 1 g/s liquefaction. More precisely, a liquefier producing 1 g/s liquid helium at 4.5 K will absorb the same power (and thus have similar size) as a refrigerator extracting about 100 W at 4.5 K, provided they both have the same efficiency with respect to the Carnot cycle. For machines with mixed refrigeration and liquefaction duties, this equivalence can be approximately verified by trading some liquefaction against refrigeration around the design point and vice versa.

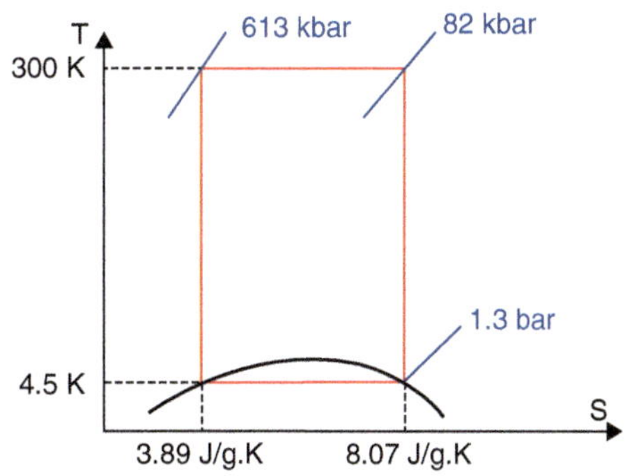

Fig. 2.16 A hypothetical Carnot cycle for helium liquefaction

2.5.3 *Real Cycles and Refrigeration Equipment*

Thus far, we have only addressed cryogenic refrigeration and liquefaction through thermodynamics, i.e., through the exchanges of mass, heat, and work at the boundaries of machines seen as "black boxes." We will now consider cycles, cooling methods, and equipment of real refrigerators.

In order to minimize the specific mechanical work requirement (and hence the size and power consumption), an efficient refrigerator should try and approximate the Carnot cycle, which is represented by a rectangle on the temperature-entropy diagram: the two isotherms are horizontal lines, while the two isentropic transforms are vertical lines. To liquefy helium, the base of the rectangle should intercept the liquid-vapor dome (Fig. 2.16).

However, superimposing this cycle on the temperature-entropy diagram of helium shows that one should operate at a high pressure of about 613 kbar, with a first isentropic compression from 1.3 bar to 82 kbar, followed by an isothermal compression. This is clearly impractical, and real helium cycles are elongated along isobar (or isochoric) lines, thus involving transforms, which require heat exchange between the high- and low-pressure streams. This heat exchange can be performed in recuperative or regenerative heat exchangers, respectively, for continuous or alternating flows. In the following, we focus on the continuous flow cycles using recuperative heat exchangers, which constitute the operating principles of large-capacity helium refrigerators and liquefiers.

Practical elementary cooling processes are shown on the temperature-entropy diagram, which is depicted in Fig. 2.17. The gas stream can first undergo quasi-isobar cooling in a counterflow heat exchanger (segment AB_1): modern refrigerators make use of brazed aluminum plate heat exchangers (Fig. 2.18). Refrigeration can be produced by adiabatic (quasi-isentropic) expansion with extraction of mechanical work (segment AB'_2): the expansion engine is a gas turbine, with the extracted power transmitted to a compressor wheel sharing a common shaft and later dissipated in a brake circuit (Fig. 2.19). A third process is isenthalpic Joule-Thomson expansion, i.e., without extraction of mechanical work, in a valve or restriction (segment AB_3).

Unfortunately, this latter process does not does work for ideal gases, the enthalpy of which is a sole function of temperature. For real gases, however, enthalpy depends both on temperature and pressure, so that isenthalpic expansion can produce warm-

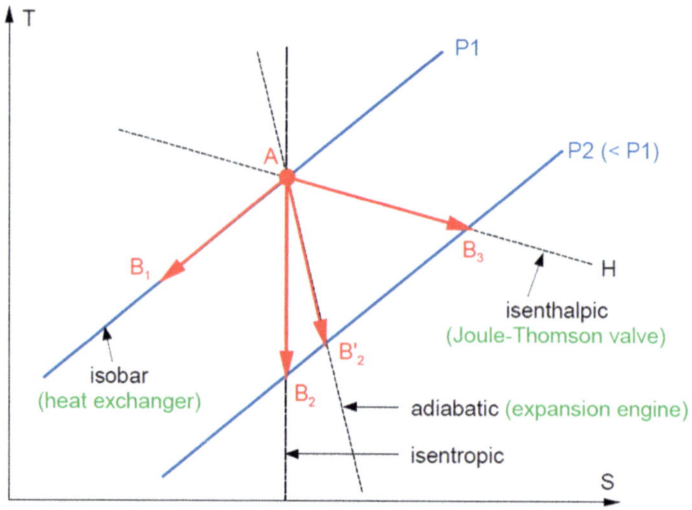

Fig. 2.17 Elementary cooling processes shown on temperature-entropy diagram

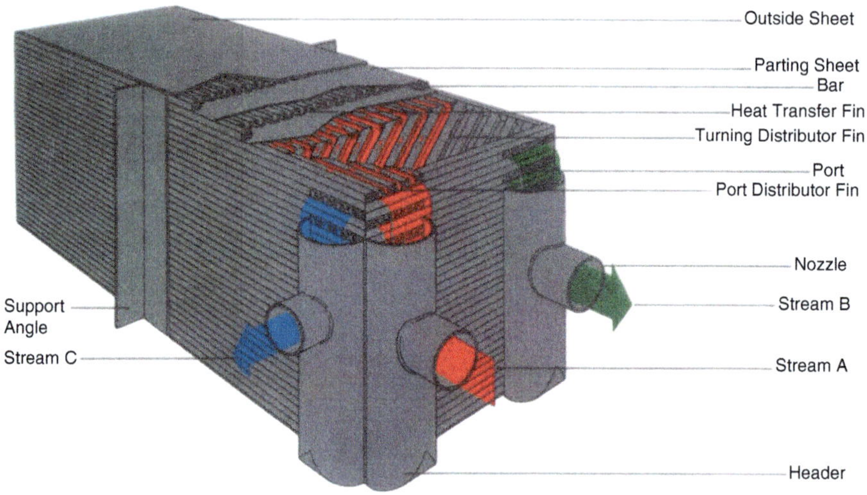

Fig. 2.18 Schematic view of a brazed aluminum plate heat exchanger

ing or cooling, depending upon the slope of the isenthalps on the diagram. In order to cool the gas stream, Joule-Thomson expansion must start below a limit called the inversion temperature. The values of inversion temperature for cryogenic fluids (Table 2.11) show that while air can be cooled from room temperature by Joule-Thomson expansion (the risk of freezing the pressure reducer on the air bottle is well known to scuba divers), helium must first be precooled down to below its inversion temperature of 43 K. The moderate downward slope of isenthalps on the temperature-entropy diagram indicates that in any case, Joule-Thomson expansion

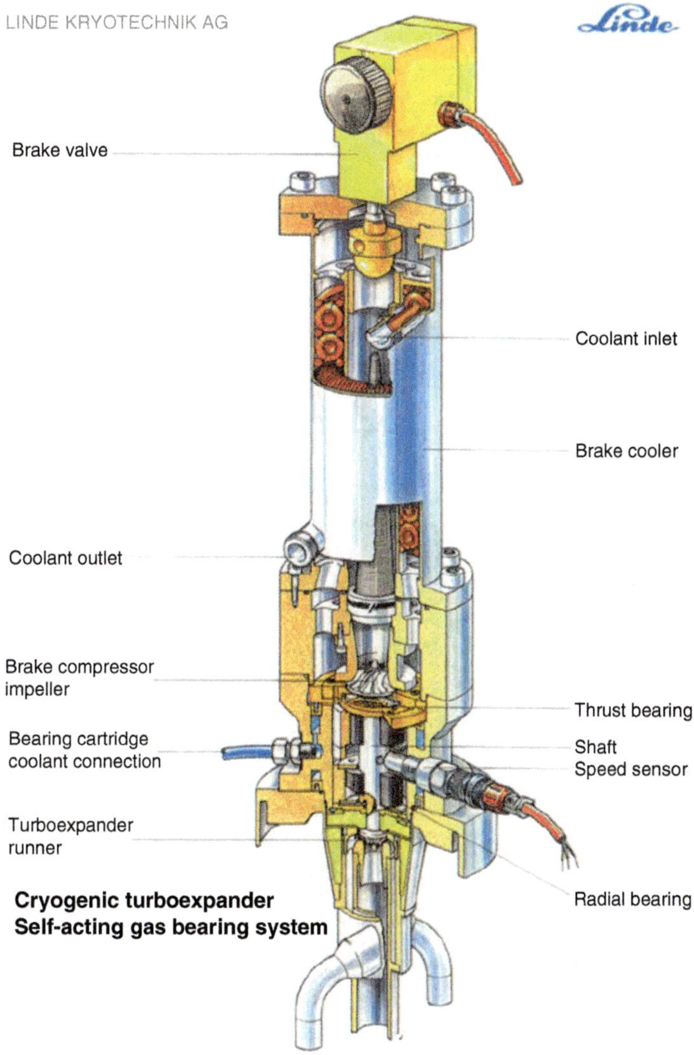

Fig. 2.19 Cryogenic turbo-expander

Table 2.11 Maximum values of Joule-Thomson inversion temperature

Cryogen	Maximum inversion temperature (K)
Helium	43
Hydrogen	202
Neo	260
Air	603
Nitrogen	623
Oxygen	761

generates substantial entropy. Its relative inefficiency with respect to adiabatic expansion is, however, accepted in view of the simplicity of its implementation, particularly when it results in partial condensation of the stream entailing two-phase flow conditions, which would be difficult to handle in an expansion turbine. In Chapter 14 of this book by Zohuri [16], we have defined the inversion temperature.

For better understanding of heat exchanger and compact heat exchanger, readers can refer to famous book by Kay and London 25 and their application of such heat exchanger in the two books by Zohuri [17–19].

These elementary cooling processes are combined in practical cycles; a common example for helium refrigeration is provided by the Claude cycle and its refinements. A schematic two-pressure, two-stage Claude cycle is shown in Fig. 2.20: gaseous helium, compressed to HP in a lubricated screw compressor, is recooled to room temperature in water coolers, dried and purified from oil aerosols down to the ppm level, before being sent to the HP side of the heat exchange line where it is refrigerated by heat exchange. This process takes place with the counterflow of cold gas returning on the low-pressure (LP) side. Part of the flow is tapped from the HP line and expanded in the turbines before escaping to the LP line. At the bottom of the heat exchange line, the remaining HP flow is expanded in a Joule-Thomson valve and partially liquefied.

Large-capacity helium refrigerators and liquefiers operate under this principle, however, with many refinements aiming at meeting specific cooling duties and improving efficiency and flexibility of operation, such as three- and sometimes four-pressure cycle process. In this case, liquid nitrogen precooling of the helium stream requires numerous heat exchangers, many turbines in series or parallel arrangements, Joule-Thomson expansion replaced by adiabatic expansion in a "wet" turbine, and cold compressors to lower the refrigeration temperature below 4.5 K. A view of such a large plant appears in Figs. 2.21, 2.22 and 2.23.

The capital cost of these complex machines is high, but scales less than linearly with refrigeration power, which favors large units. Operating costs are dominated by that of electrical energy, typically amounting to about 10% of the capital cost per year in case of quasi-continuous operation. For overall economy, it is therefore very important to seek high efficiency, which is also easier to achieve on large units. For a review of these aspects, see reference [20].

2.6 Industrial Applications

Liquefied gases, such as liquid nitrogen and liquid helium, are used in many cryogenic applications. Liquid nitrogen is the most commonly used element in cryogenics and is legally purchasable around the world. Liquid helium is also commonly used and allows the lowest attainable temperatures to be reached.

These liquids may be stored in Dewar flasks, which are double-walled containers with a high vacuum between the walls to reduce heat transfer into the liquid.

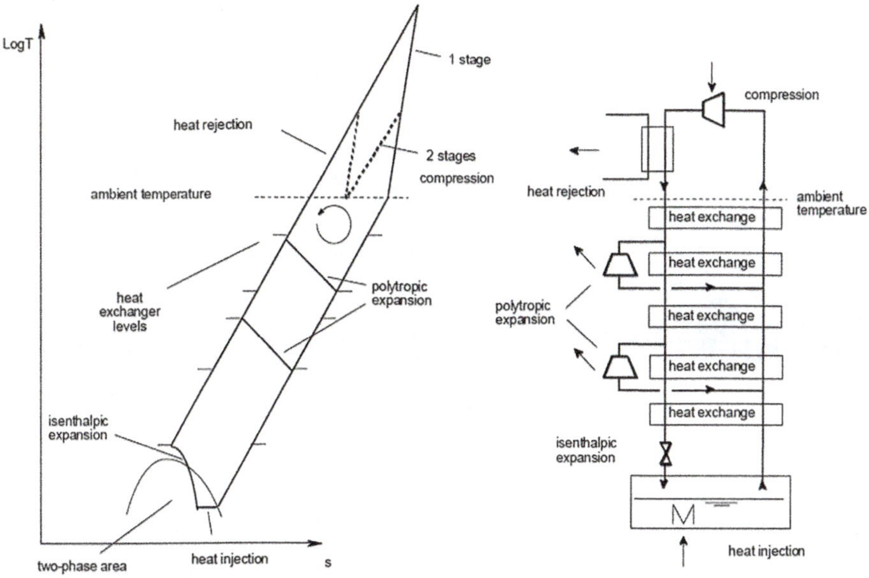

Fig. 2.20 Schematic example of two-pressure, two-stage Claude cycle: T-S diagram (left) and flow scheme (right)

Fig. 2.21 Image of compressor station at CERN

Fig. 2.22 Image of a 4.5 K cold box for the LHC (Linde) at CERN

Fig. 2.23 Front view of 4.5 K cold box for the LHC (Linde) at CERN

Typical laboratory Dewar flasks are spherical, made of glass, and protected in a metal outer container. Dewar flasks for extremely cold liquids such as liquid helium have another double-walled container filled with liquid nitrogen. Dewar flasks are named after their inventor, James Dewar, the man who first liquefied hydrogen. Thermos bottles are smaller vacuum flasks fitted in a protective casing.

Cryogenic bar code labels are used to mark Dewar flasks containing these liquids and will not frost over down to −195 °C.

Cryogenic transfer pumps are the pumps used on LNG piers to transfer liquefied natural gas from LNG carriers to LNG storage tanks, as are cryogenic valves.

2.6.1 Cryogenic Processing for Alloy Hardening

The field of cryogenics advanced during World War II when scientists found that metals frozen to low temperatures showed more resistance to wear. Based on this theory of cryogenic hardening, the commercial cryogenic processing industry was founded in 1966 by Ed Busch. With a background in the heat-treating industry, Busch founded a company in Detroit called CryoTech in 1966, which merged with 300 Below in 1999 to become the world's largest and oldest commercial cryogenic processing company. Busch originally experimented with the possibility of increasing the life of metal tools to anywhere between 200% and 400% of the original life expectancy using cryogenic tempering instead of heat-treating. This evolved in the late 1990s into the treatment of other parts.

Cryogens, such as liquid nitrogen, are further used for specialty chilling and freezing applications. Some chemical reactions, like those used to produce the active ingredients for the popular statin drugs, which reduce cardiovascular disease (CVD), must occur at low temperatures of approximately −100 °C (−148 °F). Special cryogenic chemical reactors are used to remove reaction heat and provide a low-temperature environment. The freezing of foods and biotechnology products, like vaccines, requires nitrogen in blast freezing or immersion freezing systems. Certain soft or elastic materials become hard and brittle at very low temperatures, which make cryogenic milling (cryomilling) an option for some materials that cannot easily be milled at higher temperatures.

Cryogenic processing is not a substitute for heat treatment, but rather an extension of the heating-quenching-tempering cycle. Normally, when an item is quenched, the final temperature is ambient. The only reason for this is that most heat-treaters do not have cooling equipment. There is nothing metallurgically significant about ambient temperature. The cryogenic process continues this action from ambient temperature down to −320 °F (140 °R; 78 K; −196 °C). In most instances, the cryogenic cycle is followed by a heat tempering procedure. As all alloys do not have the same chemical constituents, the tempering procedure varies according to the material's chemical composition, thermal history, and/or a tool's particular service application.

The entire process takes 3–4 days.

2.6.2 Cryogenic Fuels

Another use of cryogenics is cryogenic fuels for rockets with liquid hydrogen as the most widely used example. Liquid oxygen (LOX) is even more widely used but as an oxidizer, not a fuel. NASA's workhorse space shuttle used cryogenic hydrogen/oxygen propellant as its primary means of getting into orbit. LOX is also widely used with RP-1 kerosene, a non-cryogenic hydrocarbon, such as in the rockets built for the Soviet space program by Sergei Korolev.

Russian aircraft manufacturer Tupolev developed a version of its popular design Tu-154 and later on was known as Tu-155 with a cryogenic fuel system, known as the Tu-155. The plane uses a fuel referred to as liquefied natural gas or LNG and made its first flight in 1989.

2.6.3 Cryogenic Application in Nuclear Magnetic Resonance Spectroscopy (NMR)

Nuclear magnetic resonance (NMR) is one of the most common methods to determine the physical and chemical properties of atoms by detecting the radio frequency absorbed and subsequent relaxation of nuclei in a magnetic field. This is one of the most commonly used characterization techniques and has applications in numerous fields. Primarily, the strong magnetic fields are generated by supercooling electromagnets, although there are spectrometers that do not require cryogens. In traditional superconducting solenoids, liquid helium is used to cool the inner coils because it has a boiling point of around 4 K at ambient pressure. Cheap metallic superconductors can be used for the coil wiring. So-called high-temperature superconducting compounds can be made to superconduct with the use of liquid nitrogen (LN), which boils at around 77 K.

2.6.4 Cryogenic Application in Magnetic Resonance Image (MRI)

Cryogenics is the study and use of materials at extremely low temperatures. Such low temperatures cause changes in the physical properties of materials that allow them to be used in unusual engineering, industrial, and medical applications. For example, in the cryogenic temperature range, air becomes a liquid—or even a solid—and living tissue freezes instantly. Matter behaves strangely at the lowest temperatures of the cryogenic range. Electric currents never stop flowing, liquids run uphill, and rubber becomes as brittle as glass. In medicine, cryogenic cooling is used in some diagnostic techniques, such as magnetic resonance imaging (MRI). Cryosurgery uses liquid nitrogen to kill unhealthy tissue by freezing

Fig. 2.24 Cryogenic gases delivery truck at a supermarket

it. Cryogenics is expected to play an important role in the development of better procedures for preserving human organs for transplant.

MRI is a complex application of NMR where the geometry of the resonances is deconvoluted and used to image objects by detecting the relaxation of protons, which have been perturbed by a radio-frequency pulse in the strong magnetic field. This is mostly commonly used in health applications.

2.6.5 Cryogenic Application in Frozen Food Transport

Cryogenic gases are used in transportation of large masses of frozen food. When very large quantities of food must be transported to regions like war zones, earthquake-hit regions, etc., they must be stored for a long time, so cryogenic food freezing is used. Cryogenic food freezing is also helpful for large-scale food processing industries. See Fig. 2.24 for transportation of cryogenic gases.

Cryogenic technology gives low-temperature applications in the food sector. There is a tremendous scope for application of cryogenic technology in food processing and preservation. Cryopreservation is the only viable method available for long-term preservation of the both plant and animal origin species, such as dairy products. Cryogenic preservation of food offers great promise for the country, both for export and for domestic consumptions due to assurance of the food quality and safety, also. Most industries employ evaporative air chilling systems; preservation by cryogenic technology is less familiar in this sector. Product shrinkage, toughening and loss of tenderness, products' shelf life, microbial products, drip loss, and dehydration losses are the major quality considerations in freezing of the food products, e.g., meat products [21].

The preservation by cryogenic technology will improve the situation. Proper economic considerations including payback period and life of the system, etc. should be taken into account while selecting the cryogenic applications. The availability of indigenous cryogenic technology for food processing would ensure production of better quality products within the country and export the processed products to different countries.

In food processing, cryogenics implies use of very low-temperature materials for chilling and freezing. This type of refrigeration differs from other procedure because it does not depend on the external low-temperature production systems.

The freezing of foods and biotechnology products, like vaccines, requires nitrogen in blast freezing or immersion freezing systems. Cryogenic freezing with nitrogen is carried out by first passing the food through nitrogen vapor at about −50 °C and then freezing the food by spraying the refrigerant directly onto the food.

Fish, meat, poultry, fruit, vegetables, and bakery products can all be frozen in this way.

The extremely low-temperature characteristics of cryogenic food provide the ultimate in chilling and freezing rates.

The physical properties of the cryogenic gases provided an important tool to help the food industry to improve the plant automation, versatility, efficiency, and manufacturing cost.

2.6.6 Cryogenic Application in Forward Looking Infrared (FLIR)

Many infrared cameras require their detectors to be cryogenically cooled as it can be seen in Figs. 2.25 and 2.26.

Miniaturized versions of cryocoolers are known as micro-coolers, which have potential applications to space sensors and military airborne systems such as IR line scanners, IR search and track sensors, high-resolution thermal imaging systems, and forward looking infrared (FLIR) sensors. Micro-cooler design based on the Stirling-cycle operating principle can achieve a cooling temperature of 77 K within 3 min.

The micro-cooler configuration normally incorporates a generator to produce a compression and expansion refrigerator cycle with no valves. It is important to mention that the regenerator has a large heat capacity and acts like an efficient heat exchanger.

A micro-cooler designed for NASA thermal imaging applications consumes electrical power less than 3 W, weighs less than 15 oz., has a life expectancy of 5 years, and has demonstrated continuous operation exceeding 8000 h with no degradation in performance. High reliability of a micro-cooler requires improved materials with self-lubrication capability, elimination of gaseous contamination, clearance seals with very low friction, and a linear drive mechanism. For cryogenic temperature measurement down to 30 K, PT-100 sensor (resistance temperature detectors (RTDs)) is used. Additionally, for lower than 30 K, it is required to use silicon diode for accuracy, and there are also other cryogenic detectors which are used to detect cryogenic particles.

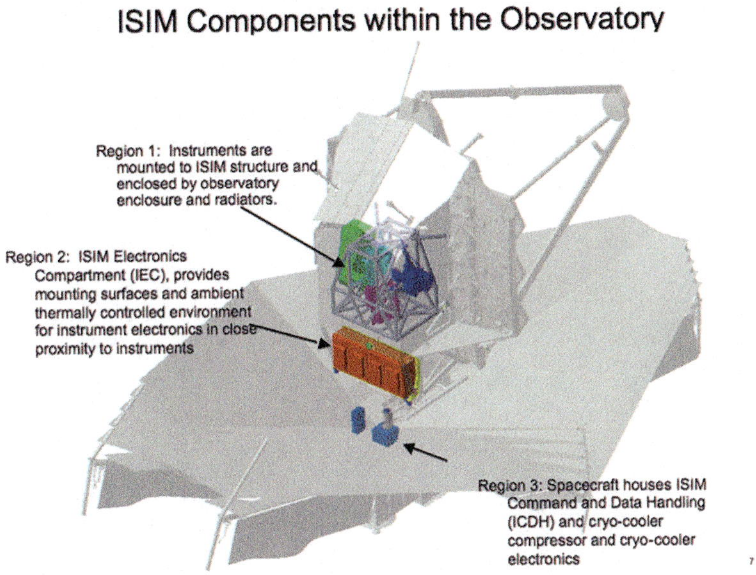

Fig. 2.25 Diagram of an infrared space telescope

Fig. 2.26 Astronomical instruments on the very large telescope are equipped with continuous flow cooling systems

Fig. 2.27 Left is Spitzer spacecraft during final testing (NASA) and right is Herschel spacecraft (ESA)

2.6.7 *Cryogenic Application in Space*

Einstein and IRAS are the first "cryogenic missions," which flew, respectively, in 1978 and 1983. *Einstein (HEAO-2)* was the second of NASA's three High Energy Astrophysical Observatories and the first X-ray telescope put into space in 1978. Among other instrument, Einstein carried a solid-state spectrometer using a Si(Li) crystal detector (range 0.5–4.5 keV) cooled at about 100 K via a solid ammonia/ methane cryostat [22, 23].

IRAS (Infrared Astronomical Satellite, launched in 1983) was the first cryogenic scientific satellite. Its mission was to map the entire sky from 8 μm to 120 μm, and it was equipped with a 0.6 m telescope cooled with liquid He to about 4 K. The focal plane assembly operated 62 photoconductive elements at 3 K.

Cosmic Background Explorer (COBE) was developed by NASA to measure the cosmic background radiation. The satellite was launched in November 1989. It carried three instruments, operating at wavelengths between 1.25 μm and 240 μm with focal planes at 1.6 K, cooled by 650 l, superfluid helium cryostat.

As we stated, the first cryogenic missions, such as IRAS, launched in 1983, COBE (COBE 1992), launched in 1989, and ISO, launched in 1995, were based on liquid He cryostats, with the bath temperature regulated by adjusting the vapor pressure. Lifetime was correspondingly limited by the amount of cryogen, typically to about 12–18 months. More recently, the same approach has been used by Spitzer (Fig. 2.27, left). Spitzer, thanks to an optimized cryogenic system (passive radiation, use of helium gas enthalpy, orbit choice), was designed to provide a minimum lifetime of 2.5 years, using only 360 l of superfluid He.

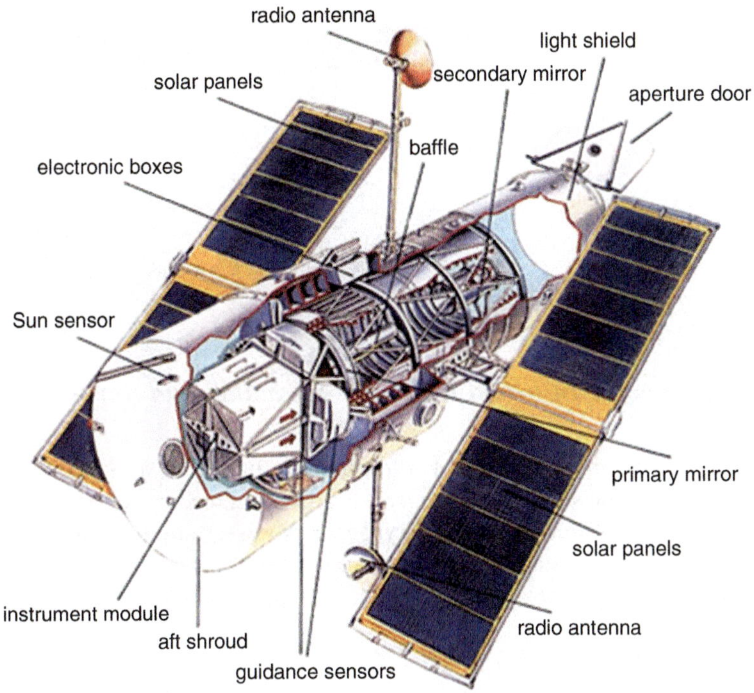

Fig. 2.28 Hubble space telescope image

The European Space Agency (ESA) is presently developing (phase C/D, 2003) two important cryogenic astronomical missions: *Planck* and *Herschel* [24]. *Planck's* main objective is to map the temperature anisotropies of the cosmic microwave background (CMB) over the whole sky, with a sensitivity $\Delta T/T = 2.10^{-6}$ and an angular resolution of 10 arc min. Such goals require bolometers operating at 0.1 K, HEMT at 20 K, and a cooled telescope (60 K). Planck's cryogenic system uses precooling to 60 K by passive radiators, cooling to 20 K with a H2 Joule-Thomson cooler (adsorption compressors), cooling to 4 K with a He Joule-Thomson cooler (mechanical compressors), and final cooling to 0.1 K with an open-loop dilution refrigerator.

ESA's Infrared Space Observatory (ISO) operated at wavelengths from 2.5 μm to 240 μm between November 1995 and May 1998 in a highly elliptical orbit. The satellite is based on a cryostat containing about 2200 liters of superfluid helium and on a 0.6 m diameter telescope. The instruments made use of different photoconductors based on InSb, Si, and Ge and operated between 1.8 K and 10 K [25].

Near Infrared Camera and Multi-Object Spectrometer (NICMOS) is a Hubble Space Telescope (HST) instrument based on three cameras designed for simultaneous operations, operating between 0.8 μm and 2.5 μm and using HgCdTe photoconductive detectors, cooled down to 50–60 K via 120 kg of solid nitrogen. See Fig. 2.28.

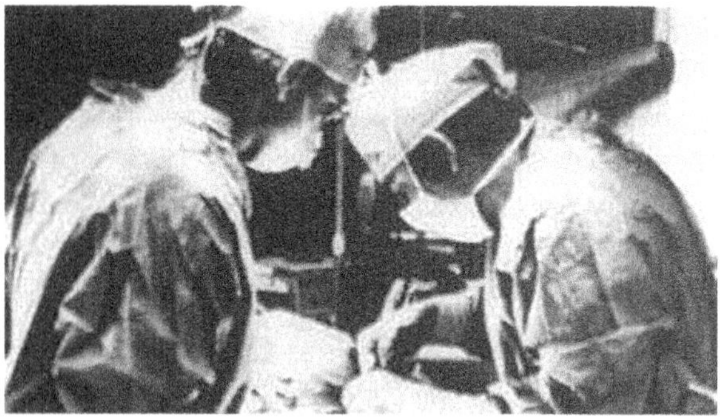

Fig. 2.29 Doctors prepare a patient for cryogenic surgery

2.6.8 *Cryogenic in Blood Banking, Medicine, and Surgery*

Certain rare blood groups are stored at low temperatures, such as −165 °C. Liquid nitrogen is one of the safest cooling agents available. In medicine, it is, used to kill unhealthy tissues by freezing them. Cryogenic processes are also used to supply "banks" storing eye corneas, blood, and sperm for future surgical procedures. Some embryos have also been frozen and stored for later implantation (surgical placement) in women.

In 1961 American surgeon, Irving S. Cooper, introduced a freezing technique called cryosurgery. Cryosurgery is relatively bloodless because the low temperatures used constrict the blood vessels, stemming the flow. Special instruments are used that have freezing tips to kill the damaged tissue and shields to protect surrounding tissue. Cooper used cryosurgery to freeze and destroy damaged tissue in the brains of patients with Parkinson's disease (a degenerative illness). Since then, cryosurgery has found many applications. It is used to repair detached retinas and to remove cataracts. It is also used to treat liver cancer and prostate cancer.

In medicine and surgery, the low-temperature and cryogenic application goes a long way. The application of ultralow temperature and cryogenics to clinical situation and underlying principles in the absence of ice have under study fall back to ancient history of humankind, which goes back to ancient Egypt 2500 years ago. In the fifth century B. C., the Greek physician Hippocrates advocated the clinical utility of cold for relieving pain in trauma and in certain diseases affecting the bones and joints. James Arnott is often described as the father of modern cryosurgery. In 1851, he achieved temperature of −24 °C by using a solution of ice and saline for the treatment of various surface conditions (see Fig. 2.29). The effects of low temperature on mammalian systems can be explained together with the effect of hypothermia by the researcher in this field of medicine [26].

Low-temperature storage of tissues and organs for transplantation in the liquid state is playing a role of life and death in today's emergency room (ER) around the world, and we can see the application of cryogenic state that is associated with the formation of ice and subsequent freezing of cells and tissues as well as organ cryopreservation.

There are two categories of the processes, which occur within living cells:

1. One is the biochemical processes, which are the distinguishing feature of living materials by using metabolic energy often involving enzymatic catalysts.
2. The second one or the others are the physical processes, which are also common in nonliving systems.

One example is the diffusion of a solute due to a concentration gradient. Both categories can be affected by temperature gradient and changes. It is observed that the biochemical processes are usually slowed to great extent upon exposure to cooling environment and vice versa. This observation can be described by some mathematical relationship, by putting it into some conceptual perspective. In other words, the starting point of cryobiology, including cryopreservation, cryosurgery, and cryogenic medicine, is the Van't Hoff's rule, which states that the change in the rate of a process such as metabolism produced by 10 °C changes the system temperature. It is called Q_{10}, which can be expressed in form of the following equation:

$$Q_{10} = \frac{\text{Reaction Rate at T} + 10}{\text{Reaction Rate at T}} \tag{2.27}$$

Usually, Q_{10} has a value of about 2 in biological systems, 2.3 in most thermochemical reactions, and 1–3 for reactions in organism up to 50 °C. For example, a reduction in human body temperature from its basal state of 37 °C results in a decline in the metabolic rate by one-half.

Another important law basic to cryobiology is the Arrhenius relationship, which can be expressed in the logarithmic form as

$$\log v = -\left(\frac{\Delta H_a}{2.303R}\right)/T \tag{2.28}$$

where v denotes the specific rate of change such as degradation in the biomaterial, R is presenting universal gas constant, and ΔH_a is heat of activation, while T stand for absolute temperature in this relationship. Or in the exponential form, Eq. 2.28 takes a new form as

$$v = C\exp\left(\frac{\Delta H_a}{2.303RT}\right) \tag{2.29}$$

where C is a constant.

When a tissue is cooled, the rate of heat transfer depends mainly on water content, blood supply, thermal conductivity of the tissue, rate of freeze, and the temperature of the refrigerant. Table 2.12 lists the extent of surface temperature reductions attainable with various refrigerants [26].

The contour of cryolesion by an open spray is rounded down to the depths of about 6 mm, but below this, it becomes more triangular shape. The method gives a more rapid drop in temperature and will freeze to a greater depth than a closed probe.

However, the shape of the cryolesion is similar for the two methods.

Table 2.12 Surface tissue temperature reductions attainable with various refrigerants [26]

Refrigerant	Temperature attainable, °C
Ice	0
Salt-ice	−20
CO^2 snow	−79
Nitrogen oxide	−75
Liquid nitrogen	−196 (spray or probe)

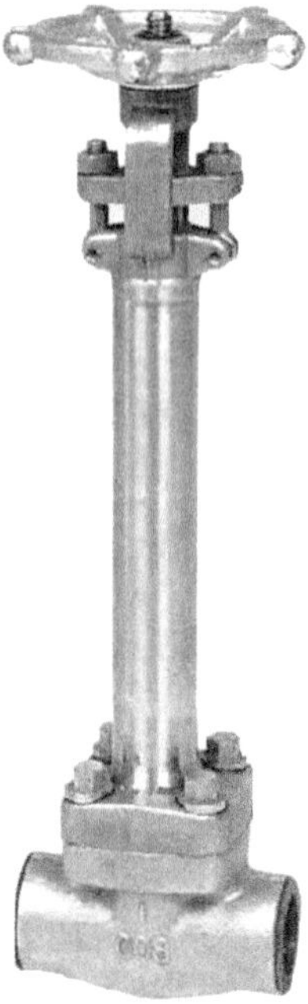

Fig. 2.30 Cryogenic valve

Cryosurgery is also widely used in the fields of dermatology, gynecology, plastic surgery, orthopedics, and podiatry. Cryosurgery has also been used successfully for more than 30 years in veterinary medicine.

Cryogens, like liquid nitrogen, are further used for specialty chilling and freezing applications. Some chemical reactions, like those used to produce the active ingredients for the popular statin drugs, must occur at low temperatures of approximately −100 °C. Special cryogenic chemical reactors are used to remove reaction heat and provide a low-temperature environment.

2.6.9 Cryogenic in Manufacturing Process

The introduction of cryogenic gases in the early 1960s as an alternative to improve freezing processes in the frozen industry was a major product quality and process improvement.

Cryogenic cooling is used to cool the tool tip at the time of machining. It increases the tool life. Oxygen is used to perform several important functions in the steel manufacturing process. See Fig. 2.30.

2.6.10 Cryogenic in Recycling of Materials

By freezing the automobile or truck tire in liquid nitrogen, the rubber is made brittle and can be crushed into small particles. These particles can be used again for other items. The theory was based on how heat-treating metal works. A heat-treated metal is cooled from a very high temperature down to room temperature causing certain strength increases in the molecular structure to occur. They theorized that continuing the temperature descent would allow for further strength increases. Using liquid nitrogen, CryoTech formulated the first early version of the cryogenic processor. Unfortunately, for the newly born industry, the results were unstable, as components sometimes experienced thermal shock when they were cooled too fast. Some components in early tests even shattered because of the ultralow temperatures. In the late twentieth century, the field improved significantly with the rise of applied research, which used new controls and technology to create more stable results.

2.7 Cryogenic Application in Research

Experimental research on certain physics phenomena, such as spintronics and magneto-transport properties, requires cryogenic temperatures for the effects to be observed. Cryogenic cooling of devices and material is usually achieved via the use of liquid nitrogen, liquid helium, or a mechanical cryocooler (which uses high-pressure helium lines). Gifford-McMahon cryocoolers, pulse tube cryocoolers, and Stirling cryocoolers are in wide use with selection based on required base temperature and cooling capacity. The most recent development in cryogenics is the use of magnets as regenerators as well as refrigerators. These devices work on the principle known as the magnetocaloric effect.

2.7.1 Research Overview

Cryocoolers are the key component of ZBO propellant storage systems. New developments are required to significantly improve the performance of coolers for the ZBO storage of liquid hydrogen. For ZBO liquid hydrogen systems, cooling powers of 1–20 W are required at 20 K. These systems require extensive development to achieve the same levels of efficiency that are reached for the higher-temperature

Fig. 2.31 High efficiency cryocooler

liquid oxygen (LOX) and methane ZBO coolers. At present, no long-lived 30 K or colder closed-cycle coolers have flown in space. Current commercial, non-flight pulse tube cryocoolers are available for temperatures down to 3 K; however, these machines are not space qualified and are inefficient.

The final development frontier for LH2 coolers is to achieve high efficiency and reliability at lower operating temperatures. Pulse tube and Stirling coolers offer the best opportunity for achieving high efficiency at these temperatures and power levels. The key to such improvements is the design of the regenerator and the selection and formation of the regenerator material. The ARC Cryogenics Group and its partners have been developing high heat capacity rare earth alloys for just this purpose.

2.7.2 *Right: Lightweight, High Efficiency Cryocooler*

Along with advancing the SOA of cryocoolers, system studies are also being conducted with GRC, JPL, and MSFC. One of the major advances in this area is the development of an analytical tool for sizing the ZBO system, including tankage, passive insulation, cryocooler, radiator, and power system mass. This model optimizes cryogenic thermal storage system performance. Tank sizes and configurations for selected mission scenarios can be easily and quickly evaluated. Topologies can be compared and trade-offs performed to arrive at a concept optimized for specific mission parameters. See Fig. 2.31.

Once the system is adequately modeled, the insulation system can be designed appropriately. New materials and integration techniques can be quantitatively analyzed with our SOA transport phenomena test facilities. Our thermal conductivity apparatus features a sample chamber within a vacuum chamber (isothermal). A hot plate is embedded in the sample to minimize alternate heat paths. In this manner, the thermal conductivity can be determined within an accuracy of a few percent. What makes the apparatus truly unique is that the thermal conductivity measurements can be made in the presence of a gas, including planetary analogues.

2.7.3 Background

Since 1976, Ames Research Center's (ARC) Cryogenics Group has provided cryogenic support for the agency's missions. For the Infrared Astronomical Satellite (IRAS) mission, ARC performed focal plane testing, thermal design, and development of a backup cryogenic valve. For the Superfluid Helium On-Orbit Transfer (SHOOT) mission, ARC developed the flight flow meters and the flight and ground operation software.

The Cryogenics Group has also established a database for cryogenic thermal contact conductance down to superfluid helium temperatures, characterized swirl in liquid helium in a rotating Dewar, measured thermal conductivity of cryogenic insulation in the presence of gases, and developed a cryogenic mirror test facility for IR astronomy.

2.7.4 Right Liquefier Demo and Cryogenic Insulation Test Facility

The Cryogenics Group has developed coolers down to 50 mK, including dilution coolers, helium-3 refrigerators, and adiabatic demagnetization coolers. Recently, ARC developed a flight-qualified pulse tube cryocooler, in collaboration with the Department of Defense (DOD). This program is for demonstration of zero boil-off (ZBO) storage, which is working with Atlas Scientific. This is also in collaboration with the Department of Energy (DOE), to develop high-performance regenerator materials to improve cryocooler efficiency. In addition, ARC has several unique capabilities for measuring the thermal conductivity of insulation materials and moisture absorption in the presence of gases, including planetary analogues within an accuracy of a few percent. See Fig. 2.32.

2.8 Cryogenic Fluid Management

Cryogenic Fluid Management (CFM) technology is an integral part of exploration systems for Earth-to-Orbit Transportation, manned missions to the Moon and Mars, planetary exploration, and In Situ Resource Utilization (ISRU). CFM also plays a key role in infrared and X-ray astronomy, biological sciences, and fundamental investigations into the origins of our universe.

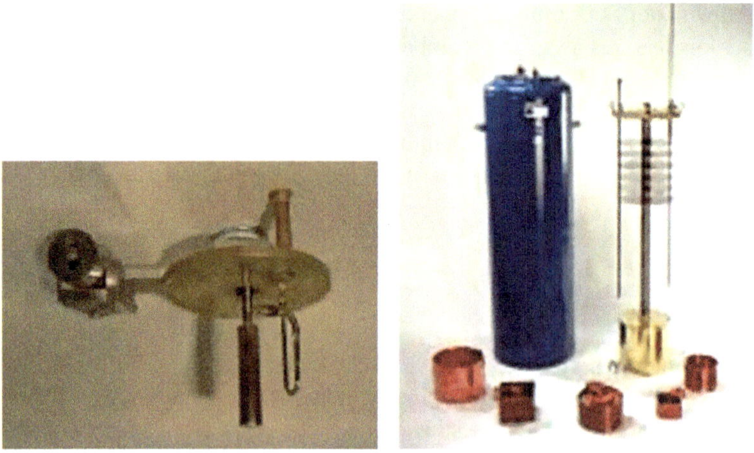

Fig. 2.32 Liquid demo and cryogenic insulation test facility

2.8.1 *Benefits*

The challenges of NASA's exploration vision require advanced Cryogenic Fluid Management technology. The exploration vision requires high-performance propulsion systems (high specific impulse) for both human and robotic missions. The vision includes in-space cryo-propulsion stages and In Situ Resource Utilization (ISRU) for cryo-propellant production and liquefaction of breathable gases. Cryogenic propellants such as oxygen, methane, and hydrogen can satisfy this requirement. The current state of the art (SOA) for cryo-propellant storage is a loss rate of 3%/month in low Earth orbit (LEO) using passive technology. Advances in passive thermal control technology might reduce losses to 1%/month, still an unacceptable rate for a 2 + year mission to Mars. By using cryocoolers to balance the entire parasitic and internally generated heat loads in the cryo-tank, no propellant will be lost, resulting in a zero boil-off (ZBO) system and eliminating the need for oversized tanks and extra propellant. Each pound of propellant tank mass saved is directly tradable for payload mass.

2.9 Conclusion

The brief introductory in this chapter has presented the basic ideas and principles of the most important aspects of cryogenics, i.e., cryogenic fluids, heat transfer, thermal design, and refrigeration. It has also provided the reader with typical numerical values of the relevant parameters, enabling him or her to perform orders of magnitude estimates and apply his or her engineering judgment. There is of course much more to say on each of these topics, some of which have significantly developed over the years and still constitute areas of technical progress. Many other subjects not addressed here also pertain to cryogenic engineering, such as materials at low temperature, storage, handling and transfer of fluids, two-phase flow and discharge, vacuum and leak-tightness technology, instrumentation (in particular thermometry), process control, impurity control, and safety. In all cases, the interested reader is referred to the selected bibliography for detailed information and to the proceedings of the cryogenic engineering conferences for recent developments.

To cool down the Large Hadron Collider (LHC) superconducting magnets at CERN, eight refrigerators of 18 kW at 4.5 K have been built (four by Air Liquide and four by Linde). The compressor station is composed of five oil-lubricated screw compressors able to compress a total mass flow of 1.6 kg/s of helium between 1 bar and 19 bar. The cold box is composed of ten heat exchangers and ten turbines. Each refrigerator cools at 1/8 of the LHC that constitute a cold mass of 4595 t over 3.3 km. A refrigerator can provide up to 240 g/s of supercritical helium at 4.5 K and 3 bar to the LHC and 250 g/s between 50 K and 75 K to cool down thermal screens.

References

1. *Oxford English Dictionary*, 2nd edn. (Oxford University Press, 1989)
2. J. Wilks, D.S. Betts, *An Introduction to Liquid Helium*, 2nd edn. (Oxford Science Publications/ Clarendon Press, Oxford, 1987)
3. B. Zohuri, *Combined Cycle Driven Efficiency for Next Generation Nuclear Power Plants: An Innovative Design Approach*, 1st edn. (Springer, Cham, 2015)
4. B. Zohuri, *Nuclear Energy for Hydrogen Generation through Intermediate Heat Exchangers: A Renewable Source of Energy*, 1st edn. (Springer, Cham, 2016)
5. A Kanni Raj, *Cryogenics: Energy Storage in Nuclear Plants* (Create Space Independent Publishing Platform, 20 Nov 2015)
6. J.M. Nash, Vortex expansion devices for high temperature cryogenics. Proc. Intersoc. Energy Convers. Eng. Conf. **4**, 521–525 (1991)
7. B. Zohuri, *Magnetic Confinement Fusion Driven Thermonuclear Energy*, 1st edn. (Springer, Cham, 2017)
8. B. Zohuri, *Inertial Confinement Fusion Driven Thermonuclear Energy*, 1st edn. (Springer, Cham, 2017)
9. B.A. Hands, *Cryogenic Engineering* (Academic Press, 1986)
10. W. Obert et al., Emissivity measurements of metallic surfaces used in cryogenic applications. Adv. Cryo. Eng. **27**, 293 (1982)
11. Ph. Lebrun, Cryogenics, key to advanced science and technology. Bull IIR, **LXXXIII**(2003–6), 4 (2003). http://www.iifiir.org/2enarticles_bull03_6.pdf
12. M.C. Jones, V. Arp, Review of hydrodynamics and heat transfer for large helium cooling systems, in *Advances in Refrigeration at the Lowest Temperatures*, IIR-IIF Commission A1-2, (Zürich, 1978) p. 41
13. Y.L. Buyanov, Current leads for use in cryogenic devices, principle of design and formulae for design calculations. Cryogenics **25**, 94 (1985)
14. Ph. Lebrun, Design of a cryostat for superconducting accelerator magnets: The LHC main dipole case, CERN-2004-008, (Geneva, 2004) p. 348
15. U. Wagner, Refrigeration, CERN-2004-008, (Geneva, 2004) p. 295
16. B. Zohuri, *Compact Heat Exchangers: Selection, Application, Design and Evaluation*, 1st edn. (Springer, Cham, 2017)

17. B. Zohuri, *Application of Compact Heat Exchangers For Combined Cycle Driven Efficiency In Next Generation Nuclear Power Plants: A Novel Approach* (Springer, Cham, 2017)
18. B. Zohuri, *Nuclear Energy for Hydrogen Generation through Intermediate Heat Exchangers: A Renewable Source of Energy* (Springer, Cham, 2016)
19. B. Zohuri, *Physics of Cryogenic An Ultra-Low Temperature Phenomena* (Elsevier, 2017)
20. S. Claudet et al., Economics of large helium cryogenic systems: Experience from recent projects at CERN. Adv. Cryo. Eng. **45B**, 1301 (2000)
21. P. Barnwal, K.K. Singh, Application of Cryogenic Technology in Food Processing and Preservation. Indian Dairy Engineers Association (IDEA) ICAR—National Dairy Research Institute (NDRI), Karnal, Haryana, India
22. B. Collaudin, N. Rando, Cryogenics **40**, 797–819 (2001)
23. R. Giacconi, et al., Astrophys. J. Part 1, **230**, 540–550 (1 June 1979)
24. B. Collaudin, T. Passvogel, Proc. SPIE, **3358** (1998)
25. M.F. Kessler et al., Astron. Astrophys. **315**, L27–L31 (1996)
26. W.J. Yang, S. Mochizuki, Low temperature and cryogenic application in medicine and surgery, in *Low Temperature and Cryogenic Refrigeration*, NATO Science Series, vol. 99 (Springer), pp. 295–308
27. R. G. Scurlock (ed.), *History and Origins of Cryogenics* (Clarendon Press, Oxford, 1992)
28. W.M. Kay, A.L. London, *Compact Heat Exchangers,* 3 Sub edn. (Krieger Publishing, Malabar, Jan 1998)

Chapter 3
Reliable Renewables with Cryogenic Energy Storage

Electricity demand varies, influenced by factors like time of the day and season. The National Grid is prepared for surges in demand, with power stations on standby ready to crank up the power. However, dealing with these peaks and troughs will become increasingly difficult as coal-fired power stations closed down and more intermittent renewable energy like wind and solar comes online. In 2015, renewables provided almost a quarter of UK electricity. The intermittent nature of green sources has seen researchers focus on trying to improve energy storage. The cryogenic energy facility stores power from renewables or off-peak generation by chilling air into liquid form. When the liquid air warms up, it expands and can drive a turbine to make electricity. The 5 MW plant near Manchester can power up to 5000 homes for around 3 h. Cryogenic storage works by using renewable or off-peak electricity to cool air down to −190 °C, which turns it into a liquid. *Intermittent supply is an ongoing problem for the development of the renewable power market. Could cryogenic energy storage prove the best way to reduce waste by saving off-peak power for later use? Birmingham University's Professor Richard Williams argues the case.*

3.1 Introduction

As eco-friendly energy generation methods such as wind and solar power continue to mature, the ultimate goal of an efficient, economically viable renewable energy network now seems tantalizingly close. The ongoing refinement of technologies like photovoltaics (PVs) and wind turbines has eased doubts that renewables can, with the right conditions and plenty of investment plugs the gap left by decommissioned fossil fuel and nuclear plants. The plant has shown that cryogenic energy system (CES) is technically achievable and meets expectations of performance.

B. Zohuri, *Hybrid Energy Systems*,
https://doi.org/10.1007/978-3-319-70721-1_3

In February 2011, the Spanish Wind Energy Association announced that between November 2012 and the end of January 2013, wind energy had produced more electricity than any other power sources in the country.

With more than 6 TWh generated by Spanish wind farms in January alone, it has become clear the scalability of renewables (or wind power at least) is no longer the problem. But like Tantalus, the tragic figure of Greek myth, the fruit of renewable power success always seems just out of reach, no matter how low-lying it seems. One of the major stumbling blocks holding back renewable energy development is the intermittent nature of its supply. If we lived in a world where the sun always shines and the wind always blows, these natural resources would likely already be our main sources of energy. In reality they are fickle forces, prone to overproducing when demand is low and failing to meet requirements when demand peaks.

As Professor Williams argues on the nature of intermitted generation issues, as such, surplus electricity generated by wind farms is often wasted, when it is not required. In the United Kingdom (UK), constraint payments of more than £26.5 m have been made for around 185 GWh of unused wind power since 2011, shifting the cost of a complex and wasteful system on to taxpayers. It is certainly true that constraint payments are also made to nonrenewable power generators, but it is still a dispiriting statistic for a power source that is shouldering the burden of increasing expectations.

"Why do we tolerate a marketplace that throws away last night's energy from wind farms that should have been captured for use today?" he wrote. "That we are subsidizing the costs of wind farms only to dump the energy they generate is clearly abhorrent. The answer is that we have no cost-effective way to provide for the storage of energy."

Along with smart upgrades such as better grid interconnection, developing a more extensive network of energy storage systems is the key to reducing waste in the energy supply chain, in general, and boosting the fortunes of intermittent renewables, in particular.

As a result, utilities and technology specialists have been experimenting with different energy storage methods for years now, with well-established hydroelectric storage and more modern battery and fuel cell systems attracting attention.

However, while these technologies have proven successful in many cases, they also have drawbacks for extensive power applications—industrial-scale batteries require costly rare-earth metals, fuel cells can be prohibitively expensive to scale up, and there are obvious geographical limitations when it comes to pump-storage hydroelectric dams.

Cryogenic energy storage (CES) is an innovative new technique of capturing and storing electricity—its developers hope it will address the niggling issues that have prevented other systems from solving the energy market's storage woes.

For reliable and efficient power generation from renewable power sources, such as solar cells and windmills, electrical energy storage (EES) systems are often mandatory. The level of power generation of such power sources is largely determined by uncontrollable environmental factors. The uses of EES systems have two major benefits for renewable power generation. First, an EES system can increase energy utilization by mitigating the temporal mismatch between the power generation and load demand.

If load demand is lower than maximum power generation capability, excess power that is not stored is wasted and cannot be used later. By storing the excess power and using it when load demand is higher than maximum power generation capability, we can reduce energy waste and fully utilize the power generation capability always. Second, an EES system improves power generation stability. Due to highly variable power generation depending on the environmental changes, instantaneous power supply-demand mismatch results in severe variations in frequency. EES systems can maintain a desired frequency by rapidly modulating its power supply or draw in response to the frequency variations.

Electrical energy storage (EES) systems provide various benefits of high-energy efficiency, high reliability, low cost, and so on, by storing and retrieving energy on demand. EES systems have a wide range of applications, such as contingency service and peak shaving for power grid, energy buffer for renewable power sources, power train in electric vehicles (EVs), and so on. Current EES systems mainly rely on a single type of energy storage technology, but unfortunately no single type of EES element technology can fulfill all the desirable characteristics, such as high-power/energy density, low cost, high cycle efficiency, and long cycle life. Hybrid electrical energy storage (HEES) systems, on the other hand, are composed of multiple, heterogeneous EES element technologies, aiming at exploiting the strengths of each technology while hiding its weaknesses. This is a practical approach to improve the performance of EES systems with currently available EES element technologies. A HEES system may achieve a combination of performance metrics that is superior to those for any of its individual energy storage elements with elaborated system design and control schemes.

3.2 Cryogenic Application in Electric Power Transmission within Big Cities

It is difficult to transmit power by overhead cables in big cities, so underground cables are used (see Fig. 3.1). However, underground cables get heated, and the resistance of the wire increases leading to waste of power. Superconductors could be used to increase power throughput, although they would require cryogenic liquids such as nitrogen or helium to cool special alloy-containing cables to increase power transmission. Several feasibility studies have been performed, and the field is the subject of an agreement within the International Energy Agency.

Bear in your mind that as part of transmission grid application, innovation and smart transmission grids system for electric power transmission is required and necessary, using cryogenic as an application.

A successful transmission automation system is the foundation for a high level of functionality and flexibility in energy usage. A smart transmission grid increases overall grid reliability and efficiency while reducing line losses and faults. It incorporates decentralized energy sources, such as offshore wind farms, and delivers

Fig. 3.1 High-power and high-voltage electricity tower distribution grids

Fig. 3.2 Illustration of smart grid transmission (Courtesy of Siemens Corporation)

electricity through the power lines on demand. An optimally engineered power transmission network must be both economically viable and physically feasible.

Siemens, for example, is the one-stop supplier of sophisticated transmission automation technology and services from individual components all the way to turnkey solutions and transmission system planning. See Fig. 3.2.

Siemens portfolio of products and solutions for power transmission covers high-voltage transmission solutions, high-voltage switching products and systems, and innovative direct current (DC) and alternating current (AC) transmission systems. Turnkey substations and power transformers along with top-notch consulting services and training from Siemens completes the spectrum.

3.3 The Basic of Cryogenic Energy Storage (CES)

It appears the UK Government is also confident in the inherent qualities of CES. The Engineering and Physical Sciences Research Council (EPSRC), the country's public research funding organization, recently announced £85 m in government funding for university research projects. Among the recipients was the University of Birmingham, which won £6 m to create a new center for cryogenic energy storage.

> The 'clever' bit is storing the cold that is released when the air expands at the end of the process.

Professor Williams, who is the lead investigator on the university's newly funded project, explains the basics of how CES works. "The process starts by using off-peak electricity—perhaps when the wind is blowing and demand is low—to refrigerate air down to −196 °C so that it becomes a cryogenic liquid, "liquid air" [or liquid nitrogen].

> This can be easily stored in containers in large quantities, until power is needed at peak times, or when the wind stops blowing. At this point, the liquid air is pumped out of the container and warms up in the ambient temperature or by using waste heat from a nearby source. This causes it to expand rapidly into a gas with 700 times the liquid volume, turning a turbine which generates the electricity.

The base efficiency of this process is relatively low at around 25%, but there are a number of ways to significantly improve performance.

"The 'clever' bit is storing the cold that is released when the air expands at the end of the process and using that to help in the refrigeration at the start of the cycle," says Williams. "This dramatically increases the efficiency of the system."

Cold recycling can reportedly increase the power efficiency of the system to around 50%, according to Highview Power Storage, a London-based CES developer that has been heavily involved in the technology's early refinement. The company adds that by using waste industrial heat from another nearby facility—a power station, say—to warm up the liquid nitrogen, efficiency could go as high as 70%.

3.4 Cryogenic Energy Storage (CES)

Cryogenic energy storage (CES) is a novel method of storing grid electricity. The idea is that off-peak or low-cost electricity is used to liquefy air (by way of a compressor, cooler, and then expander), that is then stored in an energy dense cold liquid

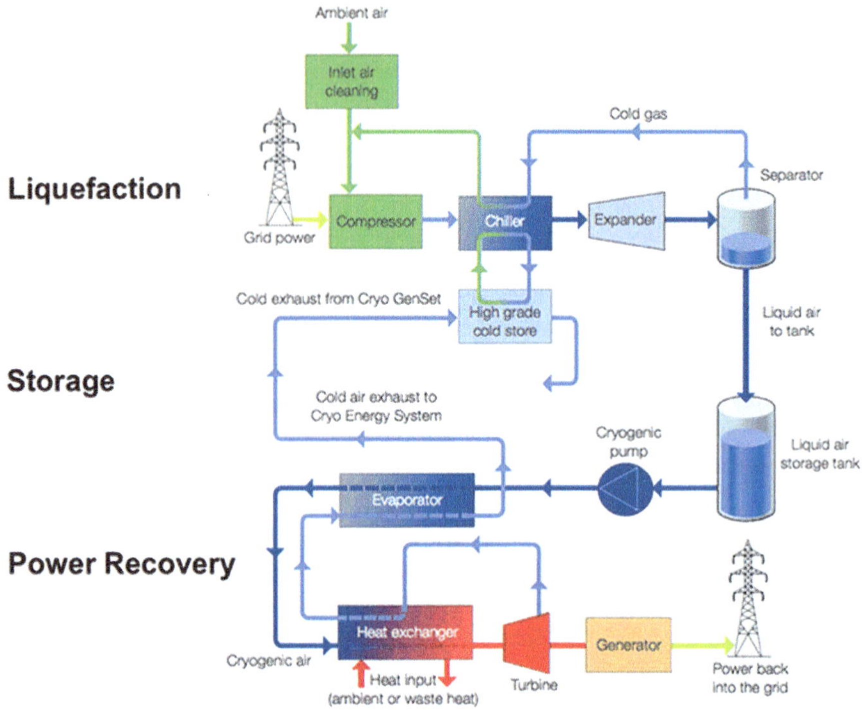

Fig. 3.3 Schematic diagram of a CES system (Courtesy of Highview Power Storage Corporation)

form. When electricity is required then, the cold liquid air is pumped to increase its pressure, superheated in a heat exchanger using either ambient or low-grade waste heat, and then expanded through a turbine generating useful electricity. This process is illustrated in Fig. 3.3.

3.5 Cryogenic Energy Storage (CES) Characteristics

The energy density for liquid air is around 100–200 W/kg, and in a recent report, on CES by the Centre for Low Carbon Futures in the United Kingdom, the cost of liquid air was estimated between \$200–530/kW (Centre for low Carbon Futures and the Liquid Air Energy Network, 2013). Proponents of CES suggest that some of the main advantages of CES are the ability to use existing gas infrastructure, its high cycling ability, the fact that it has no geographical constraints and that CES is a mechanical system, which isn't constructed with any exotic materials, unlike batteries, capacitors etc. It should be suitable for grid energy storage on a medium to large scale, for back-up generation, renewable integration and energy management.

There is also interest in liquid air as an "energy vector," which could be used in smaller scale applications in zero-emissions vehicles. Mobile refrigeration units are especially interesting as a liquid air engine could provide power and cooling.

Table 3.1 Cryogenic energy storage

Typical capacity	Typical power	Efficiency (%)	Storage duration	$/ kWh	$/ kW	Life span	Cycling capacity	Comments
Up to 15 MWh (anticipated up to 200 MWh)	100 kW to 5 MW (anticipated up to 50 MW)	Presently low, anticipated in excess of 60[a]	Mines—Days	200–530[a]	–			

[a] http://www.lowcarbonfutures.org/sites/default/files/potential-guide.pdf

At present the proven efficiency of CES is low in the prototype devices; however, Highview Power Storage expect that their second-generation device should achieve around 60% efficiency utilizing the waste heat from a "waste to power" thermal generation plant. Much of the expertise in CES is in the United Kingdom with Highview Power Storage and a new center for cryogenic energy storage at the University of Birmingham working with the Chinese Academy of Sciences. There has also been some research at the Universities of Washington and North Texas. CES is still a technology in the demonstration phase rather than being commercial though it is expected to see rapid development in the next few years. The table below illustrates some of the anticipated characteristics of CES.

Table 3.1 is presenting such characteristics.

3.5.1 Cryogenic Energy Storage (CES) a Wise Investment

> Surplus electricity generated by wind farms is often wasted when it is not required.

The process seems both innovative and technically sound, but with research centers such as the University of Birmingham, companies like Highview and the UK Government all committing resources to the technology, will it live up to its promise?

Williams, who expects the center for cryogenic energy storage's new labs to be up and running within 9 months with the goal of studying the CES process from beginning to end, believes CES's advantages make it well suited to addressing intermittent generation problems. The only by-product of the process is cold air, no rare or toxic materials are required, and the component parts of a CES unit are familiar and proven technologies, reducing the technology risk.

"Cryogenic storage systems are well-suited to capturing electricity from renewables as they can be easily scaled according to the electricity coming in by adjusting the size of the liquefaction unit, the total amount of energy that can be captured by having bigger or smaller containers, and the delivered power through the size of turbine installed," Williams says.

> The system doesn't have geographic or geological constraints, like pumped hydro storage or compressed air energy storage, so can be placed near generation or demand centers.

The strongest proof-of-concept for CES in the United Kingdom at the moment is Highview's pilot plant in Slough, which has been installed at an 80 MW biomass power station since 2010. This working unit, which has been used as a test-bed for collaborative research projects with Leeds University, hosts a 60 t liquid nitrogen tank capable of storing around 2.5 MWh of energy or enough to power 300 homes for about 8 h. The project is central to Williams and the new center's research efforts.

"The plant has shown that the CES system is technically achievable and meets expectations of performance," he says. "It has validated process modeling undertaken by Highview, giving confidence as to how a bigger scale plant will perform. The plant will be the ideal test-bed for new materials and processes developed at Birmingham. Professor Yulong Ding, currently at Leeds, who has the experience of working in cryogenic energy storage and helping develop the pilot plant, is coming to Birmingham in October as director of the center."

With its favorable economics, proven process, and unique operational advantages, CES could be the perfect technology to dominate the power market's increasingly valuable energy storage niche. In doing so, it could prove to be a decisive factor in the success of intermittent renewables.

But as Williams' attests, to capitalize on the momentum and maintain the United Kingdom's position as a world leader in the technology, the commitment to CES must grow from here.

However, in today's world renewable source of energy with possible innovative technology, cryogenic storage offers a new hope for renewable source of energy, and the world's largest cold energy storage plant is being commissioned at a site near Manchester in the United Kingdom.

The cryogenic energy facility stores power from renewable or off-peak generation by chilling air into liquid form. When the liquid air warms up, it expands and can drive a turbine to make electricity as it is demonstrated in Fig. 3.4 here.

"The capital equipment needs to be followed up with funding for skilled researchers to make use of the new facilities, support to help the scale-up of the technologies

Fig. 3.4 Liquid air reheat and expansion to gas (Courtesy of Highview Power Storage)

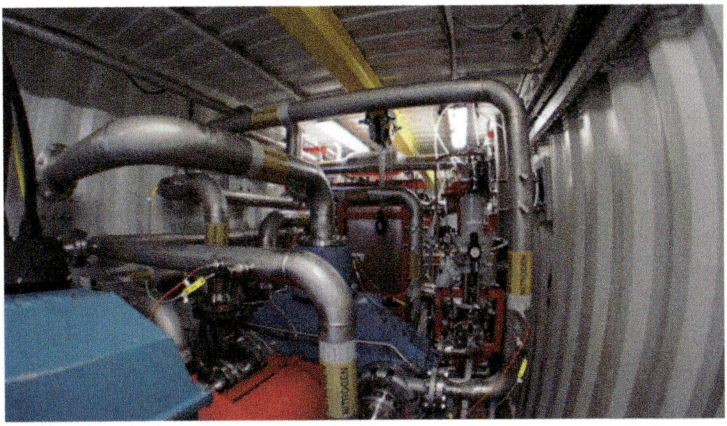

Fig. 3.5 All the component parts involved in the CES chain (Courtesy of Highview Power Storage)

being developed, and policy signals which will create the market for energy storage to draw in further private sector investment," he says. "The innovation process should be coordinated and strategic, or the advances we make in the United Kingdom will be lost again to the benefit of other countries."

Figure 3.5 is a presentation of all the component parts in the CES chain.

Note that, when the "liquid air" is reheated, it expands to a gas that drives a turbine.

Electricity demand varies, influenced by factors like time of the day and season. The National Grid is prepared for surges in demand, with power stations on standby ready to crank up the power.

However, dealing with these peaks and troughs will become increasingly difficult as coal-fired power stations close down, and more intermittent renewable energy like wind and solar comes online. In 2015, renewables provided almost a quarter of UK electricity.

This is useful for any off-peak generation; there is nothing about it which is really specific to renewables. But increasing reliance on renewables—which are more intermittent (think solar, wind) than traditional power sources—is driving increased development of such load-leveling systems, hence the headline. Also, it is very useful for capturing waste heat from power generation plants.

3.6 Cryogenic Energy Storage (CES) in Nuclear Power Plants

Recent years have seen a renewed interest in increasing nuclear power generation in both developed and developing countries due to energy security and environmental considerations. The latter is also very much associated with reducing the carbon footprint because of the highest percentage of carbon emission from the power generation sector and very limited contribution from renewable energy to total power supply.

Nuclear power plants (NPPs) feature high capital costs and low operating costs. The costs of energy from such a capital-intensive technology can be low if the facilities are operated at full capacity, and therefore NPPs have been mainly used as a baseload source of electricity production. However with increasing installations, the capacity of nuclear power may exceed the baseload of power grids. For example, in France the nuclear power contributes about 53% of the country's total installation capacity and generates about 79% of the overall electricity. In these circumstances, the excessive electricity at off-peak times has to be either exported to other countries or stored for later use. If the above measures fail to balance the generation and demand of the NPPs have to be downregulated regularly. When the NPPs operate at a part capacity, the cost of electricity production becomes very high. Furthermore, frequent changes in the load affect strongly on the aging of the equipment and the performance of the fuel, and hence causing problems in both the economic and safety aspects. See Fig. 3.6 as NPP putting electricity into grid.

Energy storage in nuclear power plants resides on a novel method of integration of nuclear power generation with cryogenic energy storage (CES) to achieve an effective time shift of the electrical power output. CES stores excess electricity in the form of cryogen (i.e., liquid air/nitrogen) through an air liquefaction process at off-peak hours and recover the stored power by expanding the cryogen at peak hours. See Fig. 3.5.

To balance the demand and supply at off-peak hours, nuclear power plants often have to be downregulated particularly when the installations exceed the baseload requirements. Part-load operations not only increase the electricity cost but also impose a detrimental effect on the safety and lifetime of the nuclear power plants. We propose a novel solution by integrating nuclear power generation with cryogenic

Fig. 3.6 Typical nuclear reactor power plant producing electricity

energy storage (CES) technology to achieve an effective time shift of the electrical power output. CES stores excess electricity in the form of cryogen (liquid air/nitrogen) through an air liquefaction process at off-peak hours and recover the stored power by expanding the cryogen at peak hours. The combination of nuclear power generation and the CES technologies provides an efficient way to use thermal energy of nuclear power plants in the power extraction process, delivering around three times the rated electrical power of the nuclear power plant at peak hours, thus effectively shaving the peak.

Simulations are carried out on the proposed process, which show that the round trip efficiency of the CES is higher than 70% due to the elevated topping temperature in the superheating process and thermal efficiency also substantially increased as well. See Fig. 3.7 as suggested concept for CES innovation [1].

As an example of this methodology, in France the nuclear power contributes about 53% of the country's total installation capacity and generates about 79% of the overall electricity. Under these conditions and circumstances, the excessive electricity at off-peak times has to be either exported to other countries or stored for later use, when the demand is asking for it. If the above measures fail to balance the generation and demand, the nuclear power plants (NPPs) have to be downregulated regularly. When the NPPs operate at a part capacity, the cost of electricity production becomes very high.

Furthermore frequent changes in the load affected strongly on the aging of the equipment and the performance of the fuel, and hence causing problems in both the economic and safety aspects [2].

One of main advantages of the CES technology is its highly efficient heat-to-power conversion in energy extraction process using cryogen itself as the working fluid [3–4]. Due to the constrained working pressure and temperature in steam generators, the thermal efficiency of pressurized water NPPs is only around 30–32%,

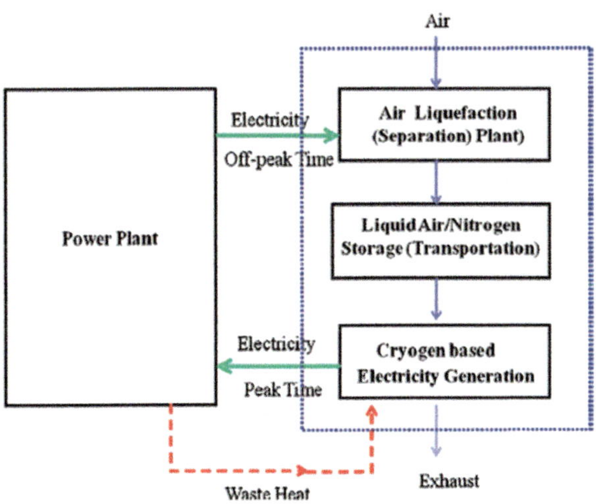

Fig. 3.7 Schematic diagram of a CES system [1]

which is much lower than that of fossil fuel-fired power plants [5]. The integration of NPPs and CES technology could increase the thermal efficiency of nuclear heat utilization at peak hours and as a result the net power output. The round trip efficiency of the energy storage could also be improved significantly. In this paper, a specific method is proposed to integrate the CES and a nuclear power plant by using pressurized water NPP as an example.

A considerable effort and research has been made to deal with load shift of NPPs, and the conventional method is pumping hydro. More recent development of new innovative approaches to the use of excess electricity for maintaining the NPPs at nearly the full load are on the way. These include steam accumulator storage, large-scale hydrogen production and storage, and geothermal heat storage. However in this section, utilization of cryogenic energy storage (CES) for the load shift of NPPs is suggested, and Fig. 3.7 is an illustration of such CES technology [1].

As Fig. 3.7 illustrates, during off-peak hours, when electricity is the cheapest and demand for electricity is the lowest, particularly during nights and weekend, liquid air/nitrogen is produced in an air liquefaction and separation plant and stored in cryogenic tanks at close to the atmospheric pressure. During peak hours, the cryogenic liquid is heated up by using the environmental heat and then superheated using other heat sources in case of availability. Boiling of the cryogenic liquid will form a high-pressure gas to drive an expander (e.g., turbine) to produce electricity [1].

One fully configured proposed system of NPPs is for pressurized water reactor (PWR) and load regulation schematically is presented in Fig. 3.6, which is illustration of a typical pressurized water nuclear power plant [1].

Note that pressurized water nuclear power reactors as generation three (GEN-III) reactor design account for a large portion of the world's nuclear power plants [6, 7]. Basic concept for these reactors to generate electricity works on the idea of that nuclear fuel in the reactor vessel is engaged in a fission chain reaction, heating primary coolant (water) in the primary coolant loop. See Fig. 3.8.

The heated primary coolant is pumped by Pump 1 into the high-pressure side of Heat exchanger 1 to transfer heat to the secondary coolant (water) in the low-pressure

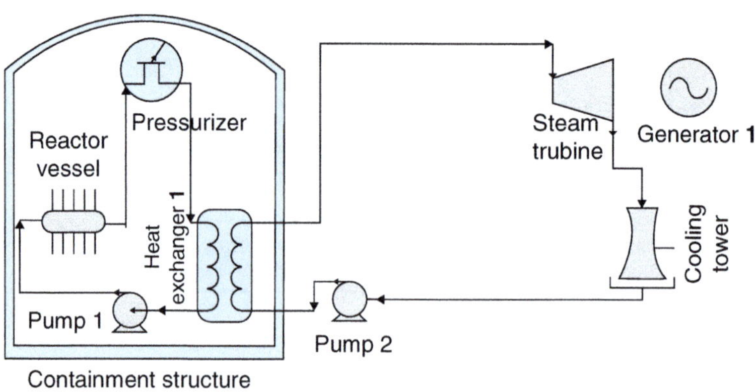

Fig. 3.8 Schematic diagram of typical pressurized water NPP [1]

side of the heat exchanger, leading to vaporization of the secondary coolant to give high-pressure steam. The high-pressure steam then expands in a steam turbine, which drives Generator 1 to produce electrical power. After the expansion in the turbine, the working fluid becomes water-steam mixture, which, upon cooling down, condenses in the cooling tower to give liquid phase water. The water is then pumped back into Heat exchanger 1 by Pump 2, completing the cycle [1].

Downregulation of the load of pressurized water NPPs in the cases of low demands includes the control assemblies being inserted into the reactor vessel and associated changes of the coolants [8]. Apart from the safety issue and lifetime reduction, such operation modes also suffer two challenges in load following. First is the limited power changing gradient which normally takes a few hours to achieve about half load. Second is that the downregulation of NPPs only balances the generation and demand at trough hours, while other plants such as gas-fired power stations have to be employed to meet the peak demands.

3.6.1 Proposed Combined Cryogenic Energy Storage (CES) in Nuclear Power Plants

An integrated NPP and CES system is suggested by Li et al. [1], which may have the potential to resolve the issues associated with the load regulation of NPPs as depicted in Fig. 3.7 schematically. As Ref. [1] describes and shows, the principle of the integrated system consists of a NPP subsystem and the CES subsystem. The NPP subsystem in the integrated system is similar to the conventional pressurized water NPP as shown in Fig. 3.8 [1].

The only difference lies in that there are two three-way valves in the secondary loop, which enables the working fluid to feed into either the steam turbine to produce electricity or Heat exchanger 4 to superheat high-pressure air in CES subsystem (see Fig. 3.9). The CES subsystem consists of an air liquefaction unit in the left part and a cryogenic energy extraction unit in the right-bottom part of Fig. 3.9. The integrated systemhas three operational modes depending on the end users' demands:

1. Energy storage mode: At trough hours when the demand is much lower than the rated power of the NPP, the NPP operates in a traditional way to drive the steam turbine to produce electricity, and the excessive power is used to drive the air liquefaction unit to produce liquid air (energy is stored in liquid air). In this process, dry air stream (2) and return gas stream (10) are mixed and compressed to an elevated pressure (5) by a two-stage compressor (Compressor 1 and Compressor 2) with intercooling (3–4) in heat exchanger (5). After rejecting heat via Heat exchanger 6 in the main cold box (Process 5–6), the high-pressure air is cooled to the lowest temperature level, followed by a near-isentropic expansion process in the so-called cryo-turbine to give liquid air. A fraction of the product is vaporized in the cryogenic tank and introduced back to the main cold box through Heat exchanger 6 and the intercooler Heat exchanger 5 to supply part of

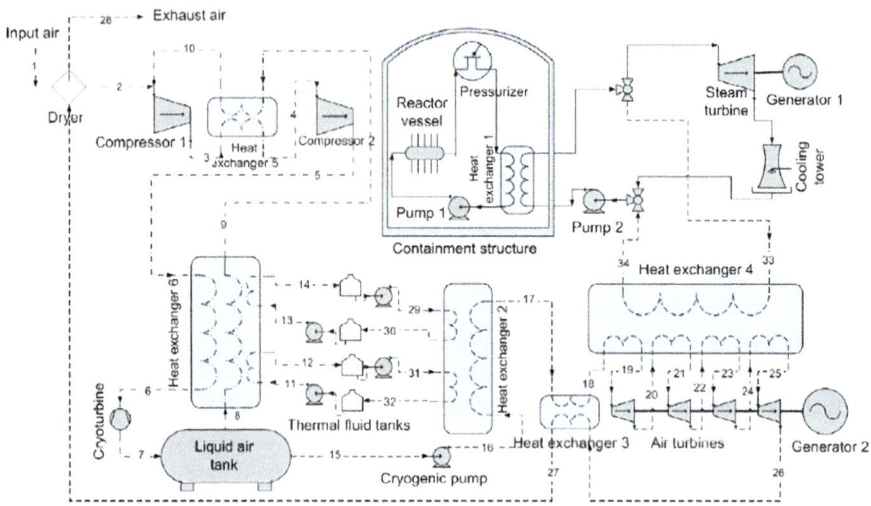

Fig. 3.9 Principle of an integrated NPP-CES system

the cold energy. The remaining cold requirement of the main cold box is met by the cold stored in thermal fluid tanks (see more details in later discussion).

2. Energy release mode: At peak hours when the demand is higher than NPP's rated power, the cryogenic energy extraction unit is turned onto produce additional power. In this process, liquid air in liquid air tank is pressurized first by the cryogenic pump. The high-pressure air transfers its cold energy to the cold storage media in Process 16–17 via Heat exchanger 2 followed by a preheating process in Heat exchanger 3 by the exhaust gas stream (Process 26–27). In the meantime, the secondary coolant in NPP is introduced into Heat exchanger 4 by switching on the three-way valves. This leads to further increase in the temperature and pressure of the air, which, upon expansion in a four-stage air turbine with interheating in Heat exchanger 4, drives Generator 2 to produce electricity. The exhaust air is used first to preheat the high-pressure air in Heat exchanger 3 and then to regenerate the desiccant in Dryer.
3. Conventional mode: At non-trough and non-peak hours, both the air liquefaction unit and cryogenic energy extraction unit are switched off so that the NPP operates in a conventional way to drive the steam turbine to produce electricity. With thermal insulation technologies, cold dissipation from liquid air tank and thermal fluid tanks is expected to be low.

More details can be found in Ref. [1], where one can see that the air liquefaction subsystem works in a similar process as the simplest Linde-Hampson liquefier (i.e., Zohuri) [9] except for taking advantages of the external cold energy in Heat exchanger 6.

References

1. Y. Li, H. Cao, S. Wang, Y. Jin, D. Li, X. Wang, Y. Ding, Load shifting of nuclear power plants using cryogenic energy storage technology. Appl. Energy **113**, 1710–1716 (2014). Journal homepage: www.elsevier.com/locate/apenergy
2. A.K. Raj, *Cryogenics Energy Storage in Nuclear Plants* (Tamil Nadu, 2015)
3. Y. Li, H. Chen, Y. Ding, Fundamentals and applications of cryogen as a thermal energy carrier: A critical assessment. Int. J. Therm. Sci. **49**, 941–949 (2010)
4. Y. Li, H. Chen, X. Zhang, C. Tan, Y. Ding, Renewable energy carriers: Hydrogen or liquid air/nitrogen? Appl. Therm. Eng. **30**, 1985–1990 (2010)
5. M.A. Rosen, Energy- and exergy-based comparison of coal-fired and nuclear steam power plants. Exergy Int. J. **1**, 180–192 (2001)
6. E.E. Michaelides, *Nuclear Power Plants: Alternative Energy Sources* (Springer, Berlin/Heidelberg, 2012), pp. 131–172
7. M.V. Kothare, B. Mettler, M. Morari, P. Bendotti, C.M. Falinower, Level control in the steam generator of a nuclear power plant. IEEE Trans. Control Syst. Technol. **8**, 55–69 (2000)
8. H. Ludwig, T. Salnikova, A. Stock, U.W. Erlangen, Load cycling capabilities of German nuclear power plants. Int. J. Nucl. Power **55**, 2–8 (2010)
9. B. Zohuri, *Physics of Cryogenic an Ultra-Low Temperature Phenomena* (Elsevier, 2017)

Chapter 4
Types of Renewable Energy

The United States currently relies heavily on coal, oil, and natural gas for its energy. Fossil fuels are nonrenewable, that is, they draw on finite resources that will eventually dwindle, becoming too expensive or too environmentally damaging to retrieve. In contrast, many types of renewable energy resources—such as wind and solar energy—are constantly replenished and will never run out. Most renewable energy comes either directly or indirectly from the sun. Sunlight, or solar energy, can be used directly for heating and lighting homes and other buildings, for generating electricity, and for hot water heating, solar cooling, and a variety of commercial and industrial uses. The sun's heat also drives the winds, whose energy is captured with wind turbines. Then, the winds and the sun's heat cause water to evaporate. When this water vapor turns into rain or snow and flows downhill into rivers or streams, its energy can be captured using hydroelectric power. Along with the rain and snow, sunlight causes plants to grow. The organic matter that makes up those plants is known as biomass. Biomass can be used to produce electricity, transportation fuels, or chemicals. The use of biomass for any of these purposes is called bioenergy.

4.1 Introduction

Most renewable energy comes either directly or indirectly from the sun. Sunlight, or solar energy, can be used directly for heating and lighting homes and other buildings, for generating electricity, and for hot water heating, solar cooling, and a variety of commercial and industrial uses.

The sun's heat also drives the winds, whose energy is captured with wind turbines. Then, the winds and the sun's heat cause water to evaporate. When this water vapor turns into rain or snow and flows downhill into rivers or streams, its energy can be captured using hydroelectric power.

B. Zohuri, *Hybrid Energy Systems*,
https://doi.org/10.1007/978-3-319-70721-1_4

Not all renewable energy resources come from the sun. Geothermal energy taps the Earth's internal heat for a variety of uses, including electric power production and the heating and cooling of buildings. And the energy of the ocean's tides comes from the gravitational pull of the moon and the sun upon the Earth.

In fact, ocean energy comes from a number of sources. In addition to tidal energy, there's the energy of the ocean's waves, which are driven by both the tides and the winds. The sun also warms the surface of the ocean more than the ocean depths, creating a temperature difference that can be used as an energy source. All these forms of ocean energy can be used to produce electricity.

Along with the rain and snow, sunlight causes plants to grow. The organic matter that makes up those plants is known as biomass. Biomass can be used to produce electricity, transportation fuels, or chemicals. The use of biomass for any of these purposes is called bioenergy.

Hydrogen also can be found in many organic compounds, as well as water. It's the most abundant element on the Earth. But it doesn't occur naturally as a gas. It's always combined with other elements, such as with oxygen to make water. Once separated from another element, hydrogen can be burned as a fuel or converted into electricity.

There are many sources of energy that are renewable and considered to be environmentally friendly and harness natural processes. These sources of energy provide an alternate "cleaner" source of energy, helping to negate the effects of certain forms of pollution. All of these power generation techniques can be described as renewable since they are not depleting any resource to create the energy. While there are many large-scale renewable energy projects and production, renewable technologies are also suited to small off-grid applications, sometimes in rural and remote areas, where energy is often crucial in human development.

So far based on existing technologies that do exist, we have listed top renewable energy source here as well as briefly explaining each one of them from top view point of view as well as what are the different types of renewable energy.

4.2 What Are the Different Types of Renewable Energies?

Renewable energy is becoming an increasingly important issue in today's world. In addition to the rising cost of fossil fuels and the threat of climate change, there have also been positive developments in this field which include improvements in efficiency as well as diminishing prices.

All of this has increased the demand for alternative energy and accelerated the transition toward cleaner, more sustainable methods of electrical power. However, it is important to note that there are many kinds—biomass, solar, wind, tidal, and geothermal power—and that each has its own share of advantages and drawbacks.

Fig. 4.1 Illustration of biomass briquettes

4.2.1 Biomass

Biomass is an industry term for getting energy by burning wood and other organic matter. Burning biomass releases carbon emissions but has been classed as a renewable energy source in the European Union (EU) and United Nation (UN) legal frameworks, because plant stocks can be replaced with new growth. Also, since the plants build themselves using carbon dioxide and release oxygen as they grow, the net balance of the carbon dioxide after the matter has burned is zero, meaning no extra carbon dioxide is added to the atmosphere. It has become popular among coal power stations, which switch from coal to biomass in order to convert to renewable energy generation without wasting existing generating plant and infrastructure. Biomass most often refers to plants or plant-based materials that are not used for food or feed and are specifically called lignocellulosic biomass. Biomass briquettes are example fuel for production of dendrothermal energy as it is illustrated in Fig. 4.1.

Biomass is organic material that comes from plants and animals, and it is a renewable source of energy.

Biomass contains stored energy from the sun. Plants absorb the sun's energy in a process called photosynthesis. When biomass is burned, the chemical energy in biomass is released as heat. Biomass can be burned directly or converted to liquid biofuels or biogas that can be burned as fuels. Examples of biomass and their uses for energy:

- Wood and wood processing wastes—burned to heat buildings, to produce process heat in industry, and to generate electricity
- Agricultural crops and waste materials—burned as a fuel or converted to liquid biofuels
- Food, yard, and wood waste in garbage—burned to generate electricity in power plants or converted to biogas in landfills
- Animal manure and human sewage—converted to biogas, which can be burned as a fuel

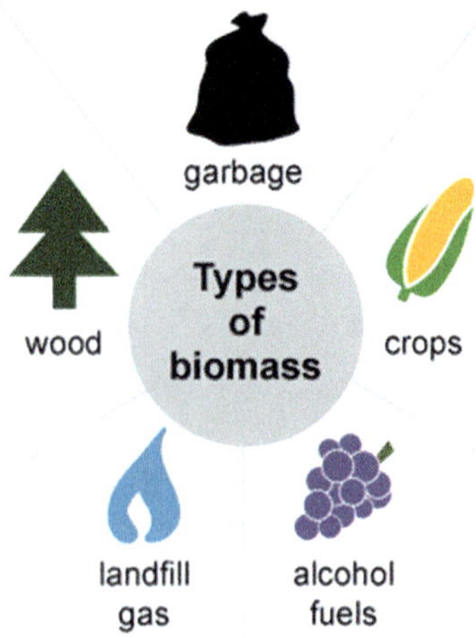

Fig. 4.2 Types of biomass illustration

Converting biomass to other forms of energy will come by burning as one way to release the energy in biomass. Biomass can be converted to other usable forms of energy such as methane gas or transportation fuels such as ethanol and biodiesel.

Methane gas is a component of landfill gas or biogas that forms when garbage, agricultural waste, and human waste decompose in landfills or in special containers called digesters.

Crops such as corn and sugar cane are fermented to produce fuel ethanol for use in vehicles. Biodiesel, another transportation fuel, is produced from vegetable oils and animal fats.

In order to calculate or estimate how much biomass is used for fuel, we can see that roughly speaking, biomass fuels provide about 5% of the primary energy used in the United States in 2016 according to account by the Department of Energy. From this overall total source of energy, about 48% was from biofuels mainly ethanol, 41% was from wood and wood-derived biomass, and about 11% was from the biomass in municipal waste. Researchers are trying to develop ways to use more biomass for fuel. See Fig. 4.2.

As we have stated so far, the most widely used form of renewable energy is biomass. Biomass simply refers to the use of organic materials and converting them into other forms of energy that can be used. Although some forms of biomass have been used for centuries—such as burning wood—other, newer methods are focused on those that do not produce carbon dioxide. The other process aspect of this renewable source of energy is photosynthesis, and in this process, plants convert radiant energy from the sun into chemical energy in the form of glucose or sugar as it is shown in its chemical reaction here. See Fig. 4.3 as well.

$$\underbrace{6H_2O}_{\text{Water}} + \underbrace{CO_2}_{\substack{\text{Carbon}\\\text{Dioxide}}} + \underbrace{\text{Radiant Energy}}_{\text{Sunlight}} \rightarrow \underbrace{C_6H_{12}O_6}_{\text{Glucose}} + \underbrace{6O_2}_{\text{Oxygen}}$$

Fig. 4.3 Demonstration of photosynthesis process for biomass

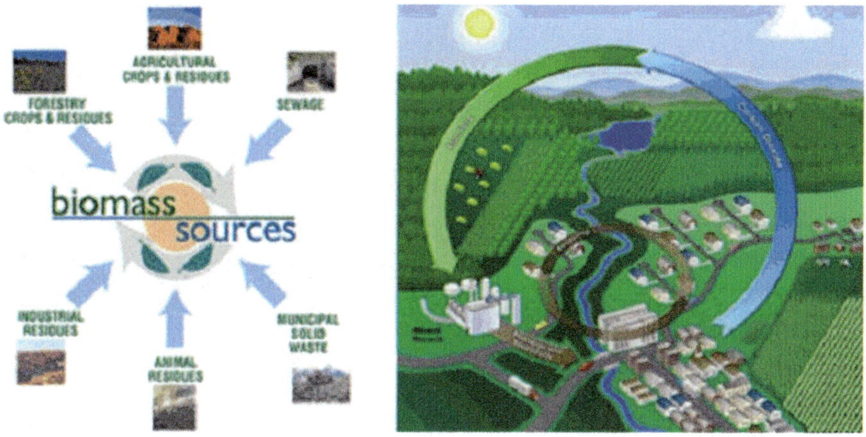

Fig. 4.4 Illustration of biomass clean burning

Figure 4.4 shows few examples that are part of clean burning biofuels that are alternative to oil and gas. Unlike fossil fuels, which are produced by geological processes, a biofuel is produced through biological processes—such as agriculture and anaerobic digestion.

Common fuels associated with this process are bioethanol, which is created by fermenting carbohydrates derived from sugar or starch crops (such as corn, sugarcane, or sweet sorghum) to create alcohol.

Another common biofuel is known as biodiesel, which is produced from oils or fats using a process known as transesterification—where acid molecules are exchanged for alcohol with the help of a catalyst. These types of fuels are popular alternatives to gasoline and can be burned in vehicles that have been converted to run on them [1].

4.2.2 Solar Power

Many power plants today use fossil fuels as a heat source to boil water. The steam from the boiling water rotates a large turbine, which activates a generator that produces electricity. However, a new generation of power plants, with concentrating

Fig. 4.5 A 25 kW dish stirling system (Courtesy of Stirling Energy Systems)

solar power systems, uses the sun as a heat source. There are three main types of concentrating solar power systems: parabolic trough, dish/engine, and power tower.

Parabolic trough systems concentrate the sun's energy through long rectangular, curved (U-shaped) mirrors. The mirrors are tilted toward the sun, focusing sunlight on a pipe that runs down the center of the trough. This heats the oil flowing through the pipe. The hot oil then is used to boil water in a conventional steam generator to produce electricity.

A dish/engine system uses a mirrored dish (similar to a very large satellite dish). The dish-shaped surface collects and concentrates the sun's heat onto a receiver, which absorbs the heat and transfers it to fluid within the engine. The heat causes the fluid to expand against a piston or turbine to produce mechanical power. The mechanical power is then used to run a generator or alternator to produce electricity. See Fig. 4.5, which shows a 25 kW Dish Stirling system catches its last rays of light at the end of the day [1].

A power tower system uses a large field of mirrors to concentrate sunlight onto the top of a tower, where a receiver sits. This heats molten salt flowing through the receiver. Then, the salt's heat is used to generate electricity through a conventional steam generator. Molten salt retains heat efficiently, so it can be stored for days before being converted into electricity. That means electricity can be produced on cloudy days or even several hours after sunset.

Solar power (aka photovoltaics) is one of the most popular, and fastest-growing, sources of alternative energy. Here, the process involves solar cells (usually made from slices of crystalline silicon) that rely on the photovoltaic (PV) effect to absorb photons and convert them into electrons. Meanwhile, solar thermal power (another form of solar power) relies on mirrors or lenses to concentrate a large area of sunlight or solar thermal energy (STE) onto a small area (i.e., a solar cell).

Fig. 4.6 The Ivanpah Solar Power Facility in California

Initially, photovoltaic power was only used for small to medium-sized operations, ranging from solar powered devices (like calculators) to household arrays. However, ever since the 1980s, commercial concentrated solar power plants have become much more common. Not only are they a relatively inexpensive source of energy where grid power is inconvenient, too expensive, or just plain unavailable; increases in solar cell efficiency and dropping prices are making solar power competitive with conventional sources of power (i.e., fossil fuels and coal).

Figure 4.6 is an illustration of solar power facility in California called Ivanpah, showing its three towers delivering concentrated solar power, and Fig. 4.7 is the presentation of a solar farm along with its reflecting mirror of photovoltaics. Figure 4.7 is the Gemasolar solar power plant, situated near Seville in Spain.

Today, solar power is also being increasingly used in grid-connected situations as a way to feed low-carbon energy into the grid. By 2050, the International Energy Agency anticipates that solar power—including STE and PV operations—will constitute over 25% of the market, making it the world's largest source of electricity (with most installations being deployed in China and India).

Solar cells convert sunlight directly into electricity. Solar cells are often used to power calculators and watches. They are made of semiconducting materials similar to those used in computer chips. When sunlight is absorbed by these materials, the solar energy knocks electrons loose from their atoms, allowing the electrons to flow through the material to produce electricity. This process of converting light (photons) to electricity (voltage) is called the photovoltaic (PV) effect.

Solar cells are typically combined into modules that hold about 40 cells; a number of these modules are mounted in PV arrays that can measure up to several meters on a side. These flat-plate PV arrays can be mounted at a fixed angle facing south, or they can be mounted on a tracking device that follows the sun, allowing them to

Fig. 4.7 The Gemasolar solar power plant, situated near Seville in Spain (Courtesy of Torresol Energy)

capture the most sunlight over the course of a day. Several connected PV arrays can provide enough power for a household; for large electric utility or industrial applications, hundreds of arrays can be interconnected to form a single, large PV system.

Thin-film solar cells use layers of semiconductor materials only a few micrometers thick. Thin-film technology has made it possible for solar cells to now double as rooftop shingles, roof tiles, building facades, or the glazing for skylights or atria. The solar cell version of items such as shingles offers the same protection and durability as ordinary asphalt shingles.

Some solar cells are designed to operate with concentrated sunlight. These cells are built into concentrating collectors that use a lens to focus the sunlight onto the cells. This approach has both advantages and disadvantages compared with flat-plate PV arrays. The main idea is to use very little of the expensive semiconducting PV material while collecting as much sunlight as possible. But because the lenses must be pointed at the sun, the use of concentrating collectors is limited to the sunniest parts of the country. Some concentrating collectors are designed to be mounted on simple tracking devices, but most require sophisticated tracking devices, which further limit their use to electric utilities, industries, and large buildings.

The performance of a solar cell is measured in terms of its efficiency at turning sunlight into electricity. Only sunlight of certain energies will work efficiently to create electricity, and much of it is reflected or absorbed by the materials that make up the cell. Because of this, a typical commercial solar cell has an efficiency of 15%—about one-sixth of the sunlight striking the cell generates electricity. Low efficiencies mean that larger arrays are needed, and that means higher cost. Improving solar cell efficiencies while holding down the cost per cell is an important goal of the PV

Fig. 4.8 Wind power farm in Denmark

industry, NREL researchers, and other US Department of Energy (DOE) laboratories, and they have made significant progress. The first solar cells, built in the 1950s, had efficiencies of less than 4% [1].

4.2.3 *Wind Power*

Wind power has been used for thousands of years to push sails, power windmills, or to generate pressure for water pumps. Harnessing the wind to generate electricity has been the subject of research since the late nineteenth century. However, it was only with major efforts to find alternative sources of power in the twentieth century that wind power has become the focal point of considerable research and development.

Compared to other forms of renewable energy, wind power is considered very reliable and steady, as wind is consistent from year to year and does not diminish during peak hours of demand. Initially, the construction of wind farms was a costly venture. But thanks to recent improvements, wind power has begun to set peak prices in wholesale energy markets worldwide and cut into the revenues and profits of the fossil fuel industry. In Denmark, for example, wind power accounts for 28% of the country's electrical production and is now cheaper than coal power; see Fig. 4.8.

According to a report issued this past March by the Department of Energy, the growth of wind power in the United States could lead to even more highly skilled jobs in many categories. Titled "Wind Vision: A New Era for Wind Power in the United States," the document indicates that by 2050, the industry could account for as much as 35% of the US' electrical production.

In addition, last year, the Global Wind Energy Council and Greenpeace International came together to publish a report titled "Global Wind Energy Outlook 2014." This report stated that worldwide, wind power could provide as much as 25–30% of global electricity by 2050. At the time of the report's writing, commercial installations in more than 90 countries had a total capacity of 318 gigawatts (GW), providing about 3% of global supply [1].

4.2.4 *Tidal Power*

Similar to wind power, tidal power is considered to be a potential source of renewable energy because tides are steady and predictable. Much like windmills, tide mills have been used since the days of ancient Rome and the middle ages. Incoming water was stored in large ponds, and as the tides went out, they turned waterwheels that generated mechanical power to mill grain.

In ocean, there are two types of energy that can be produced by the ocean: thermal energy from the sun's heat and mechanical energy from the motion of tides and waves. "Ocean thermal energy" can be converted into electricity using a few different systems that rely on warm surface water temperatures. "Ocean mechanical energy" harnesses the ebbs and flows of tides caused by the rotation of the Earth and the gravitational influence of the moon. Energy from wind-driven waves can also be converted and used to cut business electricity costs. There are also lesser-developed technologies that leverage ocean currents, ocean winds and salinity gradients as sources of power conversion.

Ocean energy is an evolving sector for alternative energy production, but with over 70% of the surface of our planet covered by the ocean, its future looks promising. Commercial and public applications for this energy resource are limited to geography and regulatory guidelines. Other practical uses for energy derived from the ocean include:

- Cold ocean water from deep below the surface can be used to cool buildings (with desalinated water often produced as a by-product).
- Seaside communities can employ the methods to tap natural ocean energy described above to supplement municipal power and energy needs.

It was only in the nineteenth century that the process of using falling water and spinning turbines to create electricity was introduced in the United States and Europe. And it has only been since the twentieth that these sorts of operations have been retooled for construction along coastlines and not just rivers.

Figure 4.9 is the illustration of an artistic concept of series of the Carnegie Wave Energy's tidal system, where buoys anchored to the sea floor and use swells to move a series of pump.

Traditionally, tidal power has suffered from relatively high cost and limited availability of sites with sufficiently high tidal ranges or flow velocities. However, many recent technological developments and improvements, both in design and turbine technology, indicate that the total availability of tidal power may be much

Fig. 4.9 Carnegie wave energy's tidal system (Courtesy of Carnegie Wave Energy)

higher than previously assumed and that economic and environmental costs may be brought down to competitive levels.

The world's first large-scale tidal power plant is the Rance Tidal Power Station in France, which became operational in 1966. And in Orkney, Scotland, the world's first marine energy test facility—the European Marine Energy Centre (EMEC)—was established in 2003 to start the development of the wave and tidal energy industry in the United Kingdom.

In 2015, the world's first grid-connected wave power station (CETO, named after the Greek goddess of the sea) went online off the coast of Western Australia. Developed by Carnegie Wave Energy, this power station operates underwater and uses undersea buoys to pump a series of seabed-anchored pumps, which in turn generate electricity [1].

4.2.5 *Geothermal*

Geothermal electricity is another form of alternative energy that is considered to be sustainable and reliable. In this case, heat energy is derived from the Earth—usually from magma conduits, hot springs, or hydrothermal circulation—to spin turbines or heat buildings. It is considered reliable because the Earth contains 1031 J worth of heat energy, which naturally flows to the surface by conduction at a rate of 44.2 terawatts (TW)—more than double humanity's current energy consumption.

Fig. 4.10 The Krafla: a geothermal power station located in Iceland (Courtesy of Wikipedia Commons/Ásgeir Eggertsson)

One drawback is the fact that this energy is diffuse and can only be cheaply harnessed in certain locations. However, in certain areas of the world, such as Iceland, Indonesia, and other regions with high levels of geothermal activity, it is an easily accessible and cost-effective way of reducing dependence on fossil fuels and coal to generate electricity. Countries generating more than 15% of their electricity from geothermal sources include El Salvador, Kenya, the Philippines, Iceland, and Costa Rica. Figure 4.10 is an artistic presentation of geothermal power station located in Iceland.

As of 2015, worldwide geothermal power capacity amounts to 12.8 GW, which is expected to grow to 14.5–17.6 GW by 2020. What's more, the Geothermal Energy Association (GEA) estimates that only 6.5% of total global potential has been tapped so far, while the IPCC reported geothermal power potential to be in the range of 35 GW to 2 TW [1].

4.3 Top Ten Renewable Energy Sources

There are many sources of energy that are renewable and considered to be environmentally friendly and harness natural processes. These sources of energy provide an alternate "cleaner" source of energy, helping to negate the effects of certain forms of pollution. All of these power generation techniques can be described as renewable since they are not depleting any resource to create the energy. While there are many large-scale renewable energy projects and production, renewable technologies are also suited to small off-grid applications, sometimes in rural and remote areas, where energy is often crucial in human development [2].

The list of these top ten renewable sources of energy is listed below, and they are quotations from Wikipedia so they fall under GNU Free Documentation License (GFDL).

4.3.1 Nuclear Power

Nuclear power is any nuclear technology designed to extract usable energy from atomic nuclei via controlled nuclear reactions. The only method in use today is through nuclear fission, though other methods might one day include nuclear fusion and radioactive decay. All utility-scale reactors heat water to produce steam, which is then converted into mechanical work for the purpose of generating electricity or propulsion. In 2007, 14% of the world's electricity came from nuclear power, with the United States, France, and Japan together accounting for 56.5% of nuclear-generated electricity. There are 439 nuclear power reactors in operation in the world, operating in 31 countries. According to the World Nuclear Association, globally during the 1980s, one new nuclear reactor started up every 17 days on average, and by the year 2015, this rate could increase to one every 5 days. According to a 2007 story broadcast on 60 min, nuclear power gives France the cleanest air of any industrialized country and the cheapest electricity in all of Europe. France reprocesses its nuclear waste to reduce its mass and make more energy. Reprocessing can potentially recover up to 95% of the remaining uranium and plutonium in spent nuclear fuel, putting it into new mixed oxide fuel. This produces a reduction in long-term radioactivity within the remaining waste, since this is largely short-lived fission products, and reduces its volume by over 90%. France is generally cited as the most successful reprocessor, but it presently only recycles 28% (by mass) of the yearly fuel use 7% within France and another 21% in Russia. See Fig. 4.11.

Proponents of nuclear energy contend that nuclear power is a sustainable energy source that reduces carbon emissions and increases energy security by decreasing dependence on foreign oil. Proponents also emphasize that the risks of storing waste are small and can be further reduced by using the latest technology in newer reactors, and the operational safety record in the Western world is excellent when compared to the other major kinds of power plants. Critics believe that nuclear power is a potentially dangerous energy source, with decreasing proportion of nuclear energy in power production, and dispute whether the risks can be reduced through new technology. Proponents advance the notion that nuclear power produces virtually no air pollution, in contrast to the chief viable alternative of fossil fuel. Proponents also point out that nuclear power is the only viable course to achieve energy independence for most Western countries. Critics point to the issue of storing radioactive waste, the history of and continuing potential for radioactive contamination by accident or sabotage, the history of and continuing possibility of nuclear proliferation, and the disadvantages of centralized electricity production.

Fig. 4.11 Typical infrastructure GEN-III nuclear power plant

4.3.2 *Compressed Natural Gas*

Compressed natural gas (CNG) is a fossil fuel substitute for gasoline, diesel, or propane fuel. Although its combustion does produce greenhouse gases, it is a more environmentally clean alternative to those fuels, and it is much safer than other fuels in the event of a spill (natural gas is lighter than air and disperses quickly when released). CNG is used in traditional gasoline internal combustion engine cars that have been converted into bi-fuel vehicles (gasoline/CNG). Natural gas vehicles are increasingly used in Europe and South America due to rising gasoline prices. In response to high fuel prices and environmental concerns, CNG is starting to be used also in light-duty passenger vehicles and pickup trucks, medium-duty delivery trucks, transit and school buses, and trains. Italy currently has the largest number of CNG vehicles in Europe and is the fourth country in the world for number of CNG-powered vehicles in circulation. Canada is a large producer of natural gas, so it follows that CNG is used in Canada as an economical motor fuel. Canadian industry has developed CNG-fueled truck and bus engines, CNG-fueled transit buses, and light trucks and taxis. Both CNG and propane refueling stations are not difficult to find in major centers. During the 1970s and 1980s, CNG was commonly used in New Zealand in the wake of the oil crises but fell into decline after petrol prices receded. See Fig. 4.12, where compressed natural gas is utilized as car fuel and replacing fossil such as gasoline.

Fig. 4.12 Utilization of compressed natural gas as fuel

4.3.3 *Biomass*

Biomass, as a renewable energy source, refers to living and recently dead biological material that can be used as fuel or for industrial production. In this context, biomass refers to plant matter grown to generate electricity or produce, for example, trash such as dead trees and branches, yard clippings, and wood chip biofuel, and it also includes plant or animal matter used for production of fibers, chemicals, or heat. Biomass may also include biodegradable wastes that can be burnt as fuel. Industrial biomass can be grown from numerous types of plants, including miscanthus grass, switch grass, hemp, corn, poplar, willow, sorghum, sugarcane, and a variety of tree species, ranging from eucalyptus to oil palm (palm oil). The particular plant used is usually not important to the end products, but it does affect the processing of the raw material. Production of biomass is a growing industry as interest in sustainable fuel sources is growing. The existing commercial biomass power-generating industry in the United States produces about 0.5% of the US electricity supply. See Fig. 4.13.

Currently, the New Hope Power is the largest biomass power plant in North America. The facility reduces dependence on oil by more than one million barrels per year and, by recycling sugar cane and wood waste, preserves landfill space in urban communities in Florida.

4.3.4 *Geothermal Power*

The Earth's heat is called geothermal energy escapes as steam at a hot springs in Nevada. Geothermal energy is the heat from the Earth. It's clean and sustainable. Resources of geothermal energy range from the shallow ground to hot water and hot rock found a few miles beneath the Earth's surface and down even deeper to the extremely high temperatures of molten rock called magma.

Fig. 4.13 Pile of recycling waste image as fuel source

Almost everywhere, the shallow ground or upper 10 ft of the Earth's surface maintains a nearly constant temperature between 50 °F and 60 °F (10 °C and 16 °C). Geothermal heat pumps can tap into this resource to heat and cool buildings. A geothermal heat pump system consists of a heat pump, an air delivery system (ductwork), and a heat exchanger-a system of pipes buried in the shallow ground near the building. In the winter, the heat pump removes heat from the heat exchanger and pumps it into the indoor air delivery system. In the summer, the process is reversed, and the heat pump moves heat from the indoor air into the heat exchanger. The heat removed from the indoor air during the summer can also be used to provide a free source of hot water.

Geothermal energy is a very powerful and efficient way to extract a renewable energy from the Earth through natural processes. This can be performed on a small scale to provide heat for a residential unit (a geothermal heat pump) or on a very large scale for energy production through a geothermal power plant. It has been used for space heating and bathing since ancient Roman times but is now better known for generating electricity. Geothermal power is cost-effective, reliable, and environmentally friendly but has previously been geographically limited to areas near tectonic plate boundaries. Recent technological advances have dramatically expanded the range and size of viable resources, especially for direct applications such as home heating. The largest group of geothermal power plants in the world is located at the Geysers, a geothermal field in California, USA. As of 2004, five countries (El Salvador, Kenya, the Philippines, Iceland, and Costa Rica) generate more than 15% of their electricity from geothermal sources. Geothermal power requires no fuel and is therefore immune to fluctuations in fuel cost, but capital costs tend to be high. Drilling accounts for most of the costs of electrical plants and exploration of deep resources entails very high financial risks. Geothermal power offers a degree

Fig. 4.14 Artistic depiction of geothermal power plant

of scalability: a large geothermal plant can power entire cities, while smaller power plants can supply rural villages or heat individual homes. Figure 4.14 is an artistic illustration of a geothermal power plant infrastructure.

Geothermal electricity is generated in 24 countries around the world, and a number of potential sites are being developed or evaluated.

In the United States, most geothermal reservoirs of hot water are located in the western states, Alaska, and Hawaii. Wells can be drilled into underground reservoirs for the generation of electricity. Some geothermal power plants use the steam from a reservoir to power a turbine/generator, while others use the hot water to boil a working fluid that vaporizes and then turns a turbine. Hot water near the surface of the Earth can be used directly for heat. Direct-use applications include heating buildings, growing plants in greenhouses, drying crops, heating water at fish farms, and several industrial processes such as pasteurizing milk.

Hot dry rock resources occur at depths of 3–5 miles everywhere beneath the Earth's surface and at lesser depths in certain areas. Access to these resources involves injecting cold water down one well, circulating it through hot fractured rock, and drawing off the heated water from another well. Currently, there are no commercial applications of this technology. Existing technology also does not yet allow recovery of heat directly from magma, the very deep and most powerful resource of geothermal energy.

Many technologies have been developed to take advantage of geothermal energy—the heat from the Earth. The National Renewable Energy Laboratory (NREL) performs research to develop and advance technologies for the following geothermal applications.

Fig. 4.15 Tesla's laboratory

4.3.5 *Radiant Energy*

This natural energy can perform the same wonders as ordinary electricity at less than 1% of the cost. It does not behave exactly like electricity, however, which has contributed to the scientific community's misunderstanding of it. The Methernitha community in Switzerland currently has five or six working models of fuel-less, self-running devices that tap this energy. Nikola Tesla's magnifying transmitter, T. Henry Moray's radiant energy device, Edwin Gray's EMA motor, and Paul Baumann's Testatika machine all run on radiant energy. This natural energy form can be gathered directly from the environment or extracted from ordinary electricity by the method called fractionation. One of the earliest wireless telephones to be based on radiant energy was invented by Nikola Tesla. The device used transmitters and receivers whose resonances were tuned to the same frequency, allowing communication between them. In 1916, he recounted an experiment he had done in 1896. He recalled that "Whenever I received the effects of a transmitter, one of the simplest ways [to detect the wireless transmissions] was to apply a magnetic field to currents generated in a conductor, and when I did so, the low frequency gave audible notes." See Fig. 4.15, which is illustration of Tesla's laboratory.

4.3.6 *Hydroelectricity Power Source*

Hydroelectricity is electricity generated by hydropower, i.e., the production of power through use of the gravitational force of falling or flowing water. It is the most widely used form of renewable energy. Once a hydroelectric complex is constructed, the project produces no direct waste. Small-scale hydro- or micro-hydropower has been an increasingly popular alternative energy source, especially in remote areas where other power sources are not viable. Small-scale hydropower systems can be installed in small rivers or streams with little or no discernible

Fig. 4.16 Water dam with hydroelectricity systems

environmental effect or disruption to fish migration. Most small-scale hydropower systems make no use of a dam or major water diversion but rather use water wheels to generate energy (Fig. 4.16).

This was approximately 19% of the world's electricity (up from 16% in 2003) and accounted for over 63% of electricity from renewable sources. While many hydroelectric projects supply public electricity networks, some are created to serve specific industrial enterprises.

Dedicated hydroelectric projects are often built to provide the substantial amounts of electricity needed for aluminum electrolytic plants, for example. In the Scottish highlands, there are examples at Kinlochleven and Lochaber, constructed during the early years of the twentieth century. The Grand Coulee Dam, long the world's largest, switched to support Alcoa aluminum in Bellingham, Washington, for America's World War II airplanes before it was allowed to provide irrigation and power to citizens (in addition to aluminum power) after the war. In Suriname, the Brokopondo Reservoir was constructed to provide electricity for the Alcoa aluminum industry. New Zealand's Manapouri Power Station was constructed to supply electricity to the aluminum smelter at Tiwai Point.

The most common type of hydroelectric power plant uses a dam on a river to store water in a reservoir. Water released from the reservoir flows through a turbine, spinning it, which in turn activates a generator to produce electricity. But hydroelectric power doesn't necessarily require a large dam. Some hydroelectric power plants just use a small canal to channel the river water through a turbine.

Another type of hydroelectric power plant—called a pumped-storage plant—can even store power. The power is sent from a power grid into the electric generators. The generators then spin the turbines backward, which causes the turbines to pump water from a river or lower reservoir to an upper reservoir, where the power is

Fig. 4.17 Wind power farm

stored. To use the power, the water is released from the upper reservoir back down into the river or lower reservoir. This spins the turbines forward, activating the generators to produce electricity.

A small or micro-hydroelectric power system can produce enough electricity for a home, farm, or ranch.

4.3.7 Wind Power

Wind power is the conversion of wind energy by wind turbines into a useful form, such as electricity or mechanical energy. Large-scale wind farms are typically connected to the local power transmission network with small turbines used to provide electricity to isolated areas. See Fig. 4.17.

Modern wind turbines tower above one of their ancestors—an old windmill used for pumping water.

We have been harnessing the wind's energy for hundreds of years. From old Holland to farms in the United States, windmills have been used for pumping water or grinding grain. Today, the windmill's modern equivalent—a wind turbine—can use the wind's energy to generate electricity.

Wind turbines, like windmills, are mounted on a tower to capture the most energy. At 100 ft (30 m) or more aboveground, they can take advantage of the faster and less turbulent wind. Turbines catch the wind's energy with their propeller-like blades. Usually, two or three blades are mounted on a shaft to form a rotor.

A blade acts much like an airplane wing, thus when the wind blows, a pocket of low-pressure air forms on the downwind side of the blade. The low-pressure air pocket then pulls the blade toward it, causing the rotor to turn. This is called lift. The force of the lift is actually much stronger than the wind's force against the front side of the blade, which is called drag. The combination of lift and drag causes the rotor to spin like a propeller, and the turning shaft spins a generator to make electricity.

Wind turbines can be used as stand-alone applications, or they can be connected to a utility power grid or even combined with a photovoltaic (solar cell) system. For utility-scale sources of wind energy, a large number of wind turbines are usually built close together to form a wind plant. Several electricity providers today use wind plants to supply power to their customers.

Stand-alone wind turbines are typically used for water pumping or communications. However, homeowners, farmers, and ranchers in windy areas can also use wind turbines as a way to cut their electric bills.

Small wind systems also have potential as distributed energy resources. Distributed energy resources refer to a variety of small, modular power-generating technologies that can be combined to improve the operation of the electricity delivery system.

Residential units are entering production and are capable of powering large appliances to entire houses depending on the size. Wind farms installed on agricultural land or grazing areas have one of the lowest environmental impacts of all energy sources. Although wind produces only about 1.5% of worldwide electricity use, it is growing rapidly, having doubled in the 3 years between 2005 and 2008. In several countries, it has achieved relatively high levels of penetration, accounting for approximately 19% of electricity production in Denmark, 11% in Spain and Portugal, and 7% in Germany and the Republic of Ireland in 2008. Wind energy has historically been used directly to propel sailing ships or converted into mechanical energy for pumping water or grinding grain, but the principal application of wind power today is the generation of electricity. As of 2008, Europe leads the world in development of offshore wind power, due to strong wind resources and shallow water in the North Sea and the Baltic Sea, and limitations on suitable locations on land due to dense populations and existing developments. Denmark installed the first offshore wind farms and for years was the world leader in offshore wind power until the United Kingdom gained the lead in October 2008. Other large markets for wind power, including the United States and China, focused first on developing their on-land wind resources where construction costs are lower (such as in the Great Plains of the United States and the similarly wind-swept steppes of Xinjiang and Inner Mongolia in China), but population centers along coastlines in many parts of the world are close to offshore wind resources, which would reduce transmission costs.

4.3.8 Solar Power

Photovoltaic (PV) solar power is harnessing the suns energy to produce electricity. One of the fastest-growing energy sources, new technologies are developing at a rapid pace. Solar cells are becoming more efficient, transportable, and even flexible, allowing for easy installation. PV has mainly been used to power small- and medium-sized applications, from the calculator powered by a single solar cell to off-grid homes powered by a photovoltaic array. The 1973 oil crisis stimulated a rapid rise in the

Fig. 4.18 Solar power panels (photovoltaic)

production of PV during the 1970s and early 1980s. Steadily falling oil prices during the early 1980s, however, led to a reduction in funding for photovoltaic R&D and a discontinuation of the tax credits associated with the Energy Tax Act of 1978. These factors are moderated growth to approximately 15% per year from 1984 to 1996. Since the mid-1990s, leadership in the PV sector has shifted from the United States to Japan and Germany. Between 1992 and 1994, Japan increased R&D funding, established net metering guidelines, and introduced a subsidy program to encourage the installation of residential PV systems. Solar installations in recent years have also largely begun to expand into residential areas, with governments offering incentive programs to make "green" energy a more economically viable option. In Canada the government offers the Renewable Energy Standard Offer Program (RESOP) (Fig. 4.18).

4.3.9 Wave Power

Wave power is the transport of energy by ocean surface waves and the capture of that energy to do useful work—for example, for electricity generation, water desalination, or the pumping of water (into reservoirs). Wave energy can be difficult to harness due to the unpredictability of the ocean and wave direction. Wave farms have been created and are in use in Europe, using floating Pelamis Wave Energy Converters. Most wave power systems include the use of a floating buoyed device and generate energy through a snaking motion or by mechanical movement from the wave's peaks and troughs. Though often co-mingled, wave power is distinct from the diurnal flux of tidal power and the steady gyre of ocean currents. See Fig. 4.19.

Fig. 4.19 Image of Pelamis wave power system

Wave power generation is not currently a widely employed commercial technology although there have been attempts at using it since at least 1890. The world's first commercial wave farm is based in Portugal, at the Aguçadora Wave Park, which consists of three 750 kW Pelamis devices. In the United States, the Pacific Northwest Generating Cooperative is funding the building of a commercial wave-power park at Reedsport, Oregon. The project will utilize the PowerBuoy technology Ocean Power Technologies which consists of modular, ocean-going buoys. The rising and falling of the waves move the buoy-like structure creating mechanical energy which is converted into electricity and transmitted to shore over a submerged transmission line. A 40 kW buoy has a diameter of 12 ft (4 m) and is 52 ft (16 m) long, with approximately 13 ft of the unit rising above the ocean surface. Using the three-point mooring system, they are designed to be installed 1–5 miles (8 km) offshore in water 100–200 ft (60 m) deep.

4.3.10 Tidal Power

Tidal energy can be generated in two ways, tidal stream generators or by barrage generation. The power created through tidal generators is generally more environmentally friendly and causes less impact on established ecosystems. Similar to a wind turbine, many tidal stream generators rotate underwater and are driven by the swiftly moving dense water. Although not yet widely used, tidal power has potential for future electricity generation. Tides are more predictable than wind energy and solar power. Historically, tide mills have been used, both in Europe and on the Atlantic Coast of the United States. The earliest occurrences date from the Middle Ages or even from Roman times. Tidal power is the only form of energy which derives directly from the relative motions of the

Fig. 4.20 Artistic illustration of deep sea farm

Earth-Moon system and to a lesser extent from the Earth-Sun system. The tidal forces produced by the Moon and Sun, in combination with Earth's rotation, are responsible for the generation of the tides. British company Lunar Energy announced that they would be building the world's first tidal energy farm off the coast of Pembrokeshire in Wales.

Figure 4.20 shows the world's first deep-sea tidal-energy farm and will provide electricity for 5000 homes. Eight underwater turbines, each 25 m long and 15 m high, are to be installed on the sea bottom off St David's peninsula. Construction is due to start in the summer of 2008, and the proposed tidal energy turbines, described as "a wind farm under the sea," should be operational by 2010.

4.4 How to Indirectly Participate in Any or All of These Sustainable Energy Solutions

Many private enterprises and public entities (e.g., government agencies and educational institutions) are looking for clean, renewable energy sources to meet their sustainable energy solution needs. The motivation can be financial, driven by regulatory mandates, a desire to be more socially responsible or all the above. Sustainable energy solutions are literally found in the air, deep underground, and in our oceans. Each of the following can be tapped directly or indirectly by organizations looking to go green.

There is another way for your organization to embrace corporate social responsibility and invest in renewable energy that does not require the construction or maintenance of any equipment. Renewable energy certificates (RECs) are tradable, non-tangible energy commodities that confirm electricity was generated by a renewable energy resource and fed into a shared power grid. A certifying agency assigns a unique identification number to each REC produced by a green energy provider.

The REC can then be sold on the open market. Electric utilities, businesses, and public entities can purchase these certificates to fulfill clean energy regulatory requirements or to otherwise reduce their environmental impact. RECs allow buyers to support renewable energy initiatives while also allowing market forces to spur the further development of green energy. Depending on where you are located, you may also be able to purchase renewable energy directly from an offsite power-generating facility.

4.5 Renewable Energy Certificates

Renewable energy certificates (RECs) provide businesses a means to support renewable energy generation and meet clean energy goals. This fact sheet answers some basic questions about RECs and REC purchasing.

Renewable energy certificates (RECs), also known as "green tags," "green certificates," and "renewable energy credits," are tradable instruments which can be used to meet voluntary renewable energy targets as well as to meet compliance requirements for renewable energy policies. A REC is a certificate that indicates the generation of 1 megawatt hour (MWh) of electricity from an eligible source of renewable power. Each REC denotes the underlying generation source, location of generation, and year of generation (aka "vintage"). RECs are often considered to represent a claim to the environmental attributes associated with renewable energy generation.

Policies that require electricity service providers to incorporate a minimum level of renewable energy in their electricity supply create compliance markets or mandatory markets for Renewable Energy Certificates. These energy policies, such as state renewable electricity standards (RES) and renewable portfolio standards (RPS), specify the eligible energy resources or technologies that can be utilized and describe how electricity service providers must comply. (For more information on RES/RPS, see Issue 5 of *The Bottom Line*) [3, 4]. Electricity service providers must produce or obtain RECs in an amount sufficient to meet their renewable energy obligations under an RES/RPS.

Voluntary markets for RECs, while not mandated by law, have developed in response to energy user preferences for green electricity. Retail, commercial, and industrial energy users can meet voluntary renewable energy goals and support the deployment of green power through the purchase of RECs.

As far as ensuring that RECs come from a reputable source is concerned and is able to meet environmental, disclosure, and accounting standard, we can state that the best way to ensure the credibility of RECs is to purchase those certified by an independent third party. The most common certification standard is Green-e®, which is administered by the Center for Resource Solutions. Green-e® certifies renewable electricity products to meet the programs strict environmental and consumer protection standards, which ensure the electricity and its associated RECs are produced by the purported renewable generation facility, delivered in the amount specified, and not claimed by more than one party.

4.5.1 Which Technologies Qualify for Certification?

Under Green-e® standards, electricity generated from the following resources qualifies for certification:

- Solar electric
- Wind
- Biomass
- Low-impact hydropower
- Biomass
- Fuel cells using renewable fuels
- Geothermal

States with RES/RPS requirements apply their own definitions of eligible technologies that produce RECs for compliance purposes.

Now the question is that are RECs and Carbon offsets the same? The answer is No. RECs and carbon are different mechanisms that accomplish different goals.

Carbon offsets allow companies to reduce their greenhouse gas (GHG) emissions liability by purchasing the emission reductions made by another entity. Each carbon offset purchased represents the equivalent of 1 t of carbon dioxide (CO_2) emissions avoided. There are multiple standards that evaluate the quality of these offsets, including the gold standard, the voluntary carbon standard, the Climate Action Reserve, and regulatory standards defined under cap-and-trade schemes.

RECs allow companies to meet renewable energy goals, support renewable power projects, and demonstrate a commitment to clean, renewable electricity. RECs are measured in terms of electricity production, and each REC represents 1 MWh of electricity produced from renewable energy resources. The standards for evaluating RECs (such as Green-e®, described above) are different than those applied to offsets.

The factors that influence REC prices can vary significantly based on a range of factors including, but not limited to, the technology, local supply and demand, and regulations in compliance markets, and each one is defined below:

- *Technology*: REC buyers exhibit preferences for specific technologies for a variety of reasons, including marketing and brand value. These preferences affect the pricing of RECs. For example, RECs from wind farms generally command a premium over RECs from landfill gas.
- *Supply and demand*: Green power purchasers sometimes prefer RECs produced locally as a way of demonstrating support for local businesses and communities. This can create higher prices in areas where production capacity is limited and local demand is high.
- *Regulations*: RECs in compliance markets are generally more expensive than in voluntary markets. Regulations that prescribe renewables volume, technology requirements, and penalties for non-compliance can significantly affect the cost of RECs in a given market. For example, New Jersey has a "solar carve out" in its RES, which requires that a percentage of renewable power come from solar resources. As a result, solar RECs in New Jersey are dramatically more expensive than solar RECs purchased from the voluntary market.

The remaining concern is that what environmental claims can be made for an on-site renewable energy project if the project owner sells the RECs? Companies that install and own on-site renewable power systems can claim the use of green power from their projects. However, companies that choose to sell the RECs from their system to improve project economics give up the right to claim they are buying the green power from the system, even though it is located at their facility. However the difference between buying "GREEN" electricity and RECs is if purchasing green electricity and purchasing RECs accomplish the same goal but follow different paths. In a green electricity transaction, the electricity service provider sources power from a renewable energy project and sells that power directly to their customer. In this transaction, the power provider passes to the customer all RECs associated with the renewable energy purchased. The electricity and RECs are bundled as a single green energy product. Renewable energy projects need to be in the same power grid as the customer in this type of green power transaction [5–7].

In a REC purchase, electricity and assurance of renewable generation are purchased separately. In this "unbundled" approach, the customer buys regular electricity from their electricity service provider but purchases the renewable energy certificates from a REC vendor. In this transaction, the renewable energy project and customer do not need to be in the same grid. An on-line resource created by the Department of Energy that lists REC vendors can be accessed at http://apps3.eere.energy.gov/greenpower/.

Some utilities also offer packages that include electricity and RECs as part of the same purchase, but the RECs are not necessarily from renewable energy projects in the utility's service area. In this case, the utility has purchased the RECs from another party and resells them to their customer. The net effect for the customer is the same as buying RECs plus regular power but with the convenience of combined billing.

Companies purchase RECs, because RECs are a flexible tool to help achieve clean energy goals and support the renewable energy market. They can be attractive to companies in regions where green pricing programs are not offered or are insufficient or where policy support for on-site projects is lacking. By purchasing RECs, businesses do not need to alter existing power contracts to obtain green power. Additionally, RECs are not limited by geographic boundaries or transmission constraints. For corporations with facilities in multiple states or energy grids, a consolidated REC procurement can be part of a strategy to meet overall clean energy goals.

4.5.2 Bottom Line on Renewable Energy Certification

Renewable energy certificates (RECs) provide businesses a means to support renewable energy generation and meet clean energy goals. This fact sheet answers some basic questions about RECs and REC purchasing.

RECs, also known as renewable energy credits, green certificates, green tags, or tradable renewable certificates, represent the environmental attributes of the power produced from renewable energy projects and are sold separately from commodity

Fig. 4.21 High pressure electrical line

electricity. RECs are attractive for federal facilities located where renewable power is not readily available. Find REC guidance in the Council on Environmental Quality's Federal Renewable Energy Certificate Guide.

Additional REC information is available on the US Environmental Protection Agency's (EPA) Green Power Partnership website, including REC price information.

The following organizations execute REC aggregations for federal agencies:

- Defense Logistics Agency Energy
- General Services Administration
- Western Area Power Administration

Under section 203(c)(3) of the Energy Policy Act of 2005 (EPAct 2005, 42 U.S.C. 1582), electricity purchased from renewable projects situated on Indian land (as defined by the Energy Policy Act of 1992, 25 U.S.C. 3501, et seq.) receives double credit toward its renewable goals. This double credit is extended in the Executive Order 13693 Implementing Instructions. As a result, purchasing renewable power that is produced on Indian land bundled with renewable energy certificates (either original or replacement) will count double for agencies pursuing their goals toward the federal 30% renewable energy by 2025 goal [8–10].

Figure 4.21 is the presentation of high-pressure electrical gridline going through the Indian land.

Section 503 of EPAct 2005 (25 U.S.C. 3502) gives federal agencies the authority to give preference to tribal businesses when purchasing electricity, energy products, or energy by-products.

4.6 Issues with Adoption Forms of Renewable Source of Energy

One problem with many forms of renewable energy is that they depend on circumstances of nature—wind, water supply, and sufficient sunlight—which can impose limitations. Another issue has been the relative expense of many forms of alternate energy compared to traditional sources such as oil and natural gas. Until very recently, running coal-fired or oil-powered plants was cheaper than investing millions in the construction of large solar, wind, tidal, or geothermal operations.

However, ongoing improvements made in the production of solar cells, wind turbines, and other equipment—not to mention improvements made in the amount of energy produced—have resulted in many forms of alternative energy becoming competitive with other methods. All over the world, nations and communities are stepping up to accelerate the transition toward cleaner, more sustainable, and more self-sufficient methods.

References

1. https://www.universetoday.com/59029/types-of-renewable-energy/
2. http://listverse.com/2009/05/01/top-10-renewable-energy-sources/
3. http://www.wri.org/publication/bottom-line-renewable-energy-certificates
4. The Bottom Line on Renewable Electricity Standards. http://www.wri.org/publication/bottom-line-renewableelectricity-standards
5. WRI's U.S. Climate Policy Resources. http://www.wri.org/climate/usclimate
6. The Bottom Line on Corporate GHG Inventories. http://www.wri.org/publication/bottom-line-corporateghg-inventories
7. Green-e® Certification. http://www.green-e.org
8. EPA Guide to Buying Green Power. http://www.epa.gov/greenpower/buygp/guide.htm
9. Evolution Markets. http://www.evomarkets.com
10. Switching to Green. http://pdf.wri.org/switching_to_green.pdf

Chapter 5
Hydrogen Energy Technology, Renewable Source of Energy

Hydrogen also can be found in many organic compounds, as well as water. It is the most abundant element on the Earth. But it does not occur naturally as a gas. It is always combined with other elements, such as with oxygen, to make water. Once separated from another element, hydrogen can be burned as a fuel or converted into electricity. A fuel cell uses the chemical energy of hydrogen or another fuel to cleanly and efficiently produce electricity. If hydrogen is the fuel, electricity, water, and heat are the only products. Fuel cells are unique in terms of the variety of their potential applications; they can provide power for systems as large as a utility power station and as small as a laptop computer.

5.1 Introduction

Hydrogen can be considered as the simplest element in existence. An atom of hydrogen consists of only one proton and one electron. It's also the most plentiful element in the universe. Despite its simplicity and abundance, hydrogen does not occur naturally as a gas on the Earth—it is always combined with other elements. Water, for example, is a combination of hydrogen and oxygen (H_2O). Hydrogen is also one of the most abundant elements in the Earth's crust. Hydrogen as a gas is not found naturally on Earth and must be manufactured. This is because hydrogen gas is lighter than air and rises into the atmosphere as a result. Natural hydrogen is always associated with other elements in compound form such as water, coal, and petroleum.

Hydrogen has the highest energy content of any common fuel by weight. On the other hand, hydrogen has the lowest energy content by volume. It is the lightest element, and it is a gas at normal temperature and pressure.

Hydrogen also can be found in many organic compounds, as well as water. It is the most abundant element on the Earth. But it does not occur naturally as a gas. It is always combined with other elements, such as with oxygen to make water. Once separated from another element, hydrogen can be burned as a fuel or converted into electricity.

B. Zohuri, *Hybrid Energy Systems*,
https://doi.org/10.1007/978-3-319-70721-1_5

Since hydrogen does not exist on Earth as a gas, it must be separated from other compounds. Two of the most common methods used for the production of hydrogen are electrolysis or water splitting and steam reforming. Steam reforming is currently the least expensive method for producing hydrogen. It is used in industries to separate hydrogen atoms from carbon atoms in methane. Because methane is a fossil fuel, the process of steam reforming results in greenhouse gas emissions which is linked to global warming. The other method for the production of hydrogen is electrolysis. Electrolysis involves passing an electric current through water to separate water into its basic elements, hydrogen and oxygen. Hydrogen is then collected at the negatively charged cathode and oxygen at the positive anode. Hydrogen produced by electrolysis is extremely pure and results in no emissions since electricity from renewable energy sources can be used. Unfortunately, electrolysis is currently a very expensive process.

Hydrogen is also found in many organic compounds, notably the *hydrocarbons* that make up many of our fuels, such as gasoline, natural gas, methanol, and propane. Hydrogen can be separated from hydrocarbons through the application of heat—a process known as *reforming*. Currently, most hydrogen is made this way from natural gas. An electrical current can also be used to separate water into its components of oxygen and hydrogen. This process is known as *electrolysis*. Some algae and bacteria, using sunlight as their energy source, even give off hydrogen under certain conditions.

There are also several experimental methods of producing hydrogen such as photoelectrolysis and biomass gasification. Scientists have also discovered that some algae and bacteria produce hydrogen under certain conditions, using sunlight as their energy source.

5.2 Hydrogen as an Energy Carrier

Hydrogen can be considered as a clean energy carrier similar to electricity. Hydrogen can be produced from various domestic resources such as renewable energy and nuclear energy. In the long-term, hydrogen will simultaneously reduce the dependence on foreign oil and the emission of greenhouse gases and other pollutants.

Hydrogen is also considered as a secondary source of energy, commonly referred to as an energy carrier. Energy carriers are used to move, store, and deliver energy in a form that can be easily used. Electricity is the most well-known example of an energy carrier.

Hydrogen as an important energy carrier in the future has a number of advantages. For example, a large volume of hydrogen can be easily stored in a number of different ways. Hydrogen is also considered as a high efficiency, low-polluting fuel that can be used for transportation, heating, and power generation in places where it is difficult to use electricity. In some instances, it is cheaper to ship hydrogen by pipeline than sending electricity over long distances by wire.

Currently, hydrogen is mainly used as a fuel in the NASA space program as illustrated in Fig. 5.1. Liquid hydrogen is used to propel space shuttle and other

Fig. 5.1 Shuttle launch

rockets, while hydrogen fuel cells power the electrical systems of the shuttle. The hydrogen fuel cell is also used to produce pure water for the shuttle crew [1].

In the future, hydrogen will join electricity as an important energy carrier, since it can be made safely from renewable energy sources and is virtually nonpolluting. It will also be used as a fuel for "zero-emissions" vehicles, to heat homes and offices, to produce electricity, and to fuel aircraft.

Hydrogen is high in energy, yet an engine that burns pure hydrogen produces almost no pollution. NASA has used liquid hydrogen since the 1970s to propel the space shuttle and other rockets into orbit. Hydrogen fuel cells power the shuttle's electrical systems, producing a clean byproduct—pure water—which the crew drinks.

In the future, hydrogen could also join electricity as an important energy carrier. An energy carrier moves and delivers energy in a usable form to consumers. Renewable energy sources, like the sun and wind, cannot produce energy all the time. But they could, for example, produce electric energy and hydrogen, which can be stored until it's needed. Hydrogen can also be transported (like electricity) to locations where it is needed.

Hydrogen has great potential as a way to reduce reliance on imported energy sources such as oil. Before hydrogen can play a bigger energy role and become a widely used alternative to gasoline, many new facilities and systems must be built [2].

As it can be seen in Fig. 5.2, future hydrogen energy infrastructure is illustrated. The hydrogen is produced through a wind electrolysis system. The hydrogen is compressed up to pipeline pressure and then fed into a transmission pipeline. The pipeline transports the hydrogen to a compressed gas terminal where the hydrogen

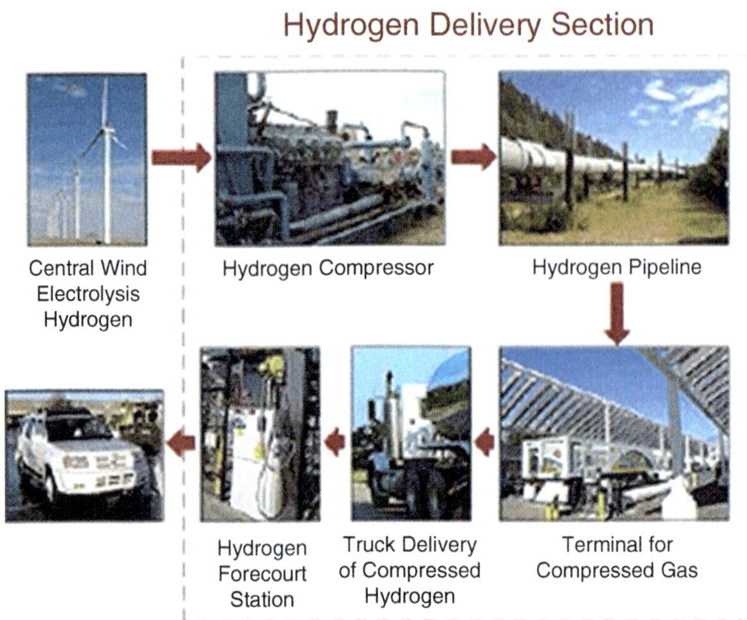

Fig. 5.2 Future hydrogen energy infrastructure (Image source US Department of Energy) [2]

is loaded into compressed gas tube trailers. A truck delivers the tube trailers to a forecourt station where the hydrogen is further compressed, stored, and dispensed to fuel cell vehicles.

5.3 Hydrogen Fuel Cell

As we stated at the beginning of this chapter, a fuel cell uses the chemical energy of hydrogen or another fuel to cleanly and efficiently produce electricity. If hydrogen is the fuel, electricity, water, and heat are the only products. Fuel cells are unique in terms of the variety of their potential applications; they can provide power for systems as large as a utility power station and as small as a laptop computer.

Figure 5.3 shows, where fuel cells directly convert the chemical energy in hydrogen to electricity, with pure water and heat as the only byproducts. Hydrogen-powered fuel cells are not only pollution-free, but a two- to threefold increase in the efficiency can be experienced when compared to traditional combustion technologies.

Fuel cells can power almost any portable devices that normally use batteries. Fuel cells can also power transportation such as vehicles, trucks, buses, and marine vessels, as well as provide auxiliary power to traditional transportation technologies. Hydrogen can play a particularly important role in the future by replacing the imported petroleum we currently use in our cars and trucks [3].

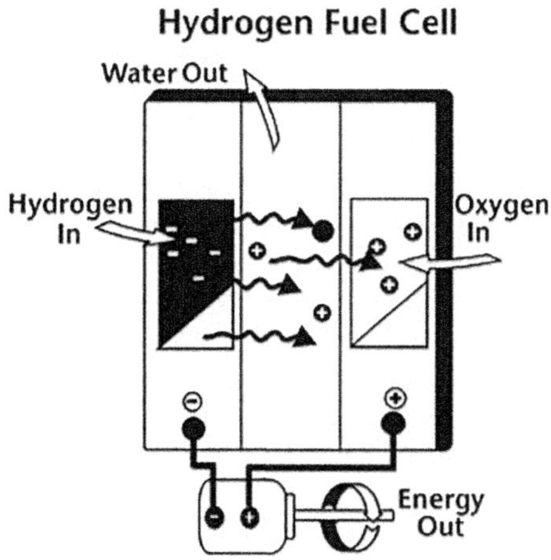

Fig. 5.3 Hydrogen fuel cell (Courtesy of US Department of Energy)

A fuel cell combines hydrogen and oxygen to produce electricity, heat, and water. As we have said, the purpose of a fuel cell is to produce an electrical current that can be directed outside the cell to do work, such as powering an electric motor or illuminating a light bulb or a city. Because of the way electricity behaves, this current returns to the fuel cell, completing an electrical circuit. (To learn more about electricity and electric power, visit "Throw The Switch" on the Smithsonian website Powering a Generation of Change.) The chemical reactions that produce this current are the key to how a fuel cell works. How the fuel cell is working as it can be observed in Fig. 5.3 is that oxygen enters the fuel cell at the cathode, and, in some cell types (like the one illustrated above), it there combines with electrons returning from the electrical circuit and hydrogen ions that have traveled through the electrolyte from the anode. In other cell types, the oxygen picks up electrons and then travels through the electrolyte to the anode, where it combines with hydrogen ions.

The electrolyte plays a key role. It must permit only the appropriate ions to pass between the anode and cathode. If free electrons or other substances could travel through the electrolyte, they would disrupt the chemical reaction.

Whether they combine at anode or cathode, together hydrogen and oxygen form water, which drains from the cell. As long as a fuel cell is supplied with hydrogen and oxygen, it will generate electricity.

Even better, since fuel cells create electricity chemically, rather than by combustion, they are not subject to the thermodynamic laws that limit a conventional power plant (see "Carnot Limit" in the glossary). Therefore, fuel cells are more efficient in extracting energy from a fuel. Waste heat from some cells can also be harnessed, boosting system efficiency still further.

Fuel cells are often compared to batteries. Both convert the energy produced by a chemical reaction into usable electric power. However, the fuel cell will produce electricity as long as fuel (hydrogen) is supplied, never losing its charge. Fuel cell research aims to lower the cost and improve the performance and durability of fuel cell technologies.

Fuel cells are a promising technology for use as a source of heat and electricity for buildings, and as an electrical power source for electric motors propelling vehicles. Fuel cells operate best on pure hydrogen. But fuels like natural gas, methanol, or even gasoline can be reformed to produce the hydrogen required for fuel cells. Some fuel cells even can be fueled directly with methanol, without using a reformer.

Fuel cells can be used in a wide range of applications, including transportation, material handling, stationary, portable, and emergency backup power applications. Fuel cells have several benefits over conventional combustion-based technologies currently used in many power plants and passenger vehicles. Fuel cells can operate at higher efficiencies than combustion engines and can convert the chemical energy in the fuel to electrical energy with efficiencies of up to 60%. Fuel cells have lower emissions than combustion engines. Hydrogen fuel cells emit only water, so there are no carbon dioxide emissions and no air pollutants that create smog and cause health problems at the point of operation. Also, fuel cells are quiet during operation as they have fewer moving parts.

Research is performed on a variety of fuel cell types—proton exchange membrane (PEMFC), alkaline membrane (AMFC), and direct-methanol (DMFC) fuel cells—which are generally differentiated by the fuel used.

Fuel cells work like batteries, but they do not run down or need recharging. They produce electricity and heat as long as fuel is supplied. A fuel cell consists of two electrodes—a negative electrode (or anode) and a positive electrode (or cathode)—sandwiched around an electrolyte. A fuel, such as hydrogen, is fed to the anode, and air is fed to the cathode. In a hydrogen fuel cell, a catalyst at the anode separates hydrogen molecules into protons and electrons, which take different paths to the cathode. The electrons go through an external circuit, creating a flow of electricity. The protons migrate through the electrolyte to the cathode, where they unite with oxygen and the electrons to produce water and heat. Learn more about [4]:

- Parts of a fuel cell
- Fuel cell systems
- Types of fuel cells

View the Fuel Cell Technologies Office's fuel cell animation to see how a fuel cell operates.

The Fuel Cell Technologies Office (FCTO) focuses on applied research, development, and innovation to advance hydrogen and fuel cells for transportation and diverse applications enabling energy security, resiliency, and a strong domestic economy in emerging technologies.

The US Department of Energy's (DOE's or the Department's) hydrogen and fuel cell efforts are part of a broad portfolio of activities to build a competitive and sustainable clean energy economy to secure the nation's energy future. Reducing greenhouse gas emissions 80% by 2050 [5] and eliminating dependence on imported fuel will require the use of diverse domestic energy sources and advanced fuels and technologies in all sectors of the economy. Achieving these goals requires a robust, comprehensive research and development (R&D) portfolio that balances short-term objectives with long-term needs and sustainability.

Fuel cells, which convert diverse fuels directly into electricity without combustion, and hydrogen, a zero-carbon fuel when produced from renewable resources, comprise key elements of the DOE portfolio. DOE's efforts to enable the widespread commercialization of hydrogen and fuel cell technologies form an integrated program—the DOE Hydrogen and Fuel Cells Program (the Program), as reflected in the Hydrogen and Fuel Cells Program Plan [6]. The Program is coordinated across the Department and includes activities in the offices of Energy Efficiency and Renewable Energy (EERE), Science, Nuclear Energy, and Fossil Energy.

As part of research and development goals, the US Department of Energy (DOE) is working closely with its national laboratories, universities, and industry partners to overcome critical technical barriers to fuel cell development. Cost, performance, and durability are still key challenges in the fuel cell industry. View related links that provide details about DOE-funded fuel cell activities.

- *Cost*—Platinum represents one of the largest cost components of a fuel cell, so much of the R&D focuses on approaches that will increase activity and utilization of current platinum group metal (PGM) and PGM-alloy catalysts, as well as non-PGM catalyst approaches for long-term applications.
- *Performance*—To improve fuel cell performance, R&D focuses on developing ion-exchange membrane electrolytes with enhanced efficiency and durability at reduced cost; improving membrane electrode assemblies (MEAs) through integration of state-of-the-art MEA components; developing transport models and in situ and ex situ experiments to provide data for model validation; identifying degradation mechanisms and developing approaches to mitigate their effects; and maintaining core activities on components, subsystems, and systems specifically tailored for stationary and portable power applications.
- *Durability*—A key performance factor is durability, in terms of a fuel cell system lifetime that will meet application expectations. DOE durability targets for stationary and transportation fuel cells are 40,000 and 5000 h, respectively, under realistic operating conditions. In the most demanding applications, realistic operating conditions include impurities in the fuel and air, starting and stopping, freezing and thawing, and humidity and load cycles that result in stresses on the chemical and mechanical stability of the fuel cell system materials and components. R&D focuses on understanding the fuel cell degradation mechanisms and developing materials and strategies that will mitigate them.

More details of technical targets and goals, as well as individual target tables, can be found in the Fuel Cells section of the Fuel Cell Technologies Office's Multi-Year Research, Development, and Demonstration Plan [4]:

- Fuel cell systems, stacks, and components for light-duty transportation applications
  - Fuel cell systems and stacks
  - proton exchange membrane fuel cell components
  - Fuel cell system humidifiers and air compressions systems

- Fuel cell transit buses
- Fuel cell backup power systems
- Fuel cell systems for stationary (combined heat and power) applications
- Fuel cell systems for portable power and auxiliary power applications

5.4 Fuel Cells

A fuel cell is a device that generates electricity by a chemical reaction. Every fuel cell has two electrodes, the anode (which is positively charged) and the cathode (which is negatively charged) [1]. The reactions that produce electricity take place at the two electrodes. Every fuel cell also has an electrolyte, which carries electrically charged particles from one electrode to the other, and a catalyst, which speeds the reactions at the electrodes [7]. Multiple fuel cells are usually assembled into a stack and generate direct current (DC).

A single fuel cell consists of an electrolyte sandwiched between two electrodes. Bipolar plates on either side of the cell help distribute gases and serve as current collectors. Hydrogen is the basic fuel for fuel cells, but fuel cells also require oxygen. Basic chemical reaction of fuel cell is given here as:

$$2H_2 + O_2 \rightarrow 2H_2O + 2e^-$$

Depending on the application, a fuel cell stack may contain a few to hundreds of individual fuel cells layered together. This "scalability" makes fuel cells ideal for a wide variety of applications, such as stationary power stations, portable devices, and transportation.

There are several kinds of fuel cells, and each operates a bit differently. But, in general terms, hydrogen atoms enter a fuel cell at the anode where a chemical reaction strips them of their electrons. The hydrogen atoms are now "ionized" and carry a positive electrical charge. The negatively charged electrons provide the current through wires to do work. If alternating current (AC) is needed, the DC output of the fuel cell must be routed through a conversion device called an inverter.

Oxygen enters the fuel cell at the cathode, and, in some cell types, it combines with electrons returning from the electrical circuit and hydrogen ions that have traveled through the electrolyte from the anode. In other cell types, the oxygen picks up electrons and then travels through the electrolyte to the anode, where it combines with hydrogen ions.

The electrolyte plays a key role. It must permit only the appropriate ions to pass between the anode and cathode. If free electrons or other substances could travel through the electrolyte, they would disrupt the chemical reaction.

Whether they combine at anode or cathode, together hydrogen and oxygen form water, which drains from the cell. As long as a fuel cell is supplied with hydrogen and oxygen, it will generate electricity.

5.4.1 Different Types of Fuel Cells

There are several kinds of fuel cells, and each operates a bit differently. But in general terms, hydrogen atoms enter a fuel cell at the anode where a chemical reaction strips them of their electrons. The hydrogen atoms are now "ionized" and carry a positive electrical charge. The negatively charged electrons provide the current through wires to do work. If alternating current (AC) is needed, the DC output of the fuel cell must be routed through a conversion device called an inverter.

The type of fuel also depends on the electrolyte. Some cells need pure hydrogen and therefore demand extra equipment such as a "reformer" to purify the fuel. Other cells can tolerate some impurities but might need higher temperatures to run efficiently. Liquid electrolytes circulate in some cells, which require pumps. The type of electrolyte also dictates a cell's operating temperature—"molten" carbonate cells run hot, just as the name implies.

Each type of fuel cell has advantages and drawbacks compared to the others, and none is yet cheap and efficient enough to widely replace traditional ways of generating power, such as coal-fired, hydroelectric, or even nuclear power plants.

While there are dozens of types of fuel cells, there are six principal kinds in various stages of commercial availability or undergoing research, development, and demonstration. These six fuel cell types are significantly different from each other in many respects; however, the key distinguishing feature is the electrolyte material.

The following list and associated images describes the five main types of fuel cells. More detailed information can be found on the web site provided by the National Museum of American History, Smithsonian Institution, at (http://americanhistory.si.edu/fuelcells/basics.htm). We have reflected on the information and description of these fuel cells from their site with courtesy.

1. *Alkali Fuel Cell*
   These fuel cells operate on compressed hydrogen and oxygen. They generally use a solution of potassium hydroxide (chemically, KOH) in water as their electrolyte. Efficiency is about 70%, and operating temperature is 150–200 °C (about 300–400 °F). Cell output ranges from 300 watts (W) to 5 kilowatts (kW). Alkali cells were used in the Apollo spacecraft to provide both electricity and drinking water. They require pure hydrogen fuel, however, and their platinum electrode catalysts are expensive. And like any container filled with liquid, they can leak. See Fig. 5.4.
2. *Phosphoric Acid Fuel Cell (PAFC)*
   These fuel cells (PAFC) use phosphoric acid as the electrolyte. Efficiency ranges from 40% to 80%, and operating temperature is between 150 °C and 200 °C (about 300–400 °F). Existing phosphoric acid cells have outputs up to 200 kW, and 11 MW units have been tested. PAFCs tolerate a carbon monoxide concentration of about 1.5%, which broadens the choice of fuels they can use. If gasoline is used, the sulfur must be removed. Platinum electrode catalysts are needed, and internal parts must be able to withstand the corrosive acid. See Fig. 5.5.

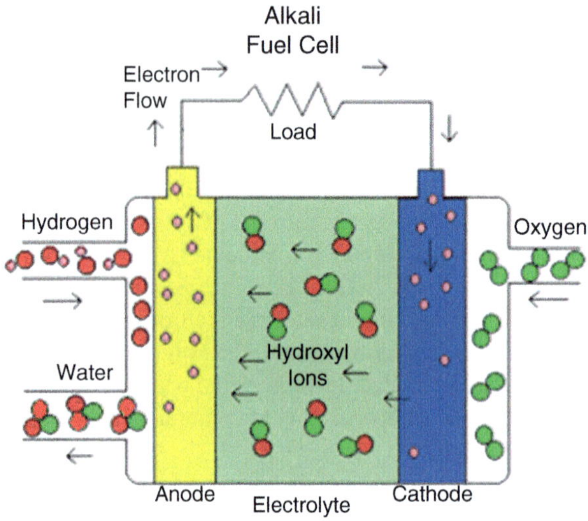

Fig. 5.4 Artistic illustration of alkali fuel cell (Courtesy of Smithsonian Institution)

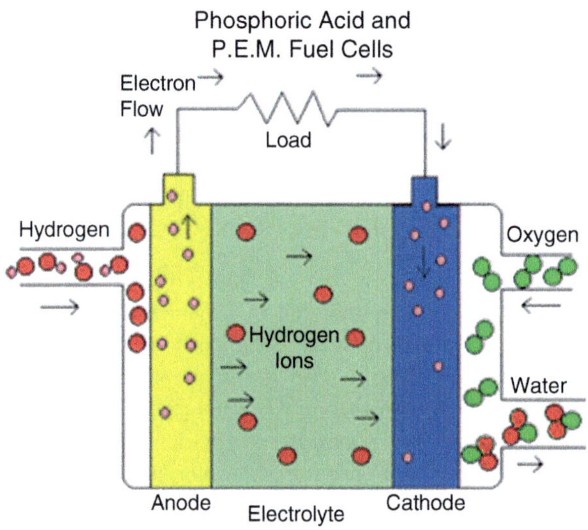

Fig. 5.5 Artistic illustration of phosphoric acid fuel cell (Courtesy of Smithsonian Institution)

3. *Molten Carbonate Fuel Cell (MCFC)*
   The fuel cells (MCFC) use high-temperature compounds of salt (like sodium or magnesium) carbonates (chemically, CO_3) as the electrolyte. Efficiency ranges from 60% to 80%, and operating temperature is about 650 °C (1200 °F). Units with output up to 2 megawatts (MW) have been constructed, and designs exist for units up to 100 MW. The high temperature limits damage from carbon monoxide "poisoning" of the cell, and waste heat can be recycled to make

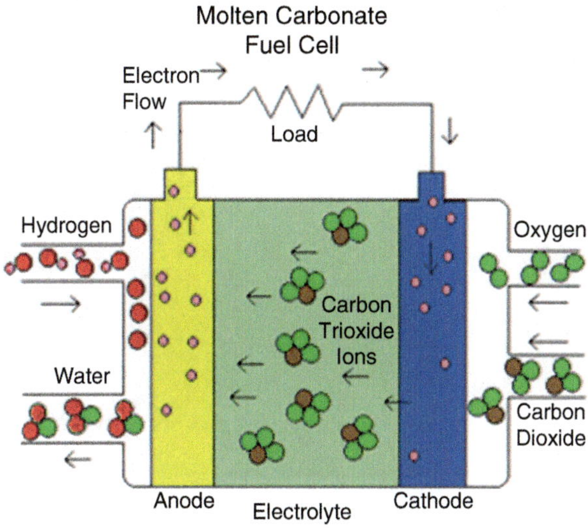

Fig. 5.6 Artistic illustration of molten carbonate fuel cell (Courtesy of Smithsonian Institution)

additional electricity. Their nickel electrode catalysts are inexpensive compared to the platinum used in other cells. But the high temperature also limits the materials and safe uses of MCFCs—they would probably be too hot for home use. Also, carbonate ions from the electrolyte are used up in the reactions, making it necessary to inject carbon dioxide to compensate. See Fig. 5.6.

Full-scale demonstration plants are now testing molten carbonate fuel cells (MCFCs). The electrolyte in an MCFC is an alkali carbonate (sodium, potassium, or lithium salts, i.e., Na_2CO_3, K_2CO_2, or Li_2CO_3) or a combination of alkali carbonates that is retained in a ceramic matrix of lithium aluminum oxide ($LiAlO_2$). An MCFC operates at 600–700 °C where the alkali carbonates form a highly conductive molten salt with carbonate ions (CO_3=) providing ionic conduction through the electrolyte matrix. Relatively inexpensive nickel (Ni) and nickel oxide (NiO) are adequate to promote reaction on the anode and cathode, respectively, at the high operating temperatures of an MCFC.

MCFCs offer greater fuel flexibility and higher fuel-to-electricity efficiencies than lower temperature fuel cells, approaching 60%. The higher operating temperatures of MCFCs make them candidates for combined cycle applications, in which the exhaust heat is used to generate additional electricity. When the waste heat is used for cogeneration, total thermal efficiencies can approach 85%.

4. *Proton Exchange Membrane (PEM)*

PEM fuel cells work with a proton exchange in the form of a thin, permeable sheet. Efficiency is about 40–50%, and operating temperature is about 80 °C (about 175 °F). Cell outputs generally range from 50 kW to 250 kW. The solid, flexible electrolyte will not leak or crack, and these cells operate at a low enough temperature to make them suitable for homes and cars. But their fuels must be purified, and a platinum catalyst is used on both sides of the membrane, raising costs. See Fig. 5.7.

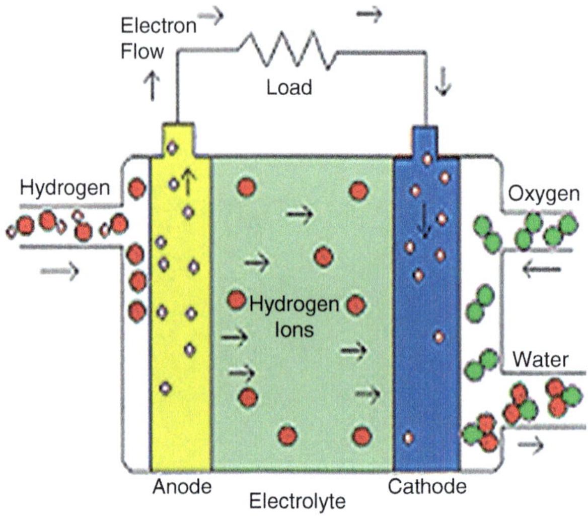

Fig. 5.7 Artistic illustration of proton exchange membrane fuel cell (Courtesy of Smithsonian Institution)

5. *Solid Oxide Fuel Cell (SOFC)*

   These cells use a hard, ceramic compound of metal (like calcium or zirconium) oxides (chemically, O_2) as electrolyte. Efficiency is about 60%, and operating temperatures are about 1000 °C (about 1800 °F). Cells output is up to 100 kW. At such high temperatures, a reformer is not required to extract hydrogen from the fuel, and waste heat can be recycled to make additional electricity. However, the high temperature limits applications of SOFC units, and they tend to be rather large. While solid electrolytes cannot leak, they can crack. See Fig. 5.8.

   Solid oxide fuel cells (SOFCs) are currently being demonstrated in sizes from 1 kW up to 250 kW plants, with plans to reach the multi-MW range. SOFCs utilize a nonporous metal oxide (usually Yttria-stabilized zirconia, Y_2O_3-stabilized ZrO_2) electrolyte material. SOFCs operate between 650 °C and 1000 °C, where ionic conduction is accomplished by oxygen ions (O=). Typically the anode of an SOFC is cobalt or nickel zirconia (Co-ZrO_2 or Ni-ZrO_2), and the cathode is strontium-doped lanthanum manganite (Sr-doped $LaMnO_3$) (Singhal 1997; Minh 1993). See Figs. 5.9 and 5.10, as well.

   SOFCs offer the stability and reliability of all-solid-state ceramic construction. High-temperature operation, up to 1000 °C, allows more flexibility in the choice of fuels and can produce very good performance in combined cycle applications. SOFCs approach 60% electrical efficiency in the simple cycle system and 85% total thermal efficiency in cogeneration applications (Singhal 1997).

   The flat plate and monolithic designs are at a much earlier stage of development typified by subscale, single-cell, and short-stack development (kW scale). At this juncture, tubular SOFC designs are closer to commercialization.

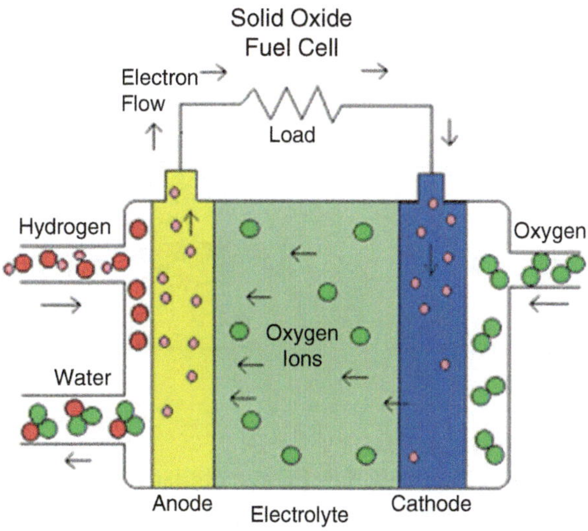

Fig. 5.8 Artistic illustration of solid oxide fuel cell (Courtesy of Smithsonian Institution)

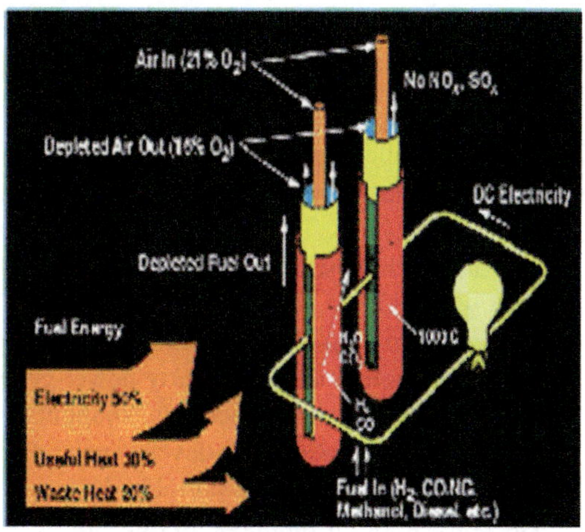

Fig. 5.9 Artistic illustration of tubular SOFC (Courtesy of Siemens Westinghouse Power Corporation)

Fig. 5.10 Illustration of planar SOFC (Courtesy of Siemens Westinghouse Power Corporation)

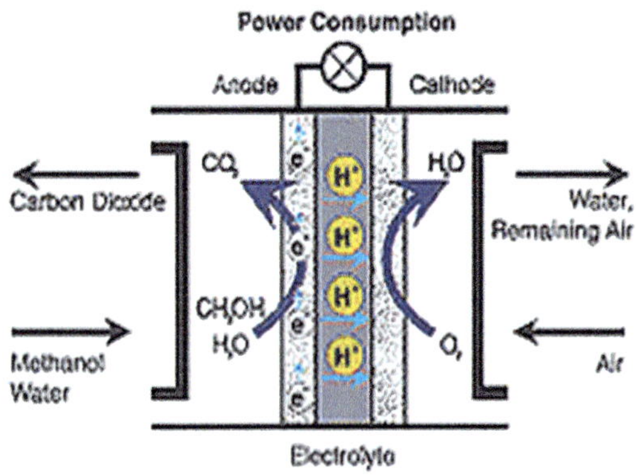

Fig. 5.11 Artistic illustration of solid oxide fuel cell (Courtesy of Smart Fuel Cell)

6. *Direct-Methanol Fuel Cell (DMFC)*
   The direct-methanol fuel cell (DMFC) is similar to the PEM cell in that it uses a polymer membrane as an electrolyte. However, a catalyst on the DMFC anode draws hydrogen from liquid methanol, eliminating the need for a fuel reformer. While potentially a very attractive solution to the issues of hydrogen storage and transportation (particularly for portable applications), the principal problem facing the commercial application of the DMFC today stems from its relatively low performance in comparison to hydrogen. See Fig. 5.11.

Since fuel cells create electricity chemically, rather than by combustion, they are not subject to the thermodynamic laws that limit a conventional power plant (known as the Carnot Limit). Therefore, fuel cells are more efficient in extracting energy from a fuel. Waste heat from some cells can also be harnessed, boosting system efficiency still further.

Fuel cells are employed in stationary power generation, portable power supply, and transportation. Small, stationary power generators provide 0.5–10 kW uninterrupted power supply to households, shopping malls, and data centers. Grid-scale fuel cell generation is also in development. Portable fuel cells are best suited for auxiliary power units (APU), portable devices, PC, smartphone, etc.

The application of fuel cells in transportation represents the future for the automotive and computation industries. Buses, light vehicles (cars), UAVs, and trains will soon be running on fuel cells.

Another great appeal of fuel cells is that they generate electricity with very little pollution—much of the hydrogen and oxygen used in generating electricity ultimately combines to form a harmless byproduct, namely, water. However, obtaining hydrogen is a challenge and can be energy intensive.

Despite its many advantages, the commercialization of fuel cell technology faces many technical and economic challenges. The durability and cost of fuel cell systems represent the biggest barriers. Fuel cells are still in the "technology development phase." Efforts are being made to reduce the cost and improve durability of fuel cells.

5.5 The Fuel Cell Technologies

The Fuel Cell Technologies Program (FCT Program), situated within EERE, addresses key technical challenges for fuel cells and hydrogen production, delivery, and storage and the institutional barriers, such as hydrogen codes and standards, training, and public awareness that inhibit the widespread commercialization of hydrogen and fuel cell technologies. The FCT Program conducts applied research, technology development, and learning demonstrations, as well as safety research, systems analysis, early market deployments, and public outreach and education activities. These activities include cost-shared, public-private partnerships to address the high-risk, critical technology barriers preventing extensive use of hydrogen as an energy carrier. Public and private partners include automotive and power equipment manufacturers, energy and chemical companies, electric and natural gas utilities, building designers, standards development organizations, other Federal agencies, state government agencies, universities, national laboratories, and other national and international stakeholder organizations. The FCT Program encourages the formation of collaborative partnerships to conduct research, development, and demonstrations (RD&D) and other activities, such as deployment, which support program goals.

The FCT Program addresses the development of hydrogen energy systems for transportation, stationary power, and portable power applications. Transportation applications include fuel cell vehicles (such as buses, automobiles, and heavy-duty vehicles), niche markets (such as lift trucks), and hydrogen refueling infrastructure. Hydrogen used for backup emergency power, commercial/industrial power and heat generation, and residential electric power generation is included in stationary power applications. Consumer electronics such as mobile phones, laptop computers, and recharging systems are among the portable power applications. The DOE is funding RD&D efforts that will provide the basis for the near-, mid-, and long-term production, delivery, storage, and use of hydrogen derived from diverse energy sources, including renewable, fossil fuels, and nuclear energy as coordinated within the Program. This document primarily describes the status, challenges, and RD&D activities of the FCT Program but also the overall DOE Hydrogen and Fuel Cells Program.

As it was stated above, fuel cell research aims to lower the cost and improve the performance and durability of fuel cell technologies. Research is performed on a variety of fuel cell types—proton exchange membrane (PEMFC), alkaline membrane (AMFC), and direct-methanol (DMFC) fuel cells—which are generally differentiated by the fuel used. They are listed below:

1. *Catalysts*—work in this area involves developing and optimizing advanced electrocatalysts and novel synthesis methods. Related projects concentrate on extended surface catalysts with reduced precious-metal loading and improved performance, durability, and activity compared to standard catalytic materials. Researchers are investigating fuel cells and electrolyzer catalysts under acidic

and alkaline conditions, with the goal of "thrifting" platinum, iridium, and their alloys (in acidic-based systems) and silver, cobalt, nickel, and their oxides/alloys (in alkaline-based systems). Also under study are support materials for catalyst dispersion, with a focus on nitrogen-doped carbon supports and corrosion-resistant, non-carbon supports.

2. *Polymer Electrolytes*—Alkaline membrane fuel cells enable the use of non-precious-metal catalysts, but they are vulnerable to ambient carbon dioxide conditions. This vulnerability decreases, however, at higher operating temperatures. Researchers are developing novel chemistries to enable higher-temperature and higher-current-density operation via the use of perfluorinated alkaline membranes. Researchers are also exploring traditional proton exchange membranes with tethered heteropoly acid functionality to allow higher-temperature, lower-humidity operation and are investigating the stability of covalently tetherable captions.
3. *Electrode Design/High-Current-Density Operation*—This crosscutting research area focuses on incorporating novel catalysts into high-performance devices and investigating the impact of low-precious-metal loading on high-current-density performance.
4. *Contaminants*—As fuel cell systems become more commercially competitive, and as automotive fuel cell research and development trends toward decreased catalyst loadings and thinner membranes, fuel cell operation becomes even more susceptible to contaminants. NREL also participates in the US Department of Energy's (DOE's) Fuel Cell Durability Working Group. Contaminants derived from fuel cell system component materials structural materials, lubricants, greases, adhesives, sealants, and hoses have been shown to affect the performance and durability of fuel cell systems. Companies are performing research to identify and quantify these system-derived contaminants and to understand the effects of system contaminants on fuel cell performance and durability. Our goal is to increase the understanding of fuel cell system contaminants and to help guide the implementation and, where necessary, development of system materials that will help enable fuel cell commercialization.

Again more details can be found in the official site of the Department of Energy office at Energy Efficiency and Renewable Energy as well as National Renewable Energy Laboratory (NREL) [8, 9].

5.6 Fuel Cell Backup Power Systems

A Department of Energy (DOE) technical target for fuel cell backup power systems is illustrated in Table 5.1. More information about targets can be found in the Fuel Cells section of the Fuel Cell Technologies Office's Multi-Year Research, Development, and Demonstration Plan [10].

Table 5.1 Technical targets: fuel cell backup power systems (1–10 kW) operating on direct hydrogen [9]

Characteristic	Units	2015 status[a]	2020 targets
Lifetime	Years	10	15
Durability[b]	Hours	8000	10,000
Energy efficiency[c]	%	50	60
Mean time between failures	Years	5	5
Ambient temperature range	°C	−10 to 40	−50 to 50
Noise	dB at 1 m	65	60
Start-up time[d]	Seconds	80	15
Availability	%	99.7	96.3
Equipment cost[e]	$/kW	6100[f]	1000
Annual maintenance cost[e]	$/kW	30	20
Annualized total cost of ownership[g]	$/kW	500	200

[a]Unless otherwise stated, status based on input from DE-FOA-0000738
[b]Time until 10% voltage degradation when operated on a backup power duty cycle
[c]Ratio of DC output energy from the power plant to the lower heating value of the input fuel (hydrogen, averaged over cycle)
[d]Time indicated is start-up time for the fuel cell. The backup power system, including hybridized batteries, is expected to provide uninterruptible power
[e]Excludes tax credits and subsidies
[f]NREL, "Current Fuel Cell System Low Volume Price by Application"
[g]Annualized cost of ownership including cost of capital equipment, installation, operation and maintenance, fuel, and fuel storage. Based on a 5 kW system with 10-year lifetime

5.7 Fuel Cell Systems for Stationary Combined Heat and Power Applications

Tables 5.2 and 5.3 list the US Department of Energy (DOE) technical targets for stationary fuel cell applications. These targets have developed with input from developers of stationary fuel cell power systems.

More information about targets can be found in the Fuel Cells section of the Fuel Cell Technologies Office's Multi-Year Research, Development, and Demonstration Plan [10].

5.8 Fuel Cell Systems for Portable Power and Auxiliary Power Applications

Tables 5.4 and 5.5 list the US Department of Energy (DOE) technical targets for fuel cell systems for portable power and auxiliary power applications.

More information about targets can be found in the Fuel Cells section of the Fuel Cell Technologies Office's Multi-Year Research, Development, and Demonstration Plan [11].

Table 5.2 Technical targets: 1–25 kW residential and light commercial combined heat and power and distributed generation fuel cell systems operating on natural gas[a]

Characteristic	Units	2015 status	2020 targets
Electrical efficiency at rated power[b]	% (LHV)	34–40	>45[c]
CHP energy efficiency[d]	% (LHV)	80–90	90
Equipment cost[e], 5-kW_{avg} system[f]	$/kW	2300–2800[g]	1500
Transient response (10–90% rated power)	Min	5	2
Start-up time from 20 °C ambient temperature	Min	10	20
Degradation with cycling[h]	%/1000 h	<2%	0.3%
Operating lifetime[i]	h	12,000–70,000	60,000
System availability[j]	%	97	99

[a]Pipeline natural gas delivered at typical residential distribution line pressures
[b]Regulated AC net/LHV of fuel
[c]Higher electrical efficiencies (e.g., 60% using SOFC) are preferred for non-CHP applications
[d]Ratio of regulated AC net output energy plus recovered thermal energy to the LHV of the input fuel. For inclusion in CHP energy efficiency calculation, heat must be available at a temperature sufficiently high to be useful in space and water heating applications. Provision of heat at 80 °C or higher is recommended
[e]Complete system, including all necessary components to convert natural gas to electricity suitable for grid connection, and heat exchangers and other equipment for heat rejection to conventional water heater, and/or hydronic or forced air heating system. Includes all applicable taxes, and mark-ups, based on projection to high-volume production (50,000 units per year)
[f]kW_{avg} is the average output (AC) electric power delivered over the life of system while unit is running
[g]Battelle preliminary 2015 cost assessment of stationary CHP systems, range represents different technologies (SOFC vs PEMFC) at manufacturing volumes of 50,000 units per year
[h]Durability testing should include effects of transient operation, start-up, and shutdown
[i]Time until >20% net power degradation
[j]Percentage of time the system is available for operation under realistic operating conditions and load profile. Unavailable time includes time for scheduled maintenance

5.9 Hydrogen Storage

The Fuel Cell Technologies Office (FCTO) is developing onboard automotive hydrogen storage systems that allow for a driving range of more than 300 miles while meeting cost, safety, and performance requirements.

Small amounts of hydrogen (up to a few MWh) can be stored in pressurized vessels at 100~300 bar or liquefied at 20.3 K (−423 °F). Alternatively, solid metal hydrides or nanotubes can store hydrogen with a very high density. Very large amounts of hydrogen can be stored in man-made underground salt caverns of up to 500,000 m^3 at 200 bar (2900 psi), corresponding to a storage capacity of 167 GWh hydrogen (100 GWh electricity). In this way, longer periods of flaws or of excess wind/PV energy production can be leveled. Even balancing seasonal variations might be possible.

Table 5.3 Technical targets[a]: 100 kW–3 MW combined heat and power and distributed generation fuel cell systems operating on natural gas[b]

Characteristic	Units	2015 status[c]	2020 targets
Electrical efficiency at rated power[d]	% (LHV)	42–47	>50[e]
CHP energy efficiency[f]	% (LHV)	70–90	90
Equipment cost, natural gas	\$/kW	1200[g]–4500[h]	1000[i]
Installed cost, natural gas	\$/kW	2400[g]–5500[h]	1500[i]
Equipment cost, biogas	\$/kW	3200–6500[j]	1400[i]
Installed cost, biogas	\$/kW	4900–8000[j]	2100[i]
Number of planned/forced outages over lifetime	–	50	40
Operating lifetime[k]	h	40,000–80,000	80,000
System availability[l]	%	95	99

[a]Includes fuel processor, stack, and ancillaries
[b]Pipeline natural gas delivered at typical residential distribution line pressures
[c]Status varies by technology
[d]Ratio of regulated AC net output energy to the lower heating value (LHV) of the input fuel
[e]Higher electrical efficiencies (e.g., 60% using SOFC) are preferred for non-CHP applications
[f]Ratio of regulated AC net output energy plus recovered thermal energy to the LHV of the input fuel. For inclusion in CHP energy efficiency calculation, heat must be available at a temperature sufficiently high to be useful in space and water heating applications. Provision of heat at 80 °C or higher is recommended
[g]M. Wei, 100 kW LTPEMFC, projection at volume of 1000 systems/year
[h]DOE Hydrogen and Fuel Cells Program Record 11,014, "Medium-scale CHP Fuel Cell System Targets"
[i]Includes projected cost advantage of high-volume production (totaling 100 MW per year)
[j]Assumed \$2500/kW higher cost to operate on biogas than on hydrogen (DOE Hydrogen and Fuel Cells Program Record 11,014, "Medium-scale CHP Fuel Cell System Targets")
[k]Time until >10% net power degradation
[l]Percentage of time the system is available for operation under realistic operating conditions and load profile. Unavailable time includes time for scheduled maintenance

5.9.1 Why Study Hydrogen Storage

Hydrogen storage is a key enabling technology for the advancement of hydrogen and fuel cell technologies in applications including stationary power, portable power, and transportation. Hydrogen has the highest energy per mass of any fuel; however, its low ambient temperature density results in a low energy per unit volume, therefore requiring the development of advanced storage methods that have potential for higher energy density.

5.9.2 How Hydrogen Storage Works

Hydrogen can be stored physically as either a gas or a liquid. Storage of hydrogen as a gas typically requires high-pressure tanks (350–700 bar [5000–10,000 psi] tank pressure). Storage of hydrogen as a liquid requires cryogenic temperatures because the boiling point of hydrogen at one atmosphere pressure is −252.8 °C. Hydrogen can also be stored on the surfaces of solids (by adsorption) or within solids (by absorption). Also see Fig. 5.12.

Table 5.4 Technical targets: portable power fuel cell systems (5–50 W/100–200 W)[a]

Characteristic	Units	2015 status	Ultimate targets
Specific power[b]	W/kg	23[h]/25[i]	45/50
Power density[b]	W/L	24[h]/30[i]	55/70
Specific energy[b, c]	Wh/kg	121[j]/450[j]	650/640
Energy density[b, c]	Wh/L	200[i]/300[i,j]	650/900
Cost[d]	$/W	15[i]/15[i]	7/5
Durability[e, f]	Hours	1500[i]/2000[i]	5000/5000
Mean time between failures[f, g]	Hours	500[i]/500[i]	5000/5000

[a]These targets are technology neutral and make no assumption about the type of fuel cell technology or type of fuel used. In addition to meeting these targets, portable power fuel cells are expected to operate safely, providing power without exposing users to hazardous or unpleasant emissions, high temperatures, or objectionable levels of noise. Portable power fuel cells are also expected to be compatible with the requirements of portable electronic devices, including operation under a range of ambient temperature, humidity, and pressure conditions, and exposure to freezing conditions, vibration, and dust. They should be capable of repeatedly turning off and on and should have turndown capabilities required to match the dynamic power needs of the device. For widespread adoption, portable power fuel cell systems should minimize life cycle environmental impact through the use of reusable fuel cartridges, recyclable components, and low-impact manufacturing techniques

[b]This is based on rated net power of the total fuel cell system, including fuel tank, fuel, and any hybridization batteries. In the case of fuel cells embedded in other devices, only device components required for power generation, power conditioning, and energy storage are included. Fuel capacity is not specified, but the same quantity of fuel must be used in calculation of specific power, power density, specific energy, and energy density

[c]Efficiency of 35% is recommended to enable high specific energy and energy density

[d]Cost includes material and labor required to manufacture the fuel cell system and any required auxiliaries (e.g., refueling devices). Cost is defined at production rates of 25,000 and 10,000 units per year for 5–50 W and 100–200 W units, respectively

[e]Durability is defined as the time until the system rated power degrades by 20%, though for some applications higher or lower levels of power degradation may be acceptable

[f]Testing should be performed using an operating cycle that is realistic and appropriate for the target application, including effects from transient operation, start-up and shutdown, and off-line degradation

[g]Mean time between failures (MTBF) includes failures of any system components that render the system inoperable without maintenance

[h]Status calculated based on commercial products from myFC at myfcpower.com/pages/jaq

[i]DOE Hydrogen and Fuel Cells Program Record 11,009

[j]Status calculated based on commercial products from ultracell at ultracell-llc.com

5.9.3 *Research and Development Goals*

FCTO conducts research and development activities to advance hydrogen storage systems technology and develop novel hydrogen storage materials. The goal is to provide adequate hydrogen storage to meet the US Department of Energy (DOE) hydrogen storage targets for onboard light-duty vehicle, material-handling equipment, and portable power applications. By 2020, FCTO aims to develop and verify onboard automotive hydrogen storage systems achieving targets that will allow hydrogen-fueled vehicle platforms to meet customer performance expectations for

Table 5.5 Technical targets: fuel cell auxiliary power units (1–10 kW) operating on ultralow-sulfur diesel fuel

Characteristic	Units	2015 status	2020 targets
Electrical efficiency at rated power[a]	% (LHV)	29[b]	40
Power density	W/L	16[b]	40
Specific power	W/kg	18[b]	45
Factory cost, system[c]	$\$/kW_e$	2100[d]	1000
Transient response (10–90% rated power)	Min	5[e]	2
Start-up time from 20 °C	Min	70[b]	30
Start-up time from standby conditions[f]	Min	–	5
Degradation with cycling[g]	%/1000 h	2.6[e]	1
Operating lifetime[g, h]	h	3000[e]	20,000
System availability[i]	%	97[e]	99

[a]Regulated DC net/LHV of fuel
[b]DESTA-Demonstration of 1st European SOFC Truck APU, Programme Review Days 2015
[c]Cost includes materials and labor to produce system. Cost defined at 50,000 unit/year production of a 5 kW system. Today's low-volume cost is expected to be higher than quoted status. Allowable cost is expected to be higher than the target for systems with rated power below 5 kW and lower than the target for systems with rated power above 5 kW
[d]Modeled cost of a 5 kW SOFC APU system produced at 50,000 units/year. F. Eubanks et al., "Stationary and Emerging Market Fuel Cell System Cost Analysis-Auxiliary Power Units," 2015 Annual Merit Review, slide 20
[e]DOE Hydrogen Program Record 11,001, "Revised APU Targets"
[f]Standby conditions may be at or above ambient temperature depending on operating protocol
[g]Durability testing should include, at minimum, daily cycles to standby condition and weekly cycles to full off condition (ambient temperature). The system should be able to meet durability criteria during and after exposure to vibration associated with transportation and highway operation, and during operation in a range of ambient temperature from −40 °C to 50 °C, a range of ambient relative humidity from 5% to 100%, and in dust levels up to 2 mg/m^3
[h]Time until >20% net power degradation
[i]Percentage of time the system is available for operation under realistic operating conditions and load profile. Scheduled maintenance does not count against system availability

range, passenger and cargo space, refueling time, and overall vehicle performance. Specific system targets include the following:

- 1.5 kWh/kg system (4.5 wt.% hydrogen)
- 1.0 kWh/L system (0.030 kg hydrogen/L)
- \$10/kWh (\$333/kg stored hydrogen capacity)

The collaborative Hydrogen Storage Engineering Center of Excellence conducts analysis activities to determine the current status of materials-based storage system technologies.

The Hydrogen Materials—Advanced Research Consortium (HyMARC) conducts foundational research to understand the interaction of hydrogen with materials in relation to the formation and release of hydrogen from hydrogen storage materials.

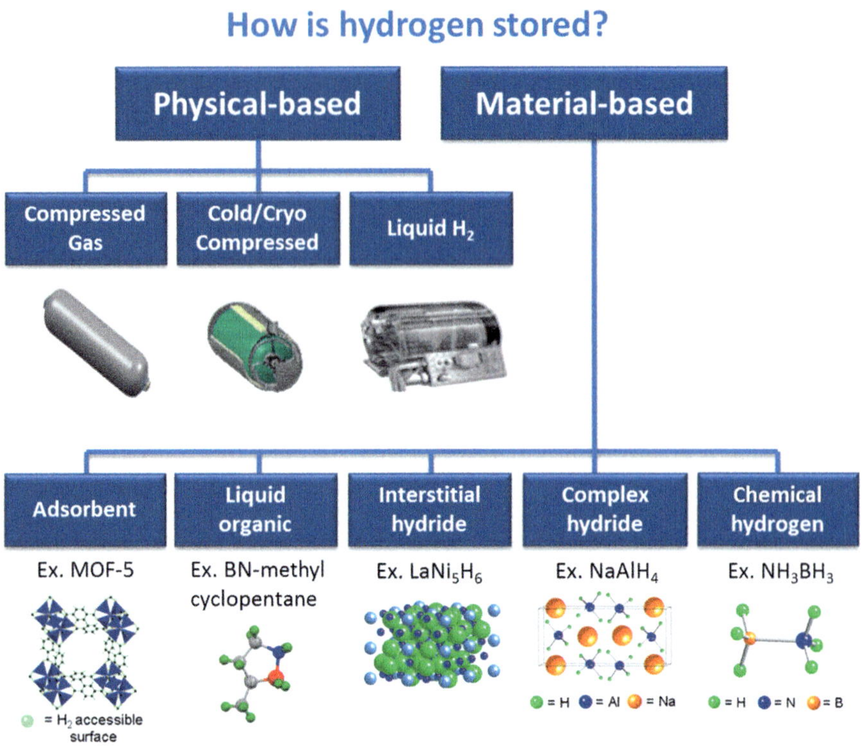

Fig. 5.12 How is hydrogen stored (Courtesy of Department of Energy)

5.9.4 Hydrogen Storage Challenges

High-density hydrogen storage is a challenge for stationary and portable applications and remains a significant challenge for transportation applications. Presently available storage options typically require large-volume systems that store hydrogen in gaseous form. This is less of an issue for stationary applications, where the footprint of compressed gas tanks may be less critical.

However, fuel cell-powered vehicles require enough hydrogen to provide a driving range of more than 300 miles with the ability to quickly and easily refuel the vehicle. While some light-duty hydrogen fuel cell electric vehicles (FCEVs) that are capable of this range have emerged onto the market, these vehicles will rely on compressed gas onboard storage using large-volume, high-pressure composite vessels. See Fig. 5.13.

The required large storage volumes may have less impact for larger vehicles, but providing sufficient hydrogen storage across all light-duty platforms remains a challenge. The importance of the 300-mile-range goal can be appreciated by looking at the sales distribution by range chart on this page, which shows that most vehicles sold today are capable of exceeding this minimum.

On a mass basis, hydrogen has nearly three times the energy content of gasoline—120 MJ/kg for hydrogen versus 44 MJ/kg for gasoline. On a volume basis,

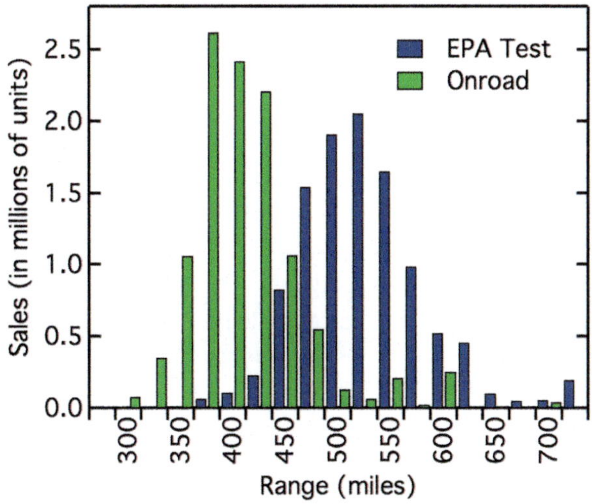

Fig. 5.13 The 2010 US light-duty vehicle sales distribution by driving range (Courtesy of Department of Energy)

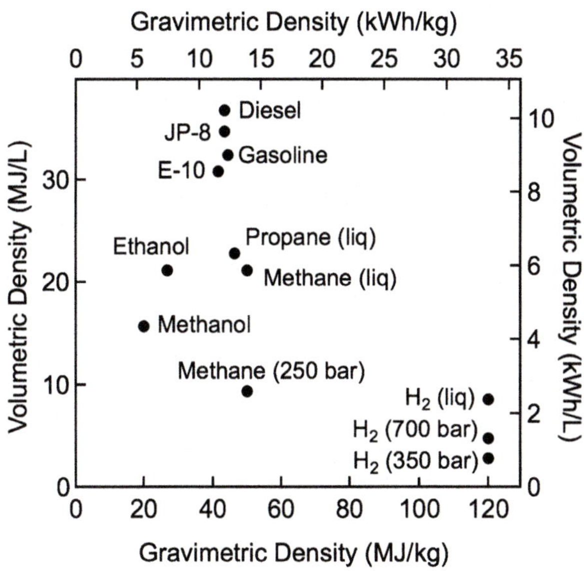

Fig. 5.14 Comparison of specific energy versus volumetric density (Courtesy of Department of Energy)

however, the situation is reversed; liquid hydrogen has a density of 8 MJ/L, whereas gasoline has a density of 32 MJ/L, as shown in the figure comparing energy densities of fuels based on lower heating values. Onboard hydrogen storage capacities of 5–13 kg hydrogen will be required to meet the driving range for the full range of light-duty vehicle platforms. See Fig. 5.14, where it compares the specific energy, (i.e., energy per mass or gravimetric density) and energy density (i.e., energy per volume or volumetric density) for several fuels based on lower heating values.

To overcome these challenges, FCTO is pursuing two strategic pathways, targeting both near-term and long-term solutions. The near-term pathway focuses on compressed gas storage, using advanced pressure vessels made of fiber reinforced composites that are capable of reaching 700 bar pressure, with a major emphasis on system cost reduction. The long-term pathway focuses on both:

1. Cold- or cryo-compressed hydrogen storage, where increased hydrogen density and insulated pressure vessels may allow for DOE targets to be met
2. Materials-based hydrogen storage technologies, including sorbents, chemical hydrogen storage materials, and metal hydrides, with properties having potential to meet DOE hydrogen storage targets

5.10 Hydrogen Energy Storage

Hydrogen is the most versatile means of energy storage—it can be produced and stored in all scales and used as a fuel, as a chemical material, or as a natural gas substitute. With its vast experience in producing and handling the molecule and the participation in leading pilot projects, Linde is among the frontrunners deploying hydrogen energy storage.

Electricity can be converted into hydrogen by electrolysis. The hydrogen can be then stored and eventually re-electrified. The round-trip efficiency today is as low as 30–40% but could increase up to 50% if more efficient technologies are developed. Despite this low efficiency, the interest in hydrogen energy storage is growing due to the much higher storage capacity compared to batteries (small scale) or pumped hydro and CAES (large scale).

Hydrogen is an energy-rich gas—this is one of the reasons why it is used as a rocket fuel. And it can be produced from a variety of feedstock—also electricity—and stored in many different ways: from a few grams in handheld cartridges to thousands of tons in an underground cavern. This gives hydrogen a unique potential to store renewable energy—both on small and very large scale. Especially for longer-term storage (weeks to months), hydrogen is today the only viable alternative in sight. And it's versatile to use: It can be converted back to power, but also be used as fuel for cars, a material for many industrial products (such as hardened fats) or even be converted to synthetic natural gas. Hydrogen makes all these markets accessible for renewable power. In other words: the wind power might end up in a margarine tub.

Hydrogen energy storage (HES) systems have been the topic of numerous studies and analyses. These systems typically involve the production of hydrogen from electricity by electrolysis, in which electrical energy is used to split water molecules into hydrogen and oxygen gas. Most electrolysis units involve alkaline or proton exchange membrane (PEM) conversion processes [12, 13]. As early as 1999, Ogden provided an overview of hydrogen infrastructure components, which included storage systems (1999), and Yang reviewed general similarities and differences between hydrogen and electricity as energy carriers (2008). Many studies of future hydrogen scenarios have been developed [14, 15], and this complementarity between hydrogen and electricity has been the focus of high-renewable scenarios developed by Barton and Gammon for the United Kingdom (2010), and more recently by Jacobson et al. [16] for California (2014). Several studies have compared hydrogen storage systems with other storage systems on the basis of cost, performance, and other attributes relevant to market viability and policy development [17–20].

In addition to numerous analytical studies, multiple grid-connected and remote demonstration projects have been executed during the past decade with approximately 80 hydrogen fueling stations currently based on electrolysis, 35 of which are located in North America (Fuel Cell Today 2013). Recently, interest has focused on power-to-gas applications, with several projects, especially in Germany, converting electrolytic hydrogen to synthetic methane (CH_4) by methanation. Methanation involves combining electrolytic hydrogen with carbon dioxide (CO_2) by a thermo-catalytic or biologic process. The concept of power-to-gas (a phrase derived from the German "Strom zu Gas") is to produce "green gas" with hydrogen from renewables and carbon dioxide from bioenergy or other sources, which allows for a significant increase in the overall utilization of renewable energy assets [21]. Power-to-gas and biogas projects in Austria, the Netherlands, Denmark, Sweden, Germany, and elsewhere were reviewed by Iskov and Rasmussen [22]. Gahleitner [23] reviewed 41 international power-to-gas projects and concluded with recommendations to improve overall system performance, develop codes and standards, and determine optimum system configurations (2013). Grond et al. [24] reviewed technologies for power-to-gas systems and concluded that these systems can provide community energy storage, time shifting/load leveling, and transmission and distribution management services (2013).

5.10.1 Hydrogen Production

Alkaline electrolysis is a mature technology for large systems, whereas PEM (proton exchange membrane) electrolyzers are more flexible and can be used for small decentralized solutions. The conversion efficiency for both technologies is about 65~70% (lower heating value). High-temperature electrolyzers are currently under development and could represent a very efficient alternative to PEM and alkaline systems, with efficiencies up to 90%. See Fig. 5.15.

5.10.2 Hydrogen Re-electrification

Hydrogen can be re-electrified in fuel cells with efficiencies up to 50% or alternatively burned in combined cycle gas power plants (efficiencies as high as 60%).

Considering other use of hydrogen can be expressed in the sense that, because of the limited round-trip efficiency, direct uses of green hydrogen are under development, e.g., as feedstock for the chemical and the petrochemical industry and as fuel for future

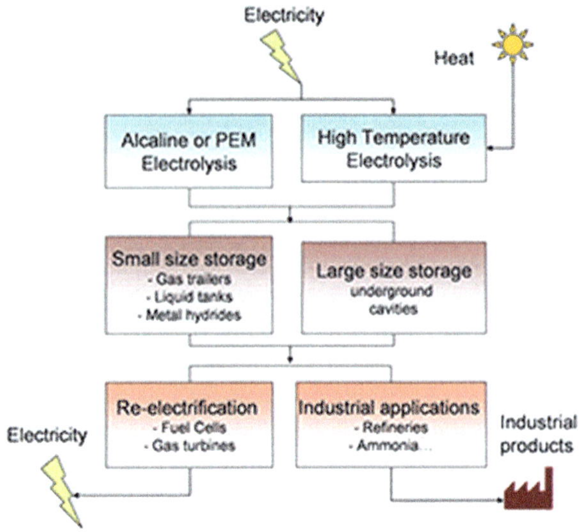

Fig. 5.15 Illustration of hydrogen production process (Courtesy of EIFER)

fuel cell cars or blending with natural gas of up to 5–15% in natural gas pipelines. Electrolytic hydrogen can also be used for the production of synthetic liquid fuels from biomass, thereby increasing significantly the efficiency of the biomass utilization.

However the deployment status is an integrated solution by several European and American companies, as the supply of electric power to small isolated sites or island is in the pipe. Demonstration projects have been performed since 2000 in Europe and the United States, and commercial products are available. Large-scale hydrogen storage in salt cavern is standard technology. To date there are two full-size hydrogen caverns in operation in Texas, USA, a third one is under construction, and three older caverns are operating at Teesside, UK.

5.11 Underground Hydrogen Storage

As part of hydrogen application as source of renewable energy during peak demand for electricity, one approach for power-to-gas is to inject hydrogen directly into natural gas pipelines rather than to undertake the additional step of methanation. This pathway was researched thoroughly in the European Union's NaturalHy project [25] and discussed by Melaina et al. [26] in the context of the US natural gas pipeline systems. In general, few changes to existing natural gas transmission or distribution pipeline networks are required if the hydrogen blend level is very low. Although industry codes and standards have become more stringent and society's tolerance for risk has decreased, for nearly a century leading up to 1950, hydrogen was a major constituent of town gas used for heating and lighting in homes, commercial buildings, and industry [27–29]. Dodds and Hawkes [30] reviewed issues related to hydrogen blending potential in the UK natural gas system and advised that early blend levels be limited to 2–3% hydrogen by volume (2014). Standards in Germany suggest up to 5%, with potential to increase to 6–20% [31]. As is evidenced by these studies, there is continued interest in pipeline material research for enabling power-to-gas. Power-to-gas projects today have a bias toward methanation, partly because of the lack of standards and pipeline-specific analysis required to approve direct injection of hydrogen. However, if suitable gas quality standards exist to facilitate direct hydrogen blending, it will likely lower the development cost for these systems. Furthermore, methanation processes are not expected to achieve 100% conversion of the input hydrogen feedstock, so the development of gas quality standards for lower levels of direct hydrogen blending is also expected to facilitate the growth of the methanation technologies.

In addition to injection into the natural gas system, underground geologic formations can be used to store large amounts of natural gas or hydrogen. This concept has several successful demonstrations and continues to attract interest in North America and Europe [32]. Salt caverns, which are currently used to store natural gas seasonally, are perhaps the best example of very large-scale hydrogen storage [33]. For example, Ozarslan [34] recently evaluated a particular large-scale solar hydrogen storage system that used salt caverns (2012).

HES units can not only increase the utilization of renewable energy resources but also have the potential to provide services to the grid. These services can be on the transmission or distribution level and enable access to additional revenue streams for HES systems. Several studies have been performed to assess the ability and value for electrolyzers, acting as demand response devices, to provide grid services [35–37]. In this respect, electrolytic hydrogen can play a role within the larger architecture of a smart grid and/or "smart gas" system by providing increased flexibility and resiliency. As is the case with other energy storage options, there are challenges to characterizing the value of these grid services to equipment owners, utilities, and electricity market operators.

Hydrogen energy storage (HES) is just beyond *electricity in, electricity out*. To understand this matter better, the reader should refer to the report by the National Renewable Energy Laboratory (NREL) published under the title of *Hydrogen Energy Storage, Grid and Transportation Services* in February 2015.

5.12 Materials-Based Hydrogen Storage

The Fuel Cell Technologies Office's (FCTO's) applied materials-based hydrogen storage technology research, development, and demonstration (RD&D) activities focus on developing materials and systems that have the potential to meet US Department of Energy (DOE) 2020 light-duty vehicle system targets with an overarching goal of meeting ultimate full-fleet, light-duty vehicle system targets.

Materials-based research is currently being pursued on metal hydride, chemical hydrogen storage, and sorbent materials.

- Metal hydride materials research focuses on improving the volumetric and gravimetric capacities, hydrogen adsorption/desorption kinetics, cycle life, and reaction thermodynamics of potential material candidates.
- Chemical hydrogen storage materials research focuses on improving volumetric and gravimetric capacity, improving transient performance, reducing release of volatile impurities, and developing efficient regeneration processes for the spent storage material.
- Sorbent materials research focuses on increasing effective adsorption temperature through increase of the dihydrogen binding energies and improving volumetric and gravimetric storage capacities through optimizing the material's pore size, increasing pore volume and surface area, and investigating effects of material densification.

A key component for advancing storage materials is the use of reliable material property measurement techniques. It is imperative to understand how the hydrogen storage properties of a material can be significantly influenced by not only individual sample characteristics—including chemical composition and distribution and microscopic and macroscopic material structure—but also pressure, temperature, and sample size. To help researchers better understand the proper measurement techniques, FCTO commissioned a best practices manual that gives a detailed overview of the recommended best practices in measuring the hydrogen storage properties of materials.

5.12.1 Technical Targets and Status

Materials-based research offers a long-term solution to the challenge of onboard automotive storage, as well as opportunities for stationary and portable power applications, with the potential to significantly reduce the required storage pressure, increase gravimetric and volumetric capacity, and reduce cost. From 2005 through 2010, the DOE Hydrogen Storage program supported three collaborative efforts—the Metal Hydride Center of Excellence, the Hydrogen Sorption Center of Excellence, and the Chemical Hydrogen Storage Center of Excellence—as well as independent projects that investigated more than 400 materials for potential use in hydrogen storage applications. Analysis activities in the Hydrogen Storage Engineering Center of Excellence (HSECoE) have determined the current status of systems using these materials. HSECoE has also developed spider charts showing three modeled systems for each material class and how they compare against all of DOE's 2020 targets.

Table 5.6 is a presentation of projects performance and cost of materials-based automotive systems compared to the 2020 and ultimate DOE targets.

The Fig. 5.16 shows hydrogen gravimetric capacity as a function of hydrogen release temperature for many of the unique hydrogen storage materials investigated by FCTO.

Table 5.6 Projected performance and cost of materials-based automotive hydrogen storage systems[a]

Storage system targets	*Gravimetric density* kWh/kg system (kg H_2/kg system)	*Volumetric density* kWh/L system (kg H_2/L system)	*Cost* \$/kWh (\$/kg H_2)
2020	1.5 (0.045)	1.0 (0.030)	\$10 (\$333)
Ultimate	2.2 (0.065)	1.7 (0.050)	\$8 (\$266)
Current status (from HSECoE)	*Gravimetric density* kWh/kg system (kg H_2/kg system)	*Volumetric density* kWh/L system (kg H_2/L system)	*Cost*[b] \$/kWh (\$/kg H_2)
Metal hydride (MH): $NaAlH_4$	0.4 (0.012)	0.4 (0.012)	\$43 (\$1430)
Sorbent: MOF-5, 100 bar, 80 K	1.3 (0.038)	0.7 (0.021)	\$15 (\$490)
Chemical hydrogen (CH) storage: off-board regenerable (AB)	1.5 (0.046)	1.3 (0.040)	\$17 (\$550)

[a]Assumes a storage capacity of 5.6 kg of usable hydrogen
[b]MH reflects status at the end of phase I; CH and sorbent reflect status at the end of phase II

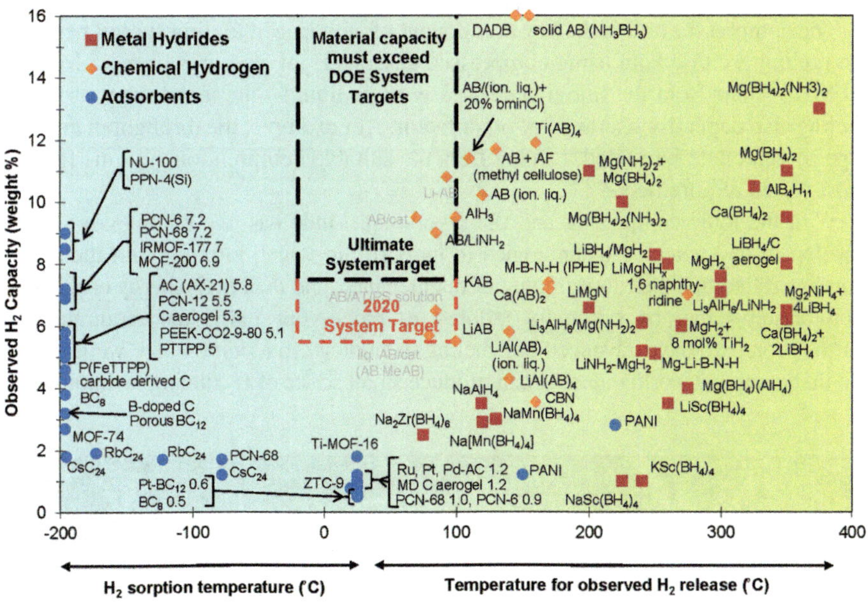

Fig. 5.16 Hydrogen gravimetric capacity as a function of hydrogen release temperature for many of the unique hydrogen storage materials (Courtesy of Department of Energy)

5.13 Industrial Application of Hydrogen Energy

We can state that hydrogen (H_2) is everywhere these days when it comes to industrial application of H_2 and claim that it is one stop shop. Hydrogen (H_2) has been deployed as an industrial gas for over 100 years, and large volumes are used across the widest range of applications every day. Hydrogen is also set to play a defining role in the much-publicized third, "green" industrial revolution. It is the most commonly occurring element in nature and—unlike fossil fuels such as crude oil or natural gas—will never run out. Like electricity, hydrogen is an energy carrier—not a source of energy. It must therefore be produced. Yet hydrogen offers several key benefits that increase its potential to replace fossil fuels. Stored hydrogen, for example, can be used directly as a fuel or to generate electricity.

Hydrogen will open up regenerative, sustainable mobility choices in our everyday lives. Hydrogen-powered vehicles have a long-distance range and can be rapidly fueled. Decades of research, development, and testing have shown that hydrogen technology is a workable, economically viable alternative suited to mass deployment. The following series of illustrations show different application of hydrogen in industry. Figure 5.17 shows a fuel cell bus at H_2 station.

The compressor unit is the key component of a hydrogen fueling station because the fueling is carried out using compressed gaseous H_2 at pressures from 35 MPa to 70 MPa. Apart from the initial state—gaseous or liquid—the technology used for fueling also depends on a range of other factors, for example, the throughput and the type of vehicle to be fueled. Linde offers the suitable compression systems for the most varied requirements.

With the ionic compressor and the cryopump, Linde has two cutting-edge, self-developed, and patented technologies in its portfolio which impress with their outstanding reliability, their low amount of maintenance, and their high energy efficiency. Both systems can be optimally tailored to meet your individual requirements. Furthermore, we have advanced our fueling technology to a point where we are now the first company worldwide that can produce small series of H_2 fueling technologies.

Fig. 5.17 A bus with fuel cell drive functionality at H_2 fuel station (Courtesy of Linde Group)

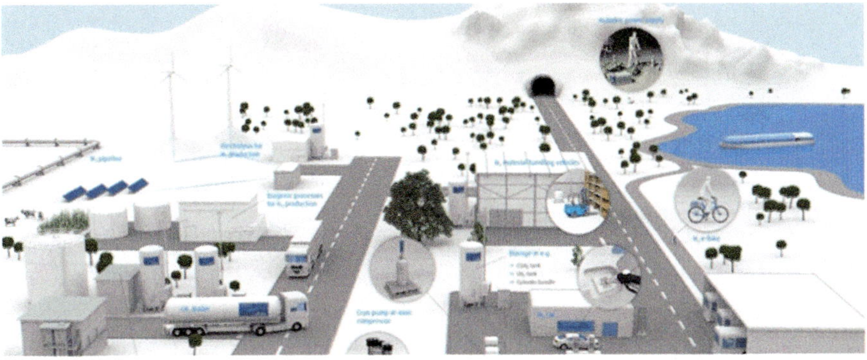

Fig. 5.18 Vehicle driven by H_2 in metropolitan area (Courtesy of Linde Group)

Fig. 5.19 Bicycle driven H_2 (Courtesy of Linde Group)

Figure 5.18 is an artistic demonstration of H_2 drive vehicles in an urban city.

Figure 5.19 shows a conceptual application of H_2 driving a bicycle.

Figure 5.20 is an overall presentation of different types of application of hydrogen in various industries.

5.14 Electrical Energy Storage

Electrical energy storage (EES) is one of the key technologies in the areas covered by the IEC. EES techniques have shown unique capabilities in coping with some critical characteristics of electricity, for example, hourly variations in demand and price. In the near future, EES will become indispensable in emerging IEC-relevant markets in the use of more renewable energy, to achieve CO_2 reduction and for smart grids.

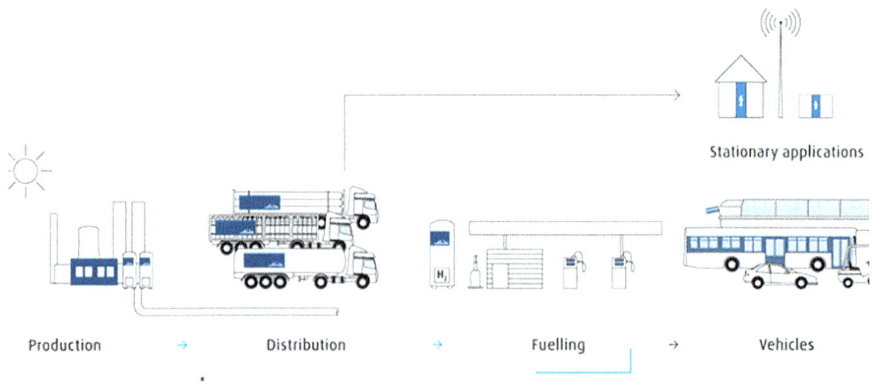

Fig. 5.20 Various industry application of hydrogen (Courtesy of Linde Group)

Historically, EES has played three main roles. First, EES reduces electricity costs by storing electricity obtained at off-peak times when its price is lower, for use at peak times instead of electricity bought then at higher prices. Secondly, in order to improve the reliability of the power supply, EES systems support users when power network failures occur due to natural disasters, for example. Their third role is to maintain and improve power quality, frequency, and voltage.

Regarding emerging market needs, in on-grid areas, EES is expected to solve problems—such as excessive power fluctuation and undependable power supply—which are associated with the use of large amounts of renewable energy. In the off-grid domain, electric vehicles with batteries are the most promising technology to replace fossil fuels by electricity from mostly renewable sources.

The smart grid has no universally accepted definition, but in general it refers to modernizing the electricity grid. It comprises everything related to the electrical system between any point of electricity production and any point of consumption. Through the addition of smart grid technologies, the grid becomes more flexible and interactive and can provide real-time feedback. For instance, in a smart grid, information regarding the price of electricity and the situation of the power system can be exchanged between electricity production and consumption to realize a more efficient and reliable power supply. EES is one of the key elements in developing a smart grid.

5.14.1 Characteristic of Electricity

Two characteristics of electricity lead to issues in its use and by the same token generate the market needs for EES. First, electricity is consumed at the same time as it is generated. The proper amount of electricity must always be provided to meet the varying demand. An imbalance between supply and demand will damage the stability and quality (voltage and frequency) of the power supply even when it does not lead to totally unsatisfied demand [38].

The second characteristic is that the places where electricity is generated are usually located far from the locations where it is consumed. Generators and consumers are connected through power grids and form a power system. In function of the locations and the quantities of power supply and demand, much power flow may happen to be concentrated into a specific transmission line, and this may cause congestion. Since power lines are always needed, if a failure on a line occurs (because of congestion or any other reason), the supply of electricity will be interrupted; also because lines are always needed, supplying electricity to mobile applications is difficult. The following sections outline the issues caused by these characteristics and the consequent roles of electrical energy storage (EES).

5.14.2 Electricity and the Roles of Electrical Energy Storages

High generation cost during peak-demand period is an essence of concern. Power demand varies from time to time (see Fig. 5.21), and the price of electricity changes accordingly. The price for electricity at peak-demand periods is higher and at off-peak periods lower. This is caused by differences in the cost of generation in each period.

During peak periods when electricity consumption is higher than average, power suppliers must complement the base-load power plants (such as coal-fired and nuclear) with less cost-effective but more flexible forms of generation, such as oil

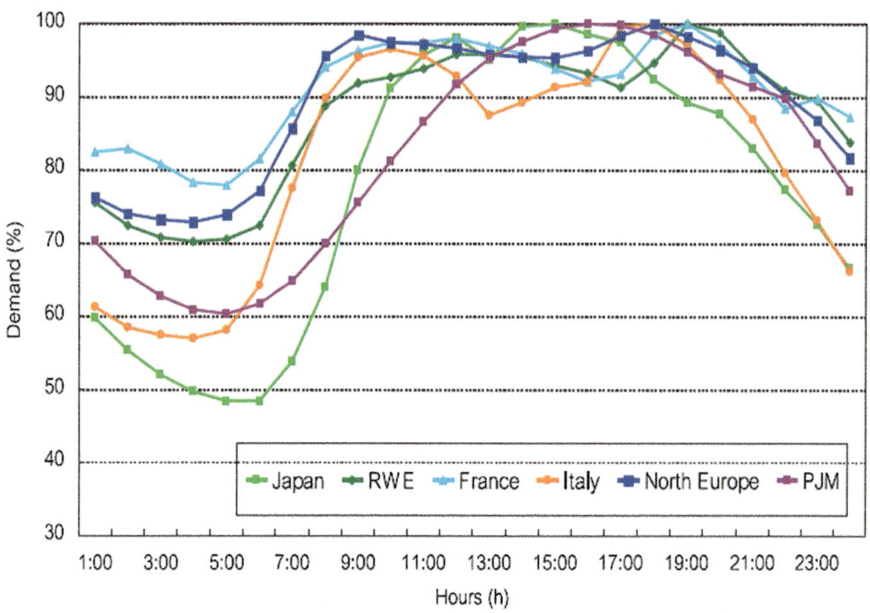

Fig. 5.21 Comparison of daily load curves (Courtesy of IEEJ—The Institute of Energy Economics, Japan, 2005)

and gas-fired generators. During the off-peak period when less electricity is consumed, costly types of generation can be stopped. This is a chance for owners of EES systems to benefit financially. From the utilities' viewpoint, there is a huge potential to reduce total generation costs by eliminating the costlier methods, through storage of electricity generated by low-cost power plants during the night being reinserted into the power grid during peak periods.

With high photovoltaic (PV) and wind penetration in some regions, cost-free surplus energy is sometimes available. This surplus can be stored in EES and used to reduce generation costs. Conversely, from the consumers' point of view, EES can lower electricity costs since it can store electricity bought at low off-peak prices, and they can use it during peak periods in the place of expensive power. Consumers who charge batteries during off-peak hours may also sell the electricity to utilities or to other consumers during peak hours.

A fundamental characteristic of electricity leads to the utilities' second issue, maintaining a continuous and flexible power supply for consumers. If the proper amount of electricity cannot be provided at the time when consumers need it, the power quality will deteriorate, and at worst this may lead to a service interruption. To meet changing power consumption, appropriate amounts of electricity should be generated continuously, relying on an accurate forecast of the variations in demand.

Power generators therefore need two essential functions in addition to the basic generating function. Firstly, generating plants are required to be equipped with a "kilowatt function," to generate sufficient power (kW) when necessary. Secondly, some generating facilities must possess a frequency control function, fine-tuning the output so as to follow minute-by-minute and second-by-second fluctuations in demand, using the extra power from the "kilowatt function" if necessary. Renewable energy facilities such as solar and wind do not possess both a kW function and a frequency control function unless they are suitably modified. Such a modification may be a negative power margin (i.e., decreasing power) or a phase shift inverter.

EES is expected to be able to compensate for such difficulties with a kW function and a frequency control function. Pumped hydro has been widely used to provide a large amount of power when generated electricity is in short supply. Stationary batteries have also been utilized to support renewable energy output with their quick response capability.

In conclusions, in respect to electrical energy storage (EES) perspective in the form of coherent picture, we can use Fig. 5.22 as an overall illustration of ESS. Note that, in the electricity market, global and continuing goals are CO_2 reduction and more efficient and reliable electricity supply and use.

Corresponding to these goals, three major drivers determining the future of EES have been identified (see Sect. 5.3): the foreseeable increase in renewable energy generation, the design and rollout of smart grids, and the future spread of dispersed generation and dispersed management of electrical energy—referred to here for simplicity as "microgrids." These drivers are only partly independent of each other: renewables clearly encourage, and simultaneously need, microgrids, and the increase in both renewables and dispersed sources demands a smarter grid. However, this paper has shown that the three drivers usefully illuminate different aspects of what will condition the future of electrical energy storage systems.

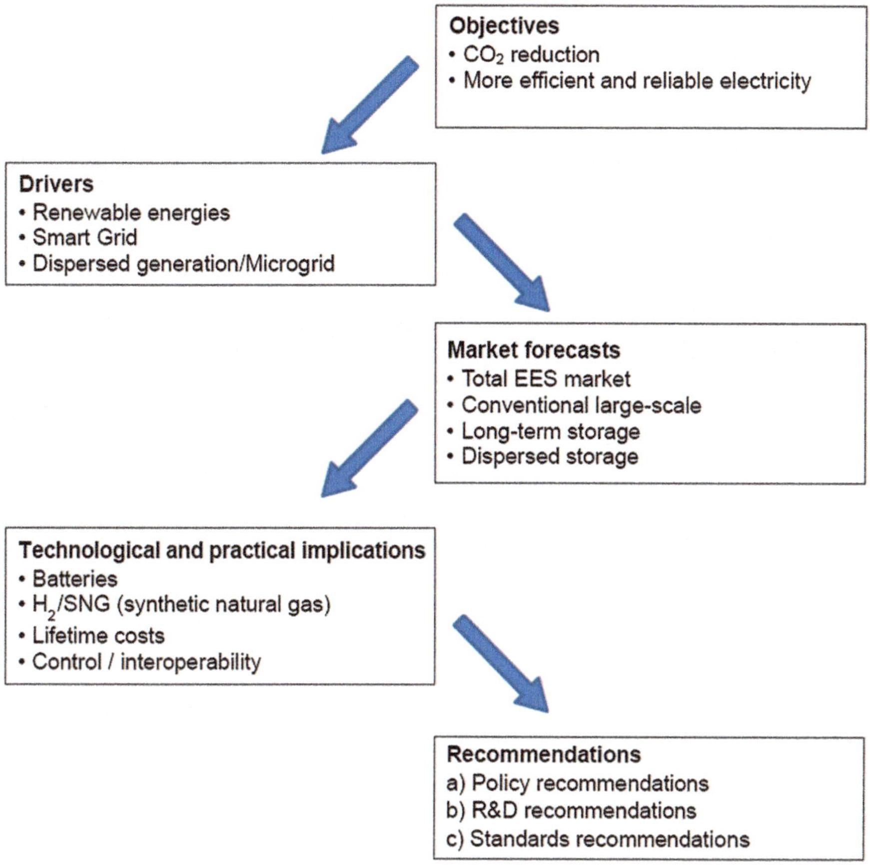

Fig. 5.22 Conclusions in the form of a logical progression (Courtesy of International Electrotechnical Commission)

The results of these drivers on future demand for EES may be conveniently divided into four market segments, the total EES market, conventional large-scale systems (e.g., pumped hydro storage (PHS)), long-term storage (e.g., H_2), and dispersed storage. How these markets are expected to develop has direct implications for which technologies will be most needed, which technology will need what type of further development, what considerations will influence roll-out and penetration, and what implementation problems may be expected. A serious analysis of these complex factors, going beyond what has already been attempted in previous sections, is not the purpose here; the four aspects listed, two technology families (batteries and H_2/SNG) and two constraints (lifetime costs and control/interoperability), seem merely to be the most important areas for future actions.

For further information refer to International Electrotechnical Commission white paper on Electrical Energy Storage [38].

5.15 Strategic Asset Management of Power Networks

Electricity networks around the world are facing a once-in-a-lifetime level of profound challenges, ranging from the massive uptake of distributed generation devices, such as rooftop solar generation, through to significant changes in the control and communications equipment used in the network itself. Power networks in developed nations are struggling with an equipment base nearing the end of its lifetime, while those in developing nations wrestle with trying to identify best practice examples on which to model their operations. Compounding these challenges, there is ever-increasing regulatory and funding pressure being placed on electricity network businesses to justify their management actions and expenditure decisions [39].

There is great variation around the world on how electricity network companies approach what are arguably their number one challenge—the design, maintenance, and operation of a large network of electrical equipment. Network companies often take quite different approaches in testing equipment, calculating the lifetime and financial costs of various equipment maintenance options, and even reporting on the performance of their system. The variety here is hardly intentional—it stems from a lack of internationally accepted global standards or guidelines on how to practice asset management in the electricity network sector.

> This current lack of international standards or guidelines on asset management for electrical networks will have a significant impact on the reliability and future viability of the electricity sector.

While standards such as the ISO 55000 series provide general guidance on best practice asset management procedures, they do not provide the industry-specific guidance that is needed given the operational methods and challenges of the electricity transmission and distribution industry.

The current situation means that:

- Network businesses around the world use different metrics to measure and report on the performance of their network. Without a commonly accepted definition of ways to calculate (for example) failure rates, it is very difficult to benchmark across organizations or jurisdictions.
- There is a lack of consensus on what are best practice methods for everything from testing the health of a particular item of equipment to prioritizing various asset management options. This makes stakeholder communication difficult (see below) and means many electricity network businesses waste time and resources developing their own methods to address a particular problem. This situation is particularly exacerbated in developing nations or in the context of relatively small organizations, who could benefit greatly by simply adopting best practice methods developed by others.
- Without worldwide standards on measuring and reporting on electricity network asset management procedures and performance, broader stakeholder engagement is very difficult. When a network business cannot benchmark its performance against peers or demonstrate that it is following industry recognized best practice,

stakeholders such as regulators or funding bodies can struggle to trust the network business's management decisions or appreciate the full depth of challenges ahead.

Electricity networks in many developed nations face the very significant challenge of an aging asset base. In many nations, electricity network rollout proceeded apace throughout the 1940s to 1980s but has slowed in recent years. Many significant items of equipment are now operating close to, or even beyond, their expected retirement age.

In many developed nations, the age of the asset base and the current slow rate of replacement mean it would take hundreds of years to renew all assets. This has significant reliability implications.

The aging equipment problem is not just one of equipment wear—it also constitutes a human resources issue, as in many cases the people with the skills and expertise to complete maintenance, or the experience needed to make asset management decisions regarding this older equipment, have retired from the industry. With an equipment fleet nearing the end of its life and a shortage of parts or people to maintain it, there are very significant implications for the reliability of electricity networks in many developed nations.

While aging equipment may not represent such a challenge in developing nations, or in others with more recently installed networks, simply understanding the optimal path forward amid a plethora of technologies, management options, and an often challenging regulatory or funding environment can be very difficult.

In conclusion, we need to consider the elaboration of detailed international standards or guidelines to introduce a common language across the electricity network business industry regarding current system performance. Metrics such as System Average Interruption Duration Index (SAIDI) and System Average Interruption Frequency Index (SAIFI) are vital to the benchmarking of electricity network performance, yet such metrics are calculated differently around the world.

5.16 Orchestrating Infrastructure for Sustainable Smart Cities

Cities are facing unprecedented challenges. The pace of urbanization is increasing exponentially. Every day, urban areas grow by almost 150,000 people, either due to migration or births. Between 2011 and 2050, the world's urban population is projected to rise by 72% (i.e., from 3.6 to 6.3 billion) and the population share in urban areas from 52% in 2011 to 67% in 2050. In addition, due to climate change and other environmental pressures, cities are increasingly required to become "smart" and take substantial measures to meet stringent targets imposed by commitments and legal obligations [40].

Furthermore, the increased mobility of our societies has created intense competition between cities to attract skilled residents, companies, and organizations. To promote a thriving culture, cities must achieve economic, social, and environmental sustainability. This will only be made possible by improving a city's efficiency, and this requires the integration of infrastructure and services. While the availability of smart solutions for cities has risen rapidly, the transformations will require radical changes in the way cities are run today.

Thus developing smart cities is not only just a process whereby technology providers offer technical solutions and city authorities procure them. Building up smart cities also requires the development of the right environment for smart solutions to be effectively adopted and used.

The development of a smart city requires participation, input, ideas, and expertise from a wide range of stakeholders. Public governance is naturally critical, but participation from the private sector and citizens of the community is equally important. It also requires a proper balance of interests to achieve the objectives of both the city and the community at large.

Reference [40] from International Electrotechnical Commission proposes a number of answers on the *what, who*, and *how* of smart city development in their executive summary, and we have quoted them here. It calls for a wider collaboration between international standardization bodies that will ultimately lead to more integrated, efficient, cheaper, and environmentally friendly solutions.

> Needs of cities differ strongly but… the main three pillars of development remain the same.

There is no single trend, solution, or specific approach for smart cities. Regional trends illustrate that there are divergent urban growth patterns among major regions with different levels of economic development. Still, significant disparities in the level of urbanization can also be observed across different countries within the same region. Nevertheless, all cities aiming to develop into smart cities have to be built on three sustainability pillars:

- *Economic sustainability*
  Cities need to provide citizens with the capacity to develop their economic potential and attract business and capital. With the global financial crisis, the economic sustainability of cities has taken center stage. The crisis has unearthed considerable weaknesses in the financial models and planning strategies of public authorities in the provision of services and in their infrastructure investments. Their financial sustainability now depends also on new financial models, as well as more efficient and better-integrated services and infrastructures.
- *Social sustainability*
  A city's attractiveness for people, business, and capital is closely related to the quality of life (QoL), business opportunities, and security and stability, which are guaranteed by social inclusiveness.
- *Environmental sustainability*
  Cities face a number of environmental sustainability challenges, generated by the city itself or caused by weather or geological events. To reduce the impact of the city on the environment resource, it is important to promote the efficient and intelligent deployment of technology and to integrate infrastructures. This process can also be developed in such a manner as to increase the resilience of the city to environmental shocks. These three pillars have one common denominator,

namely, the need to achieve more and better with less, i.e., efficiency. Efficiency must also be achieved in a manner that brings benefits and opportunities to citizens, making the city more dynamic and participatory.

5.16.1 Smart Technology Solution Create Value

Rather than being an expense, smart technology integration can create considerable opportunities for added value in any city. Technology integration helps cities to improve efficiency, enhance their economic potential, reduce costs, open the door to new business and services, and improve the living conditions of its citizens. A key condition for value creation through integration is the compatibility of technologies, which is best achieved through common and consensus-based standards that ensure interoperability.

Presently, however, smart city projects concentrate mainly on vertical integration within existing independent infrastructure and services silos, e.g., energy, transport, water, or health. A truly "smart" city requires horizontal integration as well as creating a system of systems capable of achieving considerable increases in efficiency and generating new opportunities for the city and its citizens.

5.16.2 New Approach to Smart City Solution

Cities are faced with a complex challenge, as the traditional processes of planning, procuring, and financing are not adequate for their needs. Smart cities can only exist if fundamental reforms are undertaken. Thus, there is a need for new approach, and it is necessary to design, implement, and finance smart city solution.

5.16.3 Stakeholders Are Key Drivers to Smart City Solution

A smart city cannot be imposed by decree, as the city is shaped by a large number of individual decisions and social and technological changes cannot be fully accounted for. With the present advances in telecommunications, Information and Communication Technologies (ICT) and affordable energy efficiency and energy production tools are changing the relationship between citizens and city services. Citizens are increasingly becoming providers of city services and not only users. A good plan requires the participation, input, and ideas from a wide range of stakeholders within the city. This means that city planning needs to allow for bottom-up processes of modernization. The stakeholders are:

- Political leaders, managers, and operators of the local government (city)

- The service operators—public or private: water, electricity, gas, communication, transport, waste, education, etc.
- End users and producers: inhabitants and local business representatives
- Investors: private banks, venture capitalists, pension funds, international banks
- Solution providers: technology providers, financiers, and investors

Giving to each of these groups a true stake in smart city development is important to achieve the necessary consensus for the changes. Their concerns need to be carefully considered and acknowledged, and ultimately the direction and next steps have to be collectively approved. In the absence of proper consultation, the authorities will sooner or later face considerable additional obstacles to make their vision a reality.

5.16.4 Without Integration Rising to the Level of Systems There Cannot Smart City

The transformation of a city into a smart form presents its stakeholders a wide range of challenges, including benefits and consequences when such a transformation is undertaken. A promising approach to support city planners, but also Standards Developing Organizations (SDOs), is to model a city as a collection of activity domains in an integrated virtual organization (the city), where various groups of stakeholders (local governments, public and private corporations, academia, healthcare institutions, cultural associations, religious congregations, and financial firms) participate in operating and sustaining the city as a whole. Modeling the interrelations allows identifying pain points, gaps, and overlaps in standardization and clarifying the technical needs for integration.

While the technologies to develop smart cities are mostly already readily available and improving, their deployment is hampered by technical, social, and administrative challenges. Horizontal integration of infrastructures through technology is essential to reap the benefits of innovation and the potential and necessary efficiency.

Thus, interoperability is essential; without it, city planning is marred by unexpected inefficiencies leading to suboptimal outcomes and higher costs. The planning requirements for city authorities are very complex, as there are thousands of organizations and companies working in parallel to bring on the tools, systems, and products that offer potentially affordable/sustainable solutions.

To ensure that smart integrated systems are put in place in practice, internationally agreed standards that include technical specifications and classifications in order to support interoperability (i.e., devices and systems working together) are sine qua non or essential conditions. These include technical specifications and classifications in order to support interoperability. These are metrics against which benefits can be assessed as well as best practice documents that detail controls.

5.16.5 *Horizontal and Vertical Integration a Key to Interoperability*

Electric grids, gas/heat/water distribution systems, public and private transportation systems, and commercial buildings/hospitals/homes play a key role in shaping a city's livability and sustainability. To increase their performance and efficiency, these critical city systems need to be integrated.

The successful development of a smart city will require the combining of a bottom-up systems approach with a top-down service development and a data-centric approach. Technology integration includes vertical integration from sensors to low-cost communication and real-time analysis and control and horizontal integration of historically isolated systems up to citizen-based services. Combined, this creates a system of systems.

Today's smart city projects are mainly focusing on improving the integration of historical verticals, i.e., parts of existing utilities, improving, e.g., energy efficiency or reducing water leakage. The next step is horizontal integration. Data from the different sectors can be combined to better manage the city and reduce risks. Thus, horizontal as well as vertical integration is key to creating value and interoperability.

5.16.6 *Interoperability Is the Key to Open Markets and to Competitive Solutions*

Interoperability is the key to manage systems of systems and to open markets to competitive solutions. While we are today experiencing the Internet of things (IoT) revolution (driven by the appearance of smart devices, such as wireless sensors, radio-frequency identification (RFID) tags, and Internet Protocol (IP)-enabled devices), different producers are generating technologies using their own communication specifications and data protocols.

Note that, Internet of things (IoT) market forecasts show that IoT is already making an impact on the global economy. While estimates of the economic impact during the next 5–10 years vary slightly (International Data Corporation (IDC) estimates USD 1.7 trillion in 2020 [41], Gartner sees a benefit of USD 2 trillion by that time [42], and McKinsey predicts growth of USD 4 trillion to USD 11 trillion by 2025 [43]), there seems to be a consensus that the impact of IoT technologies is substantial and growing.

Future interoperability can only be guaranteed through the existence of international standards ensuring that components from different suppliers and technologies can interact seamlessly. Continued best practice sharing and development of common standards to ensure that data can flow freely between systems are essential while maintaining the need to protect confidentiality and individual privacy.

Common terminology and procedures have to be developed in order to also ensure that organizations and businesses can efficiently communicate and collaborate, which can also be guaranteed through standards.

In addition, the multiplicity of technologies within a city now demands a top-down approach to standardization. This requires new coordination approaches between Standards Developing Organizations (SDOs), in which all the parts of the city are jointly considered by the several technical committees involved by the different organizations. This methodology is essential as systems level standards will enable the implementation and interoperability of smart city solutions.

5.16.7 Guiding Principles and Strategic Orientation

Guiding principles and strategic orientation for the International Electrotechnical Commission (IEC) and messages to other Standards Developing Organizations are indication that the electricity is core to any urban infrastructure system and the key enabler of city development. As a result, the IEC has a specific role to play in the development of a smart city's set of standards. The IEC shall call for, take initiative, invite, and strongly contribute to a more global and collaborative approach including not only international standardization organizations but also all stakeholders of the smart city landscape (city planners, city operators, etc.) and specifically the citizens.

Technology and system integration are critical to ensure interoperability, and the IEC will support active collaboration between the relevant actors as described in the following guiding principles.

The IEC shall continue to foster technology integration (electrotechnical, electronics, digital, and IT) and make sure that digital technology is fully integrated in all IEC products in a connect and share data perspective.

The IEC shall make sure digital and IT technology suppliers are actively contributing in its work. Data aspects shall become a key issue in IEC, including IoT, data analytics, data utilization, data privacy, and cyber security.

The system approach shall be accelerated as a top IEC priority taking into account flexibility, interoperability, and scalability. Value creation for users (citizens and city infrastructure and service planners and operators) will remain the main driver of standardization work.

Smart development requires solutions to be adapted to the specific needs of the city and its citizens, and standards have to be developed with this purpose in mind, removing technology barriers that prevent technology integration.

In conclusions, smart cities are necessary not only to reduce emissions but to handle the rapid urbanization growth that the world is experiencing. Inefficiencies in urban areas bring large negative environmental and social impacts. City infrastructures are the backbone of the cities, delivering the necessary services to the population and creating the conditions for citizens to develop their professional, social, and cultural activities. Infrastructures are also quintessential in guaranteeing the city's resilience to environmental risks.

Until now city infrastructures have been built independently and operated separately in parallel silos (water supply, electricity, transport). Furthermore, the citizen has mainly been a consumer of services with little direct influence on the system. In a smart city, this needs to change. First of all, efficiency requires that infrastructures are appropriately interlinked horizontally. Secondly, citizens are becoming producers and service providers. In the area of energy, individuals are starting to produce energy from renewables and thanks to the data revolution, also to deliver information and services in a number of areas. With smart systems, goods owned by citizens can be active in improving efficiency. Smart meters and electric cars can interact with the grid, data produced by the smart applications of the citizens can contribute to traffic control, improve emergency response, etc. Citizens can also use the technologies to sell new services.

This change in cities needs to be accompanied by enabling conditions, which means reforming the ways cities are governed and financed, i.e., administrative reforms and new financial systems.

However, the glue allowing infrastructures to link and operate efficiently are standards. Standards are necessary to ensure interoperability of technologies and the transfer of best practices. But standards are not yet adapted to the level of technology integration we are requiring. Standard bodies still operate in sectorial parallel silos, developing standards which are not easy to understand by nonspecialists, particularly city managers. Standards are facilitators for city planners, and they need to incorporate standards in planning and procurement. There is thus a need to reform the way standards are produced, ensuring those are adapted to the needs of the city planners and other service operators within the city.

There is a need for close collaboration between standard bodies themselves and collaboration with outside organizations, in particular the city planners.

References

1. B. Zohuri, *Physics of Cryogenic an Ultra-Low Temperature Phenomena* (Elsevier, 2017)
2. https://www.azocleantech.com/article.aspx?ArticleID=29
3. B. Zohuri, *Nuclear Energy for Hydrogen Generation through Intermediate Heat Exchangers: A Renewable Source of Energy*, 1st edn. (Springer, 2016)
4. https://energy.gov/eere/fuelcells/fuel-cells
5. http://change.gov/agenda/energy_and_environment_agenda/
6. Available at http://www1.eere.energy.gov/hydrogenandfuelcells/program_plans.html
7. X. Li, *Principles of Fuel Cells*, 1st edn. (Taylor & Francis Group, 2006). isbn:ISBN-13: 978-1591690221
8. https://energy.gov/eere/fuelcells/fuel-cell-technologies-office
9. https://www.nrel.gov/hydrogen/publications.html
10. https://energy.gov/eere/fuelcells/doe-technical-targets-fuel-cell-backup-power-systems
11. https://energy.gov/eere/fuelcells/doe-technical-targets-fuel-cell-systems-portable-power-and-auxiliary-power
12. F. Barbir, PEM electrolysis for production of hydrogen from renewable energy sources. Sol. Energy **78**(5), 661–669 (2005). https://doi.org/10.1016/j.solener.2004.09.003

13. K. Zeng, D. Zhang, Recent progress in alkaline water electrolysis for hydrogen production and applications. Prog. Energy Combust. Sci. **36**(3), 307–326 (2010)
14. D.L. Greene, P.N. Leiby, B.D. James, J. Perez, M. Melendez, A. Milbrandt, S. Unnasch, M. Hooks, *Analysis of the Transition to Hydrogen Fuel Cell Vehicles & the Potential Hydrogen Energy Infrastructure Requirements*. ORNL/TM-2008/30, ed. by S. McQueen (Oak Ridge National Laboratory, Oak Ridge, 2008). http://cta.ornl.gov/cta/Publications/Reports/ORNL_TM_2008_30.pdf. Accessed 4 Aug 2014
15. National Research Council, *Transitions to Alternative Transportation Technologies: A Focus on Hydrogen*. National Research Council of the National Academies, Committee on Assessment of Resource Needs for Fuel Cell and Hydrogen Technologies (The National Academies Press, Washington, DC, 2008)
16. M.Z. Jacobson, M.A. Delucchi, A.R. Ingraffea, R.W. Howarth, G. Bazouin, B. Bridgeland, K. Burkart, M. Chang, N. Chowdhury, R. Cook, G. Escher, M. Galka, L. Han, C. Heavey, A. Hernandez, D.F. Jacobson, D.S. Jacobson, B. Miranda, G. Novotny, M. Pellat, P. Quach, A. Romano, D. Stewart, L. Vogel, S. Wang, H. Wang, L. Willman, T. Yeskoo, A roadmap for repowering California for all purposes with wind, water, and sunlight. Energy **73**, 875–889 (2014)
17. S.M. Schoenung, W.V. Hassenzahl, *Long- Vs. Short-Term Energy Storage Technologies Analysis: A Life-Cycle Cost Study—A Study for the DOE Energy Storage Systems Program*. SAND2003-2783 (Sandia National Laboratories, Albuquerque, 2003)
18. D. Steward, G. Saur, M. Penev, T. Ramsden, *Lifecycle Cost Analysis of Hydrogen Versus Other Technologies for Electrical Energy Storage*. NREL/TP-560-46719 (National Renewable Energy Laboratory, Golden, 2009). http://www.nrel.gov/docs/fy10osti/46719.pdf. Accessed 4 Aug 2014
19. P.W. Parfomak, *Energy Storage for Power Grids and Electric Transportation: A Technology Assessment*. Tech. Rep. R42455. (Congressional Research Service, Washington, DC, 2012). http://fas.org/sgp/crs/misc/R42455.pdf. Accessed 4 Aug 2014
20. A. Oberhofer, *Energy Storage Technologies & Their Role in Renewable Integration* (Global Energy Network Institute, San Diego, 2012). http://www.geni.org/globalenergy/research/energy-storage-technologies/Energy-Storage-Technologies.pdf. Accessed 4 Aug 2014
21. M. Sterner, 100% Renewable Energy Supply for Cities and Nations. Presented at the Sustainable Energy Week, 22–26 Mar 2010. http://www.klimabuendnis.org/fileadmin/inhalte/dokumente/EUSEW2010_1.IWES_M.Sterner.pdf. Accessed 4 Aug 2014
22. H. Iskov, N.B. Rasmussen, *Global Screening of Projects and Technologies for Power-to-Gas and Bio-SNG: A Reference Report* (Danish Gas Technology Centre, Horsholm, 2013). https://www.energinet.dk/ SiteCollectionDocuments/Engelske%20dokumenter/Forskning/global_screening_08112013_final.pdf. Accessed 4 Aug 2014
23. G. Gahleitner, Hydrogen from renewable electricity: An international review of power-to-gas pilot plants for stationary applications. Int. J. Hydrog. Energy **38**(5), 2039–2061 (2013)
24. L. Grond, P. Schulze, J. Holstein, *Systems Analyses Power to Gas: Final Report—Deliverable 1: Technology Review*. GCS 13.R.23579 (DNV KEMA Energy & Sustainability (now DNV GL), Groningen, 2013). http://www.dnv.com/binaries/dnv%20kema%20%282013%29%20-%20systems%20analyses%20power%20to%20gas%20-%20 technology%20review_tcm4-567461.pdf. Accessed 4 Aug 2014
25. O. Florisson, NATURALHY: An Overview. Presented at the NATURALHY Final Public Presentation, November 19, Nederlandse Gasunie at Groningen, 2009
26. M.W. Melaina, O. Antonia, M. Penev, *Blending Hydrogen into Natural Gas Pipeline Networks: A Review of Key Issues*. NREL/TP-5600-51995 (National Renewable Energy Laboratory, Golden, 2013). http://www.nrel.gov/docs/fy13osti/51995.pdf. Accessed 6 Nov 2014
27. C.J. Castaneda, *Invisible Fuel: Manufactured and Natural Gas in America 1800–2000* (Twayne Publishers, New York, 1999)
28. J.A. Tarr, History of manufactured gas, in *Encyclopedia of Energy*, vol. 3, (Elsevier, 2004), pp. 733–743

29. M. Melaina, Market transformation lessons for hydrogen from the early history of the manufactured gas industry, in *Hydrogen Energy and Vehicle Systems*, ed. by S. E. Grasman (CRC Press, Boca Raton, 2012)
30. P. E. Dodds, A. Hawkes (eds.), *The Role of Hydrogen and Fuel Cells in Providing Affordable, Secure Low-Carbon Heat* (H2FC SUPERGEN, London, 2014)
31. R. Winkler-Goldstein, A. Rastetter, Power to gas: The final breakthrough for the hydrogen economy? Green **3**(1), 69–78 (2013). https://doi.org/10.1515/green-2013-0001
32. HyUnder, Assessment of the Potential, the Actors and Relevant Business Cases for Large Scale and Long Term Storage of Renewable Electricity by Hydrogen Underground Storage in Europe, 2014. http://www.hyunder.eu/deliverables-and-publications.html. Accessed 30 Sept 2014
33. A.S. Lord, P.H. Kobos, G.T. Klise, D.J. Borns, *A Life Cycle Cost Analysis Framework for Geologic Storage of Hydrogen: A User's Tool.* SAND20116221 (Sandia National Laboratories, Albuquerque, 2011). http://prod.sandia.gov/techlib/access-control.cgi/2011/116221.pdf. Accessed 4 Aug 2014
34. A. Ozarslan, Large-scale hydrogen energy storage in salt caverns. Int. J. Hydrog. Energy **37**(19), 14265–14277 (2012)
35. Hydrogenics, *Hydrogenics Successfully Completes Utility-Scale Grid Stabilization Trial with Ontario's Independent Electricity System Operator.* Press Release, 16 June 2011. http://www.hydrogenics.com/about-the-company/news-updates/2011/06/16/hydrogenics-successfully-completes-utility-scale-grid-stabilization-trial-with-ontario's-independent-electricity-system-operator. Accessed 4 Aug 2014
36. J. Eichman, K. Harrison, M. Peters, *Novel Electrolyzer Applications: Providing More Than Just Hydrogen.* NREL/TP-5400-61758 (National Renewable Energy Laboratory, Golden, 2014)
37. J. Judson-McQueeney, T. Leyden, C. Walker, Clean Energy Group: Resilient Power Project Webinar—Energy Storage: New Markets and Business Models, 2013. http://www.cesa.org/assets/Uploads/RPP-Webinar-Presentations- Energy-Storage-New-Markets-and-Business-Models.pdf. Accessed 4 Aug 2014
38. International Electrotechnical Commission white paper on Electrical Energy Storage at http://www.iec.ch/whitepaper/energystorage/
39. International Electrotechnical Commission white paper on Strategic Asset Management of Power Networks
40. International Electrotechnical Commission white paper on Orchestrating Infrastructure for Sustainable Smart Cities
41. International Data Corporation (IDC), *Explosive Internet of Things Spending to Reach $1.7 Trillion in 2020, According to IDC,* 2 June 2015, [Online]. Available http://www.idc.com/getdoc.jsp?containerId=prUS25658015. Accessed 22 Aug 2016
42. Gartner Inc., The Internet of Things Is a Revolution Waiting to Happen, 30 Apr 2015. [Online]. Available: http://www.gartner.com/smarterwithgartner/the-internet-of-things-is-a-revolution-waiting to happen. Accessed 22 Aug 2016
43. McKinsey Global Institute, *Unlocking the Potential of the Internet of Things*, June 2015. [Online]. Available: http://www.mckinsey.com/business-functions/business-technology/our-insights/the-internet-of-thingsthe-value-of-digitizing-the-physical-world. Accessed 22 Aug 2016
44. https://energy.gov/eere/fuelcells/doe-technical-targets-fuel-cell-systems-stationary-combined-heat-and-power

Chapter 6
Energy Storage for Peak Power and Increased Revenue

Worldwide electricity markets are changing due to decreasing prices of fossil fuels and addition of renewable generators (wind and solar). Large-scale renewables deployment collapses prices at times of high wind or solar input that limits their deployment and impacts nuclear plant revenue. These changes have reduced the demand for baseload electricity but increased the demand for dispatchable electricity—a market currently served in the United States primarily by natural gas turbines. At the same time, there is a longer-term need for dispatchable low-carbon electricity production—a replacement for fossil fuel electricity production. The changes may be challenging the economics of nuclear power today but may create new opportunities for existing and new build nuclear energy systems in the future. Heat storage coupled to electrical power plant whether gas, fissile fuel, coal, or nuclear may enable baseload reactor operation with variable electricity to the grid—heat into storage when low electricity prices and production of added electricity using stored heat when prices are high.

6.1 Introduction

To address these sources of energy challenges, scientists and engineers around the world and even universities are involved with these types of activities along with their partners at national laboratories in the United States and other countries such as the United Kingdom and France.

If we want to be completely dependent on renewable energy, then we need to find a way to store this energy for times when the sun is not shining and the wind is not blowing. Storing this energy is one of the greatest barriers to the adoption of renewable energy. However, it is an area where hydrogen can play a key role. As Jeremy Rifkin says in *The Third Industrial Revolution*, electrolyzers are the portal to the hydrogen economy. Thus, for us to be able to store energy, we need to store heat as fundamental source of energy, even for generating electricity for peak power

B. Zohuri, *Hybrid Energy Systems*,
https://doi.org/10.1007/978-3-319-70721-1_6

demand and in general for meeting the supply for the increase demand of electricity around the world at a rate of almost 17% per year.

As another source of heat and energy, light-water reactors (LWRs) are under-studied by a group of universities in support of Idaho National Laboratory (INL), and Exelon recently conducted a workshop on light-water reactor (LWR) heat storage for peak power and increased revenue on June 27–28, 2017, at Massachusetts Institute of Technology (MIT). The workshop goals were to define and understand the market, regulatory, and technical options for coupling heat storage for variable power to existing and future LWRs with recommendations for the path forward to improve LWR economics. Observations and outcomes from the workshop include [1]:

1. Nuclear reactors generate heat and thus couple efficiently to heat storage that is 10–40 times less expensive than electricity storage (pumped hydro, battery, etc.), thus potentially a lower-cost way to meet variable electricity demand. Favorable heat storage economics has resulted in concentrated solar power systems under construction to include heat storage to vary electricity production. Many of these technologies are applicable LWRs and most are applicable to other reactor types.
2. Six classes of heat storage technologies have been identified that can couple to light-water reactors: steam accumulators, sensible heat storage, cryogenic air storage, packed pebble-bed heat storage, hot-rock storage, and geothermal heat storage. Some storage technologies are ready for demonstration; others require significant R&D.
3. Heat storage systems coupled to LWRs are different from storage technologies such as batteries and pumped hydro. Batteries and pumped hydro storage have electricity input rates to storage that are near electricity output rates; thus, the strategy is buy low and sell high. With most heat storage systems, there are separate capital costs associated with heat input, storage, and heat-to-electricity production.
4. Accumulators and some other heat storage technologies have very low costs for heat addition to storage. The profitable strategy may be to send steam to storage 6 h per day when prices are the lowest and produce added electricity 18 h per day to minimize the cost of the more expensive heat-to-electricity component of the storage system. For many existing reactors up to 20% of the steam would go to storage when prices are low. The maximum power output would increase by less than 5% to avoid major upgrades of the turbine hall. When viewing the nuclear plant as a black box, the addition of storage would appear to have increased its "baseload" capacity by a few percent and dramatically increased the capability to rapidly go down and back up in power. Inside the plant, the reactor is operating at full capacity.
5. Other technologies such as nuclear geothermal inject hot water underground and use a geothermal power system for electricity production. Because of the extremely low cost of storage, such systems may enable seasonal energy storage, provide assured generating capacity, and provide the option for a strategic multi-year heat reserve—the low-carbon equivalent to a strategic oil reserve.

The business case is central. Five years ago, coupling heat storage to a LWR reactor would not have been economic. The changes in the electricity markets indicate that such an option may now be economical in some markets. As the markets continue to change, the economic case improves.

There is a need for demonstration projects to address institutional issues, to provide technology demonstrations for the near-term options, and to collect sufficient information to determine the economics.

However, what has been changed in nuclear industry in recent years to make a slow-growing technology regardless of innovative approach to LWR is a new generation that is called GEN-IV, and among this generation is small modular reactor (SMR), and these changes are described and summarized in the following paragraph.

Electricity markets are changing because of low-cost natural gas (United States and Canada) and the addition of intermittent renewable generators (wind and solar). This has reduced the demand for baseload electricity. At the same time, there is an increased demand for dispatchable electricity—a market currently served in the United States primarily by natural gas turbines—to a smaller extent by pumped hydroelectricity and to a very limited extent by batteries. These changes are hurting the economics of nuclear power but may create new opportunities for nuclear energy systems with heat storage to enable baseload reactor operation with variable electricity to the grid.

In a low-carbon world, the energy sources are nuclear, wind, and solar. These technologies have high capital costs and low operating costs. If these energy production facilities are operated at half capacity, the production cost of energy approximately doubles. Because energy is about 8% of the global economic output, increases in energy costs have large impacts on the United States and global standards of living. Equally important, the uneven distribution of renewables has serious geopolitical implications.

6.2 Variable Electricity and Heat Storage

In this section, we will discuss and expand on variable electricity and steam from salt-, helium-, and sodium-cooled baseload reactors with gas turbines and heat storage. Advances in utility natural gas-fired air-Brayton combined cycle technology are creating the option of coupling salt-, helium-, and sodium-cooled nuclear reactors to nuclear air-Brayton combined cycle (NACC) power systems. NACC may enable a zero-carbon electricity grid and improve nuclear power economics by enabling variable electricity output with baseload nuclear reactor operations. Variable electricity output enables selling more electricity at times of high prices. Peak power is achieved using stored heat or auxiliary fuel (natural gas, biofuels, hydrogen). A typical NACC cycle includes air compression, heating compressed air using nuclear heat and a heat exchanger, sending air through a turbine to produce electricity, reheating compressed air, sending air through a second turbine, and exhausting to a heat recovery steam generator (HRSG). In the HRSG, warm air produces steam that is used to produce added electricity. For peak power production, auxiliary heat (natural gas, stored heat) is added before the air enters the second turbine to raise air temperatures and power output. Like all combined cycle plants, water cooling requirements are dramatically reduced relative to other power cycles because much of the heat rejection is in the form of hot air.

Historically nuclear plants have been designed to produce baseload electricity using steam turbines—the traditional power cycle of the utility industry. Changing markets and changing technologies suggest that air-Brayton power cycles may become the preferred power cycle technology for higher-temperature reactors: fluoride-salt-cooled high-temperature reactors (FHRs), sodium-cooled fast reactors (SFRs), and high-temperature gas-cooled reactors (HTGRs). The basis for this conclusion is described herein.

6.3 Implications of Low-Carbon Grid and Renewables on Electricity Markets

In a free market, the price of electricity varies with time. Figure 6.1 shows the market price of electricity versus the number of hours per year electricity can be bought at different prices in California (blue bars). Power plants with the lowest operating costs come are dispatched first. As the price of electricity rises, power plants with higher operating costs come on line. There are near-zero and negative prices for a significant number of hours per year when electricity production exceeds demand and electricity generators pay the grid to take electricity. This is a consequence of two effects.

- *Renewables subsidies*. Production tax credits provide revenue for wind and solar plants to produce output independent of electricity demand. An owner of a wind or solar facility will sell electricity into the grid as long as the price paid to the grid to take electricity when there is excess production is less than the subsidy [2].
- *Operational constraints*. Nuclear and fossil plants cannot instantly shut down and restart. They pay the grid at times of negative prices to remain online and thus be able to sell electricity a few hours later at high prices.

The addition of significant non-dispatchable wind or solar changes the shapes of the price curve. The addition of a small amount of solar is beneficial because the electricity is added at times of peak demand. However, as additional solar is added, it drives down the price of electricity in the middle of sunny days. Each solar owner will sell electricity at whatever price exists above zero.

This implies that when 10–15% of the total electricity demand is met by solar in California, the output from solar systems during midday for parts of the year will exceed electricity demand, the price of electricity will collapse to near or below zero, and the revenue to power plants at these times will collapse to near zero. Each incremental addition of solar at this point lowers the revenue for

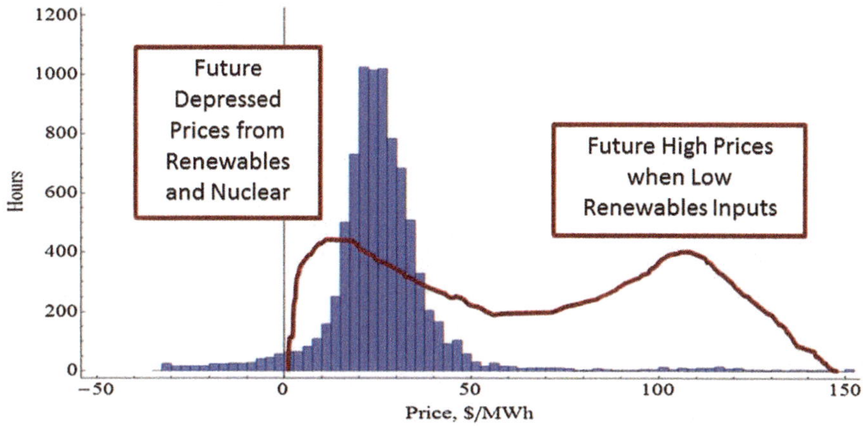

Fig. 6.1 Distribution of electrical prices (bar chart), by duration, average over CAISO (California) hubs (July 2011–June 2012), and notational price curve (red line) for future low-carbon grid

existing solar electricity producers. The percentage solar is the percentage of all electricity produced by solar zero in the middle of the night and exceeding electricity demand initially in June in the middle of sunny days. Relatively small fractions of solar have large impacts on prices in the midday but no impact at night when there is no solar.

The same effect occurs as one adds wind capacity but wind input is more random. As wind penetrates the market, it drives the price of electricity down on days with high wind conditions and low electricity demand. Recent studies have estimated this effect in the European market [3, 4]. If wind grows from providing 0% to 30% of all electricity, the average yearly price for wind electricity in the market would drop from 73 €/MWe (first wind farm) to 18 €/MWe (30% of all electricity generated). There would be 1000 h per year when wind could provide the total electricity demand, the price of electricity would be near zero, and 28% of all wind energy would be sold in the market for prices near zero.

The same will occur with nuclear but only when nuclear provides ~70% of the total electricity demand. This is because nuclear plants run at baseload and baseload is about 70% of electricity demand.

In a fossil fuel-dominated system, one does not see near-zero prices because fossil plants have low capital costs and high fuel costs. Fossil plants will shut down when electricity prices go below the costs of the fossil fuels. With renewables and nuclear, prices can approach zero for a significant number of hours per year. Without massive subsidies that increase with renewables penetration, this revenue collapse limits solar to ~10% of electricity production and wind to ~20% of electricity production. This also implies that in the long term, the price of electricity at times of low renewable input will rise. If other types of power plants operate half the time because they do not generate electricity at times of high renewable inputs, replacement plants will not be built unless there is a rise in the prices of electricity when renewable energy sources are not producing electricity. The red line in Fig. 6.1 is a notational price curve one is expected to get if there is large-scale use of renewables with more hours of low-priced electricity (high wind or high solar) and more hours of high-priced electricity with fewer hours of mid-priced electricity. Recent studies on the German grid have reached similar conclusions [5].

The changing shape of the price curve encourages technologies with low capital costs and high operating costs to provide electricity at times of low renewables inputs. The net result is that large-scale wind and solar with existing technologies can result in increased use of fossil fuels to provide electricity at times of low solar or wind conditions. Studies by the State of California [6] and Google [7] have come to similar conclusions.

6.4 Strategies for a Zero-Carbon Electricity Grid

There are multiple strategies for a zero-carbon grid, each with specific advantages and disadvantages. Traditionally storage is proposed to meet peak electricity demand; but there are major constraints in the context of a zero-carbon electricity grid. Pumped hydro coupled with nuclear can meet variable electricity demands because the pumped hydro can be charged at night at times of low electricity demand. However, the parallel strategy does not work for systems with large solar or wind inputs. There can be extended periods of low wind or solar that will deplete any storage device, and thus storage in systems with large wind or solar components requires backup electricity generation.

There are incentives in coupling storage with nuclear reactors because storing heat is generally cheaper than storing electrons for peak electricity production. Nuclear reactors produce heat and thus can couple efficiently to these systems. Another paper at this conference discusses thermal storage systems coupled to LWRs [8].

There are seasonal variations in electricity demand. Today the only proposed technology for seasonal storage in a low-carbon grid is hydrogen; but the round-trip efficiency (electricity to hydrogen to electricity) is less than 50%.

6.5 Nuclear Air-Brayton Combined Cycle Strategies for Zero-Carbon Grid

In the last 20 years, there have been dramatic improvements in utility gas turbines. Combined cycle efficiency is now ~60%. The cooling water requirements are 40% of a LWR because much of the heat is rejected as hot air. For nuclear systems, gas turbines can be run in baseload and peak mode as discussed in the next section. Furthermore, most of the research and development on power cycles worldwide is associated with these power cycles. These dramatic improvements require rethinking what types of power cycles should be coupled to higher-temperature nuclear power reactors.

We describe two sets of studies that couple gas turbines to nuclear reactors. The next section discusses coupling fluoride-salt-cooled high-temperature reactors (FHRs) to gas turbines, followed by a section on coupling sodium-cooled fast reactors (SFRs) to gas turbines. This would not have been a good idea 20 years ago—the technology was not ready.

6.6 Salt-Cooled Reactors Coupled to NACC Power System

In the 1950s the United States initiated the Aircraft Nuclear Propulsion program to develop a jet-powered nuclear bomber. To meet the required temperatures for the jet engine, the United States began the development of the molten salt reactor (MSR) where the low-pressure liquid fluoride salt coolant was developed to provide high-temperature heat for the jet engines. The development of the intercontinental ballistic missile resulted in the program being cancelled, but two test reactors were successfully built.

The program then started the development of the MSR as a power reactor with a steam cycle. The gas turbine technologies of the 1960s could not meet utility requirements. Advances in utility gas turbines over 50 years have now reached the point where it is practical to couple a salt-cooled reactor to a commercial stationary combined-cycle gas turbine.

One advanced salt-cooled reactor system has been proposed to integrate power production with high-temperature heat storage: the fluoride-salt-cooled high-temperature reactor (FHR) with nuclear air-Brayton combined cycle (NACC) and firebrick resistance-heated energy storage (FIRES). The FHR is a new reactor concept [9] that combines:

1. A liquid salt coolant.
2. Graphite-matrix coated-particle fuel originally developed for high-temperature gas-cooled reactors (HTGRs).
3. An NACC power cycle adapted from natural gas combined cycle plants.
4. FIRES. The FHR concept is a little over a decade old and has been enabled by advances in gas turbine technology and HTGR fuel.

The use of solid fuel avoids some of the complications of liquid fuel reactors. The reactor delivers heat to the power cycle between 600 and 700 °C. The Chinese plan to build an FHR test reactor by 2020.

A point design for a commercial FHR has been developed with a baseload output of 100 MWe [10]. The power output was chosen to match the capabilities of the GE 7FB gas turbine—the largest rail transportable gas turbine made by General Electric. FHRs with higher output could be built by coupling multiple gas turbines to a single reactor or using larger gas turbines. The development of an FHR will require construction of a test reactor—this size commercial machine would be a logical next step after a test reactor. This point design describes the smallest practical FHR for stationary utility power generation. The market would ultimately determine the preferred reactor size or sizes.

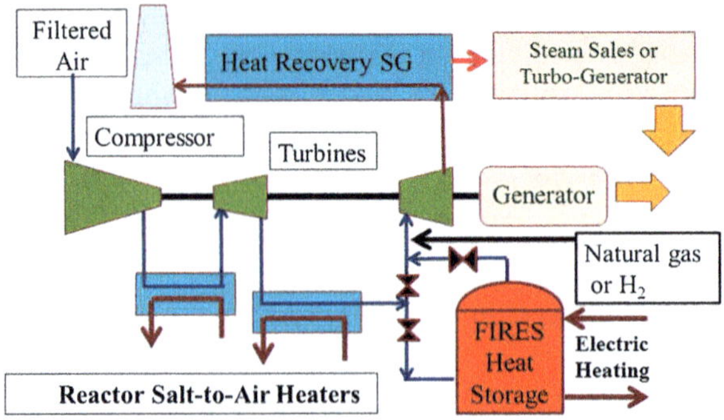

Fig. 6.2 Nuclear air-Brayton combined cycle (NACC) with firebrick resistance-heated energy storage (FIRES)

The FHR is coupled to a NACC with FIRES (Fig. 6.2). In the power cycle, external air is filtered, compressed, heated by hot salt from the FHR while going through a coiled tube air heat (CTAH) exchanger, sent through a turbine producing electricity, reheated in a second CTAH to the same gas temperature, and sent through a second turbine producing added electricity. Warm low-pressure air flow from the gas turbine system exhaust drives a heat recovery steam generator (HRSG), which provides steam to either an industrial steam distribution system for process heat sales or a Rankine cycle for additional electricity production. The air from the HRSG is exhausted up the stack to the atmosphere. Added electricity can be produced by injecting fuel (natural gas, hydrogen, etc.) or adding stored heat after nuclear heating by the second CTAH. This boosts temperatures in the compressed gas stream going to the second turbine and to the HRSG.

The incremental natural gas, hydrogen, or stored heat-to-electricity efficiency is 66.4%—far above the best stand-alone natural gas plants because the added heat is a topping cycle. For comparison, the same GE 7FB combined cycle plant running on natural gas has a rated efficiency of 56.9%. The reason for these high incremental natural gas or stored heat-to-electricity efficiencies is that this high-temperature heat is added on top of "low-temperature" 670 °C nuclear heat (Fig. 6.3). A modular 100 MWe FHR coupled to a GE 7FB modified gas turbine that added natural gas or stored heat produces an additional 142 MWe of peak electricity.

The heat storage system consists of high-temperature firebrick heated to high temperatures with electricity at times of low or negative electric prices. The hot firebrick is an alternative to heating with natural gas. The firebrick, insulation systems, and most other storage system components are similar to high-temperature industrial recuperators. The round-trip storage efficiency from electricity to heat is ~66%—based on ~100% efficiency in resistance electric conversion of electricity to heat and 66% efficiency in conversion of heat to electricity. That efficiency will be near 70% by 2030 with improving gas turbines.

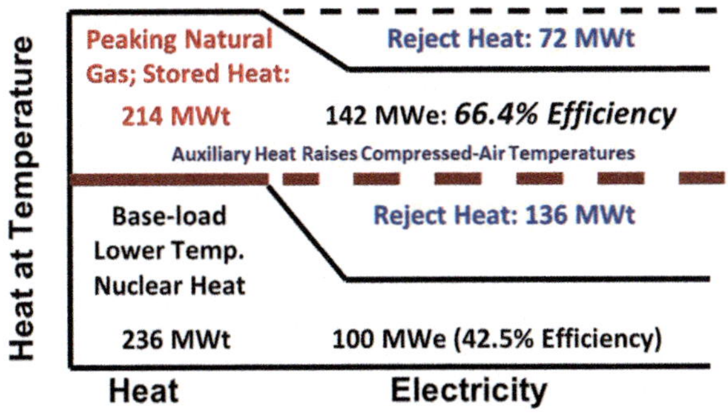

Fig. 6.3 Heat and electricity balance for NACC

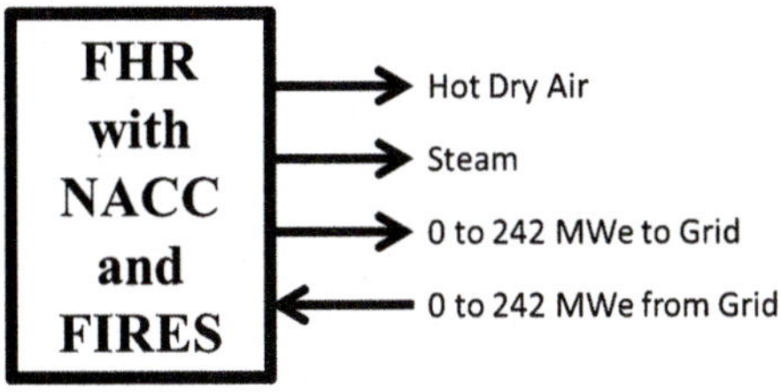

Fig. 6.4 Capability of modular FHR with NACC and FIRES

The plant output is shown in Fig. 6.4. When electricity prices are low (less than the price of natural gas), electricity from the FHR is sent to FIRES. In addition, up to 242 MWe of electricity is bought from the grid. The buy capability of the FHR matches the sell capability and thus does not require upgrades to the grid. Because electricity is used to heat the firebrick, firebrick can be heated to 1800 °C to minimize the quantity of firebrick required. The hot compressed gas from FIRES is lowered to the turbine limits by either steam injection or mixing with lower-temperature compressed air.

In the existing Texas and California grids, the revenue for an FHR with NACC is 50% higher than a baseload nuclear plant because of the capability to produce more electricity at times of peak demand. This is revenue after subtracting the cost of natural gas for peak electricity production. At natural gas prices three to four times the current low prices (natural gas prices in Europe and Asia), the FHR with NACC revenue will double a baseload nuclear plant.

With FIRES the economics are expected to be dramatically better in the German electricity market today and the expected California market by 2020. In those markets, there will be sufficient renewables to drive electricity prices to zero for significant periods of time. It enables replacement of "expensive" natural gas with cheaper electricity. FIRES enables buying massive quantities of electricity when the price is low. Unlike batteries and other electricity storage devices, resistance heaters are inexpensive, and thus the system can absorb massive quantities of low-price electricity even if available for short periods of time.

6.7 Sodium-Cooled Reactors Coupled to NACC Power System

Sodium-cooled reactors are operated within 550 °C temperature range, and since an NACC system looks quite good for a salt-cooled reactor, it is worth considering what it might do for a sodium-cooled reactor. With some modifications, it appears that it could be competitive with systems that have been built [11]. A computer model was built based on standard techniques for analyzing Brayton and Rankine systems [12–14]. System performance was optimized by varying the turbine outlet temperatures for a fixed turbine inlet temperature. A second parameter that can usually be varied to obtain optimum performance is the peak pressure in the steam cycle. For most of the cases considered here, this was held constant at 12.4 MPa (1800 psi). Fairly detailed design was attempted for the heat exchangers involved in the system as they tend to dominate system size. The techniques and data were extracted from the text by Kays and London [15].

Consider a baseline system composed of a compressor and two turbines in the standard topping cycle and a steam bottoming cycle with three turbines and two reheat cycles. For a reactor outlet temperature of 550 °C (823 K) and a turbine inlet temperature of 510 °C (783 K), the standard NACC system will achieve an efficiency of 26.2%, at a compressor pressure ratio of 3.1, hardly a competitive number. But if a 95% effective recuperator is added to the system taking the hot exhaust from the HRSG and using it to preheat the compressed gas going into the sodium to air heat exchanger, the efficiency goes up to 39.9% and the compressor pressure ratio drops to 1.8.

Since the high-pressure water in the bottoming cycle must be heated and the heating of the air in the air compressor increases the work required, it is possible to split the compressor and add an intercooler that heats the high-pressure water in the bottoming cycle and cools the output from the first part of the compressor. If this is done, the efficiency goes to 40.3% and the overall compressor pressure ratio goes to 2.0. A system diagram is provided in Fig. 6.5.

The ratio of the work in the first half of the compressor to that in the second half is about 1.3:1. For all of these calculations, a fairly standard speak steam pressure of 12.4 MPa (1800 psi) was chosen. The efficiency of the system can be increased by increasing the peak pressure in the steam cycle. This moves the evaporation temperature to a higher level and allows more heat in the HRSG exhaust to preheat the air going into the sodium to air heat exchangers. If the peak steam pressure is increased to 20 MPa (2900 psi), the overall cycle efficiency increases to 41.6% with a compressor pressure ratio of 2.1. A final option that was considered was to include two reheat cycles and three turbines in gas turbine. With a recuperator and an intercooler, this gave an efficiency of 40.2%. Therefore, it is probably not worth this added complexity.

Table 6.1 provides a summary of these calculations. The turbine exit temperature was varied to optimize the efficiency of the cycle in all cases. The least accurate number of course is the system volume calculated to estimate

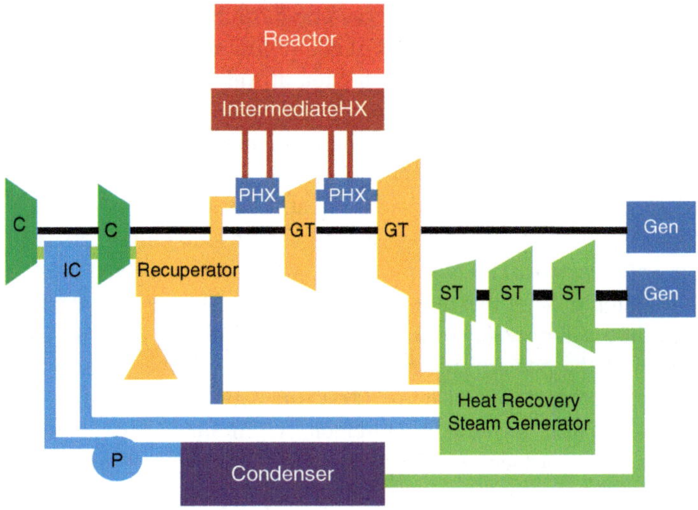

Fig. 6.5 System layout with recuperator and intercooler (*C* compressor, *GT* gas turbine, *ST* steam turbine, *PHX* primary heat exchanger, *IC* intercooler, *P* pump)

Table 6.1 Summary of performance calculations

	TET	CPR	Eff	MdA	WAR	VOL
Baseline	690	3.1	26.2	771	0.036	265
Baseline w/R	740	1.8	39.9	806	0.055	433
Baseline w/R&I	730	2.0	40.3	798	0.051	425
Baseline w/R&I @ 20 MPa	745	2.1	41.6	801	0.047	445
3 Turbines w/R&I	755	1.8	40.2	751	0.060	438

TET turbine exit temperature (K), *CPR* compressor pressure ratio, *MdA* mass flow rate of air (Kg/s), *WAR* water to air mass flow ratio, *VOL* system volume in cubic meters

relative sizes. However, since system size is dominated by heat exchangers, the relative numbers are probably fairly accurate.

The lower-temperature sodium reactor can also be augmented by a combustion chamber that burns the air after it has been heated in a nuclear heat exchanger. At this point, the boundaries on what can be done are not well defined.

Because the peak temperatures in the nuclear cycle are significantly lower than in current combustion systems, the nuclear systems do not stress turbine blade technologies. Therefore the nuclear-heated systems do not require the use of turbine blade cooling. Thus, it is reasonable to assume that the combustion augmentation will be used with uncooled turbine blades. This sets an upper limit on turbine inlet temperatures of about 1300 K. That limit is essentially what had limited the power augmentation for a 100 MWe baseload system to 142 MWe of added peak electricity in the case of a salt-cooled reactor. Choosing this same augmentation (142 MW) for a sodium-cooled reactor will require the turbine inlet temperatures to reach 1010 K, and the efficiency of burning the added gas is approximately 57.5%. This is an efficiency that is comparable to what gas combustion systems currently achieve. However note that there is almost 300 K of additional temperature increase available if desired. Thus, the combustion augmentation of sodium systems could be greater than for higher-temperature system. To take this analysis one step further requires more detailed design calculations which are being pursued at this time.

Another interesting aspect of the NACC systems is that they require significantly less circulating water to get rid of waste heat. Consider a 40% efficient system producing 100 MW(e). A closed-loop system will have to dump 150 MW(t) to the circulating water system. The NACC system with recuperator and intercooler will only have to dump 64% of that amount or 96 MW(t). The steam cycle for this type of NACC produces about 64% of the power at an efficiency of 40%. The air cycle produces 36% of the power. This ratio shifts in favor of the air-Brayton cycle as turbine inlet temperatures increase. At 700 °C the two systems produce about equal amounts of power. The decreased requirement for circulating water should make a larger fraction of the surface of the earth available for power systems installations.

It is also possible to eliminate the steam bottoming cycle entirely at the cost of about 2% in efficiency. This would completely eliminate any dependence on cooling water and allow reactor power system to be placed almost anywhere.

6.8 Power Cycle Comparisons

The efficiency of NACC power systems continues to increase with increased turbine inlet temperatures. For the foreseeable future, there does not appear to be a limitation to using off-the-shelf materials as it is not likely that a reactor heated system will exceed 1300 K turbine inlet temperature. A comparison of the cycle efficiencies for several cycles that have been proposed for the next-generation nuclear plant [16–18] is presented in Fig. 6.6. The calculations for the NACC systems are based on the system described in Fig. 6.5 with a peak steam pressure of 12.4 MPa. The data for the other systems was extracted from the referenced publications.

Around 500 °C the NACC cycle falls short of the supercritical steam and the supercritical CO_2 cycles by 4.1% and 2.6%, respectively. So it will not be competitive on a pure efficiency basis. However, by connecting to the large base of components available for gas turbines and reducing the circulating water requirement, it still offers significant advantages. At the 700 °C point, the NACC system has an efficiency that is about 1.4% less than a supercritical CO_2 system. However, if the steam bottoming cycle can be operated at 20 MPa rather than 12.4 MPa, the efficiency of the NACC system surpasses that of the supercritical CO_2 system by about 0.5%. Thus, it is clear that NACC systems can be competitive for most of the next-generation nuclear plants that have been proposed.

NACC systems can also be applied to high-temperature gas-cooled reactor (HTGR) systems. The major difficulty will be minimizing the size of the helium-to-air

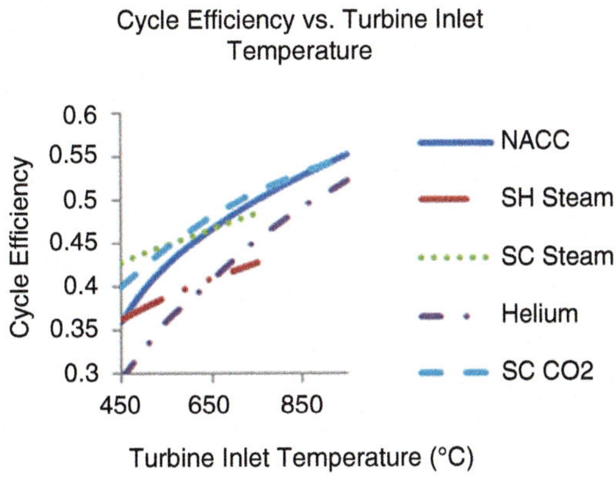

Fig. 6.6 Cycle efficiencies for various advanced cycles [18]

heat exchangers. In addition there is a matching problem with the reheat cycles in the gas turbines. Since the core of an HTGR is very large to provide a good heat transfer area and the temperature increase across the core is usually several hundred degrees, the helium temperature exiting the helium-to-air heat exchangers will be too high to reenter the reactor core directly. Therefore, some efficient technique will be required to use this extra heat before the helium is returned to the reactor core. This is not an insurmountable problem but will require a slightly different NACC topology for a HTGR-like system.

6.9 Summary

The electricity market is changing and will continue to change with deregulation, emphasis on a low-carbon grid, and addition of renewables. The change in the market requires rethinking how to best use nuclear energy to meet the needs of the electricity grid and the need for variable power.

There have been major advances in gas turbines and most of the world's R&D on advanced power cycles is to improve gas turbine technology. Those advances now make it feasible to couple NACC to power reactors. The characteristics of NACC allow baseload reactor operation with variable electricity output-improving economics and enabling a zero-carbon electricity grid. There are significant challenges but large incentives for such power cycles.

NACC systems can be applied to most of the proposed next-generation systems. Their strongest competitor in terms of cycle efficiency is the supercritical CO_2 system. NACC systems will match or better the efficiency of these systems at or above 700 °C. Furthermore, NACC systems have the competitive advantage of a large customer base for system hardware, significantly reduced circulating water requirement for rejecting waste heat, and much greater efforts to improve the technology relative to other power cycles.

References

1. Center for Advanced Nuclear Energy Systems, MIT-ANP-TR-170, Massachusetts Institute of Technology, Cambridge, MA, July 2017
2. F. Huntowisk, A. Patterson, M. Schnitzer, *Negative Electricity Prices and the Production Tax Credit: Why Wind Producers Can Pay Us to Take Their Power—And Why that Is a Bad Thing*, The Northbridge Group, 14 Sept 2012
3. L. Hirth, The market value of variable renewables, the effect of solar wind power variability on their relative prices. Energy Econ. **38**, 218–236 (2013)
4. L. Hirth, The optimal share of variable renewables: How the variability of wind and solar power affects their welfare-optimal development. Energy J. **36**(1) (2015)
5. H. Poser et al., *Development and Integration of Renewable Energy: Lessons Learned from Germany*, Finadvice, FAA Financial Advisory AG, Adliswil, July 2014
6. California Council on Science and Technology, *California Energy Futures – The View to 2050: Summary Report*, Apr 2011
7. R. Konningstein, D. Fork, What it would really take to reverse climate change, *IEEE Spectrum*, 11 Nov 2014. http://spectrum.ieee.org/energy/renewables/what-it-would-really-take-to-reverse-climate-change
8. C. Forsberg, E. Schneider, Variable Electricity from Base-Load Nuclear Power Plants Using Stored Heat, Paper 15125, ICAPP 2015
9. C. Forsberg et al., *Fluoride-Salt-Cooled High-Temperature Reactor (FHR) Commercial Basis and Commercialization Strategy*. MIT-ANP-TR-153, Massachusetts Institute of Technology, Cambridge, MA, Dec 2014
10. C. Andreades et al., *Technical Description of the "Mark 1" Pebble-Bed Fluoride-Salt-Cooled High-Temperature Reactor (PB-FHR) Power Plant*, UCBTH-14-002, University of California at Berkeley, Sept 2014
11. A.E. Waltars, A.B. Reynolds, *Fast Breeder Reactors* (Pergamon Press, New York, 1981)
12. D.G. Wilson, T. Korakianitis, *The Design of High-Efficiency Turbomachinery and Gas Turbines*, 2nd edn. (Prentice Hall, Upper Saddle River, 1998)
13. P.P. Walsh, P. Fletcher, *Gas Turbine Performance* (Blackwell Science, ASME, Fairfield, 1998)
14. M.M. El-Wakil, *Powerplant Technology* (McGraw-Hill, New York, 1984)
15. W.M. Kays, A.L. London, *Compact Heat Exchangers* (McGraw Hill, New York, 1964)
16. U. Oka, S. Koshizuka, Design Concept of Once-Through Cycle Supercritical-Pressure Light Water Cooled Reactors, SCR-2000, *Proceedings of the First International Symposium on Supercritical Reactors*, Tokyo, 2000
17. V. Dostal, P. Hejzlar, M.J. Driscoll, The supercritical carbon dioxide power cycle: Comparison with other advanced cycles. Nucl. Technol. **154**, 283–301 (2006)
18. B. Zohuri, *Combined Cycle Driven Efficiency for Next Generation Nuclear Power Plants: An Innovative Design Approach* (Springer, Cham, 2015)

Chapter 7
Fission Nuclear Power Plants for Renewable Energy Source

In this chapter, the history of nuclear power plant is presented, and need for nuclear power plant for production of electricity is argued. The foundation has been structured and consequently some technology of ongoing research that makes production of electricity from the nuclear power plant more cost-effective. For the nuclear reactors to be more comparative with fossil and gas fuel power plants, they need to be as efficient as the traditional power plants are, when it comes to output thermal efficiency. As this chapter suggests, utilizing the combined cycles to drive and produce electricity via nuclear fuel makes more sense to own them from return on investment, total cost of ownership, and efficient for their owners, namely, electricity companies. Results of modeling a combined cycle Brayton-Rankine power conversion system are presented in this chapter and based model reactor for this purpose was chosen to be the molten salt reactor type. The Rankine bottoming cycle appears to offer significant advantages over the recuperated Brayton cycle. The overall cycle in the modeling for purpose of writing this chapter was optimized as a unit, and lower pressure Rankine systems did seem to be more efficient. The combined cycle requires a lot less circulating water for a heat dump than current power plants. The Rankine bottoming cycle appears to offer significant advantages over the recuperated Brayton cycle as part of computer simulation written by this author. The overall cycle was optimized as a unit, and lower pressure Rankine systems seem to be more efficient. The combined cycle requires a lot less circulating water for a heat dump than current power plants. Molten salt reactors and lead-cooled reactors have been extended to temperatures typical of liquid metal-cooled fast reactors. A split compressor with intercooler has been added to the Brayton cycle that serves as a "feed water heater" for the steam bottoming cycle as part of computer simulation. Adding this component to the system provides about a 3% increase in efficiency. For the purpose of simplicity, an in-house computer code was developed, where the consideration was on the steady-state situation; however a more sophisticated computer code is needed, which will take the transient approach under consideration for better accuracy and optimization of these analysis.

B. Zohuri, *Hybrid Energy Systems*,
https://doi.org/10.1007/978-3-319-70721-1_7

7.1 Introduction

To address these sources of energy challenges, scientists and engineers around the world and even universities are involved with these types of activities along with their partners at national laboratories in the United States and other countries such as the United Kingdom and France.

If we want to be completely dependent on renewable energy, then we need to find a way to store this energy for times when the sun is not shining and the wind is not blowing. Storing this energy is one of the greatest barriers to the adoption of renewable energy. However, it is an area where hydrogen can play a key role. As Jeremy Rifkin says in *The Third Industrial Revolution*, electrolyzers are the portal to the hydrogen economy. Thus, for us to be able to store energy, we need to store heat as fundamental source of energy, even for generating electricity for peak power demand and in general to meet the supply for increase demand of electricity around the world at rate of almost 17% per year.

As another source of heat and energy, light-water reactors (LWRs) are understudied by group of universities in support of Idaho National Laboratory (INL), and Exelon recently conducted a workshop on light-water reactor (LWR) heat storage for peak power and increased revenue on June 27–28, 2017, at Massachusetts Institute of Technology (MIT). The workshop goals were to define and understand the market, regulatory, and technical options for coupling heat storage for variable power to existing and future LWRs with recommendations for the path forward to improve LWR economics. Observations and outcomes from the workshop include [1]:

1. Nuclear reactors generate heat and thus couple efficiently to heat storage that is 10–40 times less expensive than electricity storage (pumped hydro, battery, etc.), thus potentially a lower-cost way to meet variable electricity demand. Favorable heat storage economics has resulted in concentrated solar power systems under construction to include heat storage to vary electricity production. Many of these technologies are applicable LWRs, and most are applicable to other reactor types. Heat storage is therefore the alternative energy storage strategy for a low-carbon electricity grid-one suitable to coupling to LWRs.
2. Six classes of heat storage technologies have been identified that can couple to light-water reactors: *steam accumulators*, *sensible heat storage*, *cryogenic air storage*, *packed pebble-bed heat storage*, *hot-rock storage,* and *geothermal heat storage*. Some storage technologies are ready for demonstration; others require significant research and development (R&D) [1].
3. Heat storage systems coupled to LWRs are different from storage technologies such as batteries and pumped hydro. Batteries and pumped hydro storage have electricity input rates to storage that are near electricity output rates; thus the strategy is buy low and sell high. With most heat storage systems, there are separate capital costs associated with heat input, storage, and heat-to-electricity production.

4. Accumulators and some other heat storage technologies have very low costs for heat addition to storage. The profitable strategy may be to send steam to storage 6 h per day when prices are the lowest and produce added electricity 18 h per day to minimize the cost of the more expensive heat to electricity component of the storage system. For many existing reactors, up to 20% of the steam would go to storage when low prices. The maximum power output would increase by less than 5% to avoid major upgrades of the turbine hall. When viewing the nuclear plant as a black box, the addition of storage would appear to have increased its "base-load" capacity by a few percent and dramatically increased the capability to rapidly go down and back up in power. Inside the plant the reactor is operating at full capacity.
5. Other technologies such as nuclear geothermal inject hot water underground and use a geothermal power system for electricity production. Because of the extremely low cost of storage, such systems may enable seasonal energy storage, provide assured generating capacity, and provide the option for a strategic multi-year heat reserve—the low-carbon equivalent to a strategic oil reserve.

The business case is central. Five years ago, coupling heat storage to a LWR reactor would not have been economic. The changes in the electricity markets indicate that such an option may now be economical in some markets. As the markets continue to change, the economic case improves.

There is a need for demonstration projects to address institutional issues, to provide technology demonstrations for the near-term options, and to collect sufficient information to determine the economics.

However, what has been changed in nuclear industry in recent years to make a slow-growing technology regardless of innovative approach to LWR is a new generation that is called GEN-IV, and among this generation is small modular reactor (SMR), and these changes are described and summarized in the following paragraph.

Electricity markets are changing because of low-cost natural gas (the United States and Canada) and the addition of intermittent renewable generators (wind and solar). This has reduced the demand for base-load electricity. At the same time, there is an increased demand for dispatchable electricity—a market currently served in the United States primarily by natural gas turbines, to a smaller extent by pumped hydroelectricity and to a very limited extent by batteries. These changes are hurting the economics of nuclear power but may create new opportunities for nuclear energy systems with heat storage to enable base-load reactor operation with variable electricity to the grid.

In a low-carbon world, the energy sources are nuclear, wind, and solar. These technologies have high capital costs and low operating costs. If these energy production facilities are operated at half capacity, the production cost of energy approximately doubles. Because energy is about 8% of the global economic output, increases in energy costs have large impacts on the United States and global standards of living. Equally important, the uneven distribution of renewables has serious geopolitical implications.

7.2 Electricity Markets

Discussing about this subject, we need to ask ourselves that what has changed. We can start with the fact that mankind has had the same energy policies for 300,000 years—meet variable energy demands by throwing a little more carbon on the fire. While the technology has changed from the cooking fire to the gas turbine, the economics have not. The cost of the cooking fire (stone or brick) and the gas turbine is low. Most of the labor and capital resources are gathering the fuel (wood, natural gas, etc.) and bringing it to the fire. These are low-capital-cost and high-operating-cost technologies. As a consequence, it is economical to produce variable energy to match variable energy needs by operating the fire at part load.

In a low-carbon world, the energy sources are nuclear, wind, and solar. These technologies have high capital costs and low operating costs. If these energy production facilities are operated at half capacity, the bus-bar cost of electricity approximately doubles. Because energy is about 8% of the global economic output, increases in energy costs have large impacts on the United States and global standards of living. Equally important, the uneven distribution of renewables has serious geopolitical implications.

The differences between fossil energy technologies (low capital cost, high operating cost) and low-carbon technologies (high capital cost, low operating cost) have major impacts on electricity prices as seen in deregulated electricity markets. In these markets, electricity generators bid a day ahead to provide electricity to the grid. The grid operator accepts the lowest bids to meet electricity demands. All of the winning bids are paid the electricity price ($/MWh) of the highest-priced winning electricity bid required to meet the electricity demand for that hour. Nuclear, wind, and solar bid their marginal operating costs which are near zero. Fossil plants bid their marginal costs that are close to the cost of fossil fuels that they burn.

In a market with nuclear and fossil plants, the fossil plants set the hourly price of electricity. If one adds large quantities of solar or wind, their low operating costs set market prices at times of high wind or solar production. Figure 7.1 shows the impact of solar additions between 2012 and 2017 on California electric prices on a spring day with high solar input and low electricity demand. Electricity prices collapse at times of high solar production. In this specific example, the prices have gone negative because of government subsidies that allow the solar producer to pay the grid to take electricity to collect production tax credits. The price increases as the sun goes down because of lower solar electricity production, and peak demand occurs in the early evening.

Recent electricity market analysis for California indicates that California invested heavily in solar power in the past few years to the point that it has surplus of electricity so much that other states such as Arizona are sometimes paid to take it.

On 14 days during March of 2017, Arizona utilities got a gift from California and that was *Free Solar Power*.

In fact even, better than free solar energy. California produced so much solar power on those days that it paid Arizona to take excess electricity its residents were not using to avoid overloading its own power lines.

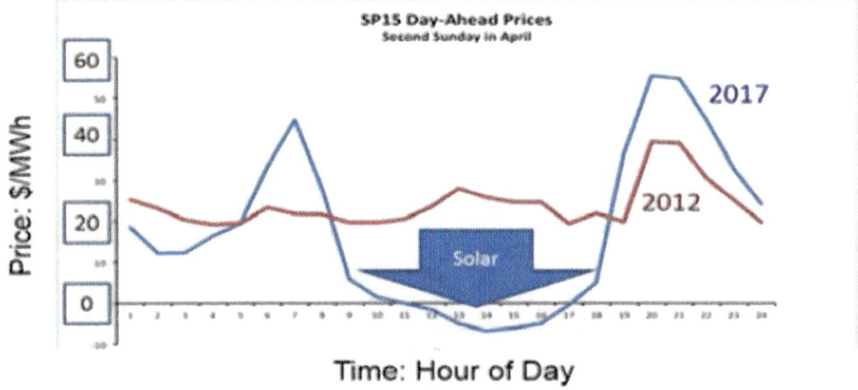

Fig. 7.1 Impact of added solar on California electricity prices for second Sunday in April 2012 and 2017

It happened on 8 days in January and 9 in February as well. All told, those transactions helped save Arizona electricity customers millions of dollars this year, though grid operators declined to say exactly how much. And California also has paid other states to take power.

The number of days that California dumped its unused solar electricity would have been even higher if the state had not ordered some solar plants to reduce production—even as natural gas power plants, which contribute to greenhouse gas emissions, continued generating electricity.

Solar and wind power production were curtailed a relatively small amount—about 3% in the first quarter of 2017—but that is more than double the same period last year. And the surge in solar power could push the number even higher in the future.

Why does not California, a champion of renewable energy, use all the solar power it can generate?

The answer, in part, is that the state has achieved dramatic success in increasing renewable energy production in recent years. But it also reflects sharp conflicts among major energy players in the state over the best way to weave these new electricity sources into a system still dominated by fossil fuel-generated power.

Today, Arizona's largest utility, Arizona Public Service, is one of the biggest beneficiaries of California's largesse because it is next door and the power can easily be sent there on transmission lines.

On days that Arizona is paid to take California's excess solar power, Arizona Public Service says it has cut its own solar generation rather than fossil fuel power. So California's excess solar is not reducing greenhouse gases when that happens.

That is a good deal for Arizona, which uses what it is paid by California to reduce its own customers' electricity bills. Utility buyers typically pay an average of $14–$45/MWh for electricity when there is not a surplus from high solar power production. See Fig. 7.2.

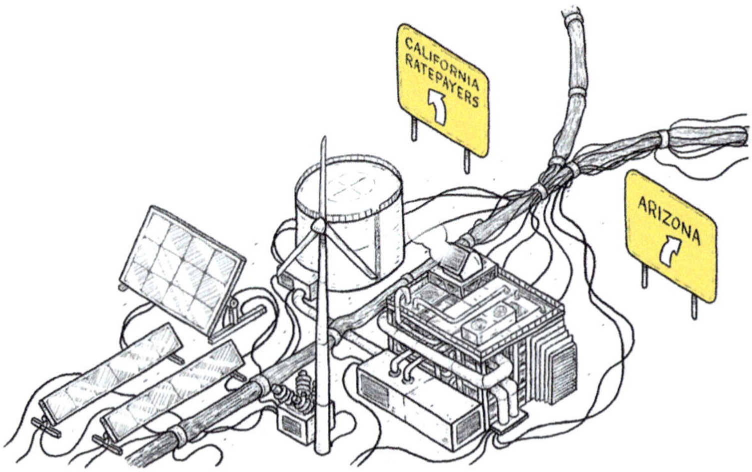

Fig. 7.2 Artistic illustration of combined solar and wind energy power

Behind the rapid expansion of solar power is its plummeting price, which makes it highly competitive with other electricity sources, in part that stems from subsidies, but much of the decline comes from the sharp drop in the cost of making solar panels and their increased efficiency in converting sunlight into electricity.

In 2010, power plants in the state generated just over 15% of their electricity production from renewable sources. But that was mostly wind and geothermal power, with only a scant 0.5% from solar. Now overall amount has grown to 27%, with solar power accounting for 10% or most of the increase. The solar figure does not include the hundreds of thousands of rooftop solar systems that produce an additional 4 percentage points, a share that is ever growing.

A key question in the debate is when California will be able to rely on renewable power for most or all of its needs and safely phase out fossil fuel plants, which regulators are studying.

The answer depends in large part on how fast battery storage improves, so it is cheaper and can store power closer to customers for use when the sun is not shining. Solar proponents say that the technology is advancing rapidly, making reliance on renewables possible far sooner than previously predicted, perhaps two decades or even less from now—which means little need for new power plants with a life span of 30–40 years.

The average cost of solar power for residential, commercial, and utility-scale projects declined 73% between 2010 and 2016. Solar electricity now costs 5–6 cents/kWh—the amount needed to light a 100W bulb for 10 h—to produce or about the same as electricity produced by a natural gas plant and half the cost of a nuclear facility, according to the US Energy Information Administration.

The same effect occurs with wind as shown in Fig. 7.3 in Iowa. Wind has a multiday cycle on the Great Plains and thus the daily prices of electricity vary.

As matter of fact in Iowa, in addition to federal programs, the state of Iowa encourages development of renewable electricity sources through a 1 cent/kWh tax credit.

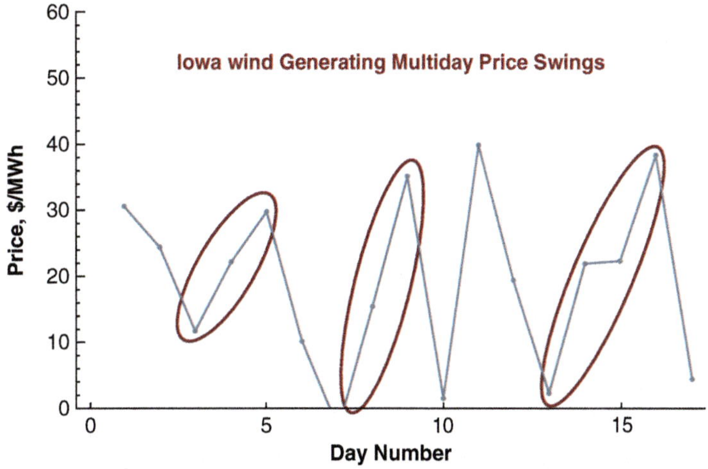

Fig. 7.3 Impact of wind on daily West-Iowa electricity prices in April 6–22, 2014

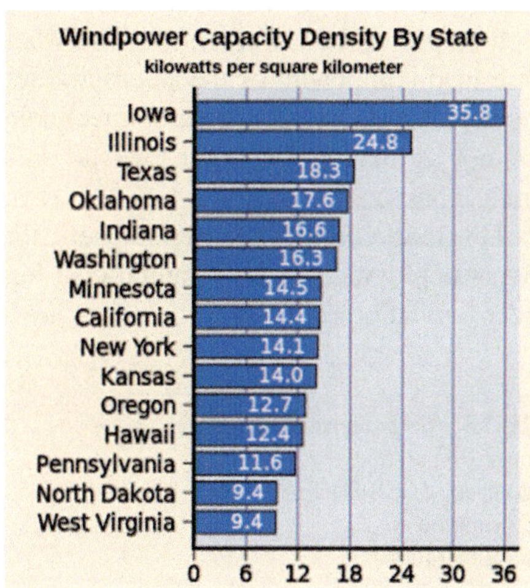

Fig. 7.4 Density of installed generation capacity (Courtesy of AWEA 2013 4th quarter Market Report State Areas from US Census Bureau)

Also, generation equipment and facilities receive property tax breaks, and generation equipment is exempt from sales tax.

In 2010 and in 2009, Iowa led the United States in the percentage of electrical power generated by wind, at 15.4% and 14.2% [2]. This was up from 7.7% in 2008, as there was a large increase in the installed capacity in 2008 [3]. Some of the wind power generated electricity is sold to utility companies in nearby states, such as Wisconsin and Illinois. See Fig. 7.4, which is the illustration of wind power capacity by state.

Wind farms are most prevalent in the north and west portion of Iowa. Wind maps show the winds in these areas to be stronger on average, making them better suited for the development of wind energy. Average wind speeds are not consistent from month to month. Wind maps show wind speeds are on average strongest from November to April, peaking in March. August is the month with the weakest average wind speeds. On a daily cycle, there is a slight rise in average wind speeds in the afternoon, from 1 to 6 p.m. Estimates by the National Renewable Energy Laboratory (NREL) indicate that Iowa has potentially 570,700 MW of wind power using large turbines mounted on 80 m towers [4]. Iowa ranks seventh in the country in terms of wind energy generation potential due to the strong average wind speeds in the midsection of the United States. The Iowa Environmental Mesonet distributes current weather and wind conditions from approximately 450 monitoring stations across Iowa, providing data for modeling and predicting wind power. Figure 7.5 is pie chart for Iowa electricity generation sources by the types of the plant.

All high-capital-cost low-operating-cost technologies will collapse the price of electricity at certain times if deployed on a sufficiently large scale.

The value of the product goes down with increased deployment. This price collapse occurs as solar provides ~15% of total electricity demand, wind provides ~30% of total electricity demand, or nuclear provides ~70% of total electricity demand when fossil fuels provide the remainder of the electricity. The low solar fraction reflects high output in the middle of the day whereas the high nuclear fraction reflects the base-load component of the electricity demand. Price collapse economically limits the deployment of all low-carbon technologies with deployment of any low-carbon technology making the other low-carbon technologies less economic—overlapping price collapse.

This market effect has two impacts. First, the deployment of these technologies favors deployment of low-capital-cost high-operating-cost fossil plants to provide electricity at other times when prices are higher. Second, this change in the market

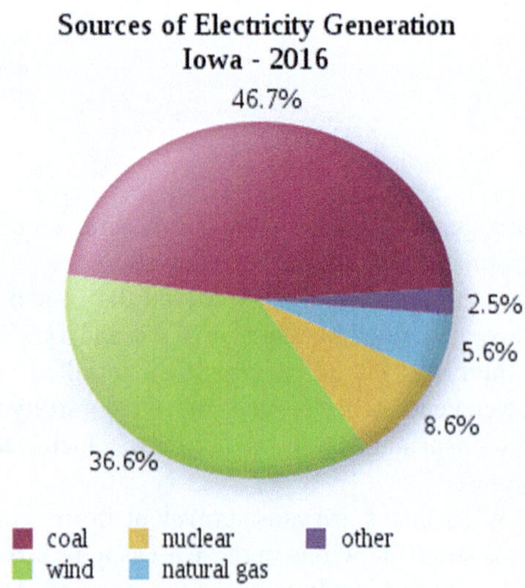

Fig. 7.5 Iowa electricity generation sources (Courtesy of US Energy Information Administration)

creates the economic incentive to deploy energy storage systems to consume low-price energy (raise its price) and provide energy at times of higher demand.

The storage times in a market with large quantities of solar generation (daily cycle) are different than the storage times in a market dominated by wind (multiday cycle). The variation of electricity demand is different across the country with large differences due to different climates. One does not expect that there will be a "single" economically optimum storage solution. The optimal storage solution will vary with location.

There are three electricity markets in which energy storage has the potential to increase revenue for the owner of an existing or new plant—each with different characteristics. These three are listed here and more details can be found in Ref. [1]:

1. *Energy markets*: Energy markets pay per unit of electricity delivered to the grid.
2. *Capacity markets*: There are two strategies to assure sufficient generating capacity to meet demand:
   (a) To have no capacity market and allow energy prices to climb to very high levels ($100 s/MWh or more) at times of scarcity.
   (b) Strategy for the grid to have contracts for assured electricity supply even during cloudy day for Solar radiation source or lack blowing wind for Wind source of energy
3. *Auxiliary service market*: This refers to other electricity grid services such as frequency control, black start (start after power outage), and reserves for rapid response grid emergencies such as another electrical generator failing.

More details can be found in Ref. [1] by interested reader to gain more knowledge in this area.

7.2.1 Global Electricity Consumption Set to Explode

In an article that was published by Jacqueline Holman on April 17, 2009, in Creamer Media's Engineering News, she presented the debate of global growth of electricity under article subject of "Global Electricity Consumption Set to Explode by 2030 [5]."

She has stated that growth partnership company Frost & Sullivan reports that, at a global level, electricity consumption is forecast to increase from 10,543 billion kWh in 1990 to over 30,116 billion kWh in 2030 at a compound annual growth rate of 3.6%.

The largest regions in terms of electricity consumption are North America and Europe. The company states that electricity consumption will continue to grow in these regions but at a lower level than in other regions, resulting in the forecast that their share of the overall market will decline.

Frost & Sullivan's 360° perspective of the African power and energy industry report states that electricity consumption is forecast to grow fastest in the developing countries and regions, such as China, India, the Middle East, Latin America, Africa, and the rest of Asia.

7.2.1.1 Market Drivers

She also continues to argue that market drivers are Substantial investment in the power sector is needed globally to substantiate growth in the oil exploration, commodity mining and reconstruction industries, that is stated by Frost & Sullivan. Additionally, the United Nations predicts that the world population will rise from 6.4 billion in 2004 to 8.1 billion by 2030, implying that there will be a global increase in electricity demand.

The company adds that the trend toward urbanization and migration to cities in many developing countries exerts increased pressure on the power grid and forces governments to invest in the electricity sector to meet growing demand.

"The power infrastructure is undeveloped in many countries, so governments have initiated rural electrification projects to ensure that a higher percentage of the population has access to electricity," states Frost & Sullivan.

7.2.1.2 Market Restraints

Electricity production requires high capital outlay, which many developing countries lack. Frost & Sullivan says that it is estimated that generating 1000 MW costs over $1 billion, which means that erecting power plants is costly. Substantial assistance is required from international institutions and the private sector to build a power station, and low electricity tariffs often reduce a power company's revenue streams.

The company comments that bureaucracy, corruption, burdensome approval procedures, rigid labor laws, and lengthy delays in bidding and construction processes affect investment in the power industry in many countries.

"Political problems and violence also affect the power industry in some countries by discouraging meaningful private sector investment," reports Frost & Sullivan.

7.2.1.3 Market Issues

Frost & Sullivan reports that the global energy market is challenged by three interlocking issues that make current energy markets attractive but also complex. These are increasing fuel costs, environmental issues, and growth in the global energy demand.

The company has predicted a timeline for the significant market issues from 2010 to 2030.

It predicts that 2010–2020 will see oil passing $100 a barrel, continued acceleration of coal-fired power, rising world oil prices dampening the demand for liquid fuel, the acceleration of natural gas, and the resurgence of nuclear energy. From 2020 to 2030, it expects that growth of natural gas will be kept in check by high gas prices and coal will fuel almost 45% of energy consumption, following growth in India and China.

Global energy demand is likely to grow strong in all regions, and later growth may be dominated by India and China. There could be a long-running boom in Russia after deceleration and maturation takes place in the European Union. From 2020 onward, developing economies will account for over 40% of electricity demand, and global energy demand will almost double its 2000 level.

Concerning environmental issues, Frost & Sullivan predicts that China is likely to overtake the United States as the world's largest emitter of carbon dioxide (CO_2). This would be followed by an accelerated investment in carbon capture and renewable energy usage in Europe increasing to 20%. Nearing 2025, stationary fuel cells will probably contribute to electricity generation, and by 2030 global CO_2 emissions are likely to reach 40 Gt, compared to the 8.38 Gt emitted from the burning of fossil fuels in 2006.

7.2.1.4 Challenges

Challenges facing the global power and energy market include environmental concerns and permits, workforce scarcity, raw materials pricing and availability, the availability of natural gas, increasing natural gas prices, and forging capacity and prices.

Frost & Sullivan states that a shortage of skilled labor and engineers is creating a bottleneck where a large amount of engineering input and customization is required. Experienced and trained engineers, as well as specialized skills from a small number of equipment manufacturers, are required to design, construct, and commission a large plant.

The construction of fuel-fired power plants is significantly delayed by the increase of pricing and delivery times of raw materials. These factors hinder the ability to estimate the scheduling and costs during the engineering phase, thereby significantly increasing the exposure to risk.

The relatively low availability of natural gas and security of supply from countries such as Russia is a challenge, as it is causing some concern about relying on natural gas as a source of fuel. The increasing price of gas is also causing some companies to reconsider investments in combined cycle gas turbine plants owing to higher running costs [5].

7.3 California and Others Are Getting It Wrong

Environmental groups are even more critical. They contend that building more fossil fuel plants at the same time that solar production is being curtailed shows that utilities—with the support of regulators—are putting higher profits ahead of reducing greenhouse gas emissions.

"California and others have just been getting it wrong," said Leia Guccione, an expert in renewable energy at the Rocky Mountain Institute in Colorado, a clean power advocate. "The way [utilities] earn revenue is building stuff. When they see a need, they are perversely [incentivized] to come up with a solution like a gas plant."

Regulators and utility officials dispute this view. They assert that the transition from fossil fuel power to renewable energy is complicated and that overlap is unavoidable.

They note that electricity demand fluctuates—it is higher in summer in California, because of air conditioning, and lower in the winter—so some production capacity inevitably will be underused in the winter. Moreover, the solar power supply fluctuates as well. It peaks at midday, when the sunlight is strongest. Even then it isn't totally reliable.

Because no one can be sure when clouds might block sunshine during the day, fossil fuel electricity is needed to fill the gaps. Utility officials note that solar production is often cut back first because starting and stopping natural gas plants are costlier and more difficult than shutting down solar panels.

Eventually, unnecessary redundancy of electricity from renewables and fossil fuel will disappear, regulators, utilities, and operators of the electric grid say.

"The gas-fired generation overall will show decline," said Neil Millar, executive director of infrastructure at CAISO, the California Independent System Operator, which runs the electric grid and shares responsibility for preventing blackouts and brownouts. "Right now, as the new generation is coming online and the older generation hasn't left yet, there is a bit of overlap."

Utility critics acknowledge these complexities. But they counter that utilities and regulators have been slow to grasp how rapidly technology is transforming the business. A building slowdown is long overdue, they argue.

Despite a growing glut of power, however, authorities only recently agreed to put on hold proposals for some of the new natural gas power plants that utilities want to build to reconsider whether they are needed.

Again as it was stated at the beginning of this chapter, the key question in all these debates are when California and others will be able to rely on renewable energy power for most or all of their needs and safely phase out fossil fuel plants, which regulators are analyzing and studying in steps toward "Decarbonization," which is an omission of CO_2 emission. See Appendix B of this book.

7.4 Outlook for Power Generation

The consultancy states that fossil fuel power is likely to continue to dominate and will grow dramatically in the next 20 years.

Global primary energy demand could increase by over 50% between now and 2030, while world electricity demand could increase by 60% over the same period. Frost & Sullivan explains that over 70% of this increase comes from developing countries, led by China and India.

CO_2 emissions could reach as much as 40 Gt in 2030, a 55% increase on the current level.

7.5 Why We Need Nuclear Power Plants

Some scientists are calling this source of energy as 100% renewable energy, and of course environmentalists arguably are saying that is a wrong approach, just because in the core of these plants there exists uranium or plutonium as fuel when we are talking about fission-type nuclear power plants that exist in grid today and produce electricity to the net. However, on the other side of the spectrum where researchers and scientist at national laboratories and universities around the globe are working toward fusion program to achieve a breakeven passionately argue that nuclear power plants of fusion type are totally clean so long as the source of energy comes in form of two hydrogen isotopes such as deuterium (D) and tritium (T) as source of fusion reaction and driving energy from it.

This is a dream that is too far away from reality of today's need and demand for electricity yet is not out of scope of near future. Physics of plasma [6] for driving energy via inertial confinement fusion (ICF) [7] or magnetic confinement fusion (MCF) [8] are in agreement with such innovative approaches.

The story of why we need nuclear power plant as source of energy was first published by *CityLab* and then was reproduced *Climate Desk Collaboration* and again here as part of this chapter and section.

Renewable energy has had a busy year. California and New York have adopted ambitious plans calling for 50% renewable energy by 2030. A group of Stanford and Berkeley scientists has put forth an even bolder vision—encouraging all 50 states to run on wind, water, and solar by 2050, without any nuclear energy or biofuels in the picture. New York City Mayor Bill de Blasio has announced his intention to go fully renewable with the city government's power, too.

A world without any fossil fuel energy would be a much cleaner place for both people and the environment. Right now renewable energy accounts for just 13% of all US electricity. A significant increase in that share would lead to a major reduction in air pollution and its attendant diseases, not to mention the costs of climate change-induced flooding or wildfires. The lives, time, and property saved could be put to work tackling other social problems.

But it is not entirely clear that a US energy grid based on 100% renewables is the best way to achieve a zero-carbon future. On the contrary, there's a strong environmentalist case for approaching that goal with caution. Limiting a zero-carbon future to wind, water, and solar means greater costs of storing this energy, discarding other existing zero-carbon sources like nuclear, and generally blanketing the earth with panels and turbines as a means to save it. The merits of total transformation were touched upon in Chap. 1 of this book was presented as their renewable energy roadmap study by Stanford Professor Mark Jacobson and his team, where they used US Energy Information Administration (EIA) data to project of "Business as Usual (BAU)." They then compiled state-by-state energy portfolios needed to meet that projected demand through expanded wind, water, and solar energy generation, and the result was depicted in Fig. 1.18 as well as rest of the story in that chapter.

Considering all the above, there are countries around the world that actively promoting the Fission Nuclear Energy as source of electricity production such as for example, India.

Prime Minister Narendra Modi signed an agreement early June of 2017 with President Vladimir Putin for two more units of nuclear plants at Kudankulam. Earlier, the government of India announced an agreement with international companies to set up 7000 MW of nuclear plants for Rs. 70,000 Crore on Indian domestic technology, although its wisdom has been questioned by many commentators.

In contrast as part of push for renewables source of energy, the government of India launched the National Solar Mission in 2009 with a target of setting up of 20,000 MW of solar plants by 2022. It was recognized that solar power plants would need subsidy through a guaranteed price via feed-in tariff (FIT) at which solar electricity would be purchased.

7.6 Is Nuclear Energy Renewable Source of Energy

Assuming for the time being we are taking fission reaction as foundation for present (GEN-III) and future (GEN-IV) nuclear power reactors, as nuclear energy source to somewhat degree, we can argue it is a clean source of energy.

Although nuclear energy is considered clean energy, its inclusion in the renewable energy list is a subject of major debate. To understand the debate, we need to understand the definition of renewable energy and nuclear energy first. However, until we manage through future technology of these fission reactors to manage to bring down the price electricity per kWh driven by fusion energy down to the point of those by gas or fossil fuels, there is no chance to push these reactors beyond GEN-III [9].

However, efforts toward reduction price of electricity driven by nuclear fission power plants, especially using some innovative design of GEN-IV plants with high-temperature baseline in conjunction with some thermodynamics cycles such as Brayton and Rankine, are on the way by so many universities and national laboratory such as the Idaho National Laboratory and universities such as the MIT, UC Berkeley, and University of New Mexico as well as this author [10].

Renewable energy is defined as an energy source/fuel type that can regenerate and can replenish itself indefinitely. The five renewable sources used most often are biomass, wind, solar, hydro, and geothermal.

Nuclear energy on the other hand is a result of heat generated through the fission process of atoms. All power plants convert heat into electricity using steam. At nuclear power plants, the heat to make the steam is created when atoms split apart—called fission. The fission releases energy in the form of heat and neutrons. The released neutrons then go on to hit other neutrons and repeat the process, hence generating more heat. In most cases the fuel used for nuclear fission is uranium.

One question we can raise here in order to further understand whether or not we need present nuclear technology as a source of energy is that:

What is the difference between clean energy and renewable energy? Put another way, why is nuclear power in the doghouse when it comes to revamping the nation's energy mix? [11]. See Fig. 7.6.

The issue has come to the forefront this week during the debate over the Waxman-Markey energy and climate bill and its provisions for a national renewable energy mandate.

Simply put that republicans have tried—and failed—several times to pass amendments that would christen nuclear power as a "low-emissions" power source eligible for all the same government incentives and mandates as wind power and solar power [11].

Fig. 7.6 Nuclear energy in doghouse [10]

Many environmental groups are fundamentally opposed to the notion that nuclear power is a renewable form of energy—on the grounds that it produces harmful waste byproducts and relies on extractive industries to procure fuel like uranium.

Even so, the nuclear industry and pronuclear officials from countries including France have been trying to brand the technology as renewable, on the grounds that it produces little or no greenhouse gases. Branding nuclear as renewable could also enable nuclear operators to benefit from some of the same subsidies and friendly policies offered to clean energies like wind, solar, and biomass.

So far, however, efforts to categorize nuclear as a renewable source of power are making little headway.

The latest setback came in around August of 2009, when the head of the International Renewable Energy Agency (IRENA)—an intergovernmental group known as IRENA that advises about 140 member countries on making the transition to clean energy—dismissed the notion of including nuclear power among its favored technologies.

"IRENA will not support nuclear energy programs because it's a long, complicated process, it produces waste and is relatively risky," Hélène Pelosse, its interim director general, told in general."

Energy sources like solar power, Ms. Pelosse said, are better alternatives—and less expensive ones, "especially with countries blessed with so much sun for solar plants," she said it in 2009.

7.6.1 Argument for Nuclear as Renewable Energy

Most supporters of nuclear energy point out the low carbon emission aspect of nuclear energy as its major characteristic to be defined as renewable energy. According to nuclear power opponents, if the goal to build a renewable energy infrastructure is to lower carbon emission, then there is no reason for not including nuclear energy in that list [11].

But one of the most interesting arguments for including nuclear energy in the renewable energy portfolio came from Bernard L Cohen, former professor at University of Pittsburg. Professor Cohen defined the term "indefinite" (time span required for an energy source to be sustainable enough to be called renewable energy) in numbers by using the expected relationship between the sun (source of solar energy) and the earth. According to Professor Cohen, if the uranium deposit could be proved to last as long as the relationship between the Earth and Sun is supposed to last (5 billion years), then nuclear energy should be included in the renewable energy portfolio [12].

In his paper Professor Cohen claims that using breeder reactors (nuclear reactor able to generate more fissile material than it consumes), it is possible to fuel the earth with nuclear energy indefinitely. Although the amount of uranium deposit available could only supply nuclear energy for about 1000 years, Professor Cohen believes that actual amount of uranium deposit available is way more than what is considered extractable right now. In his arguments, he includes uranium that could be extracted at a higher cost, uranium from the sea water and also uranium from eroding earth crust by river water. All of those possible uranium resources if used in a breeder reactor would be enough to fuel the earth for another 5 billion years and hence renders nuclear energy as renewable energy [12].

7.6.2 Argument for Nuclear as Renewable Energy

One of the biggest arguments against including nuclear energy in the list of renewable is the fact that uranium deposit on earth is finite, unlike solar and wind. To be counted as renewable, the energy source (fuel) should be sustainable for an indefinite period of time, according to the definition of renewable energy.

Another major argument proposed by the opponents of including nuclear energy as renewable energy is the harmful nuclear waste from nuclear power reactors. The nuclear waste is considered as a radioactive pollutant that goes against the notion of a renewable energy source [11]. Yucca Mountain is one of the examples used quite often to prove this point. Most of the opponents in the United States also point at the fact that while most renewable energy source could render the US energy independent, uranium would still keep the country energy dependent as United States would still have to import uranium [11].

7.6.3 Conclusion

It seems like at the heart of debate lies the confusion over the exact definition of renewable energy and the requirements that need to be met in order to be one. The recent statement by Helene Pelosi, the interim director general of the International Renewable Energy Agency (IRENA), saying IRENA will not support nuclear energy programs because it is a long, complicated process and it produces waste and is relatively risky, proves that their decision has nothing to do with having a sustainable supply of fuel [13]. And if that's the case, then nuclear proponents would have to figure out a way to deal with the nuclear waste management issue and other political implications of nuclear power before they can ask IRENA to reconsider including nuclear energy in the renewable energy list [14].

One more strong argument against fission nuclear power plants as source of renewable energy comes from Dr. James Singmaster in August 3, 2009 and has been republished here as follow:

> The basic problem of the climate crisis is the ever-expanding overload of heat energy in the closed biosphere of earth. Temperatures going up indicate the increasing heat energy overload. Everyone reading this should check out Dr. E. Chaisson's article titled "Long-Term Global Warming from Energy Usage" in EOS, Trans. Amer. Geophys. Union, V. 89, No. 28, Pgs. 253–4(2008) to learn that nuclear energy, be it fission or fusion, being developed should be dropped with money put into it being put to developing renewable energy supplies using the sun, wind and hydrogen.
>
> The hydrogen needs to be generated from splitting water using sunlight with the best one or two of seven catalysts reported in the last 2 years. Or with excess solar or wind collection generating electricity, that could be used to generate hydrogen by electrolysis of water.
>
> There is no way that nuclear power can avoid releasing trapped energy to increase the energy overload, so it should be forgotten.
>
> To remove some of the energy as well as some of the carbon overload in the biosphere, we need to turn to pyrolysis of massive ever-expanding organic waste streams to remake charcoal that will be removing some of both overloads. It will require using renewable energy and the pyrolysis process expels about 50% of the carbon as small organic chemicals that can be collected, refined and used foe fuel that is a renewable one. For more about using pyrolysis, search my name on GreenInc blog or google it for other blog comments on pyrolysis.
>
> Dr. J. Singmaster

References

1. Center for Advanced Nuclear Energy Systems, MIT-ANP-TR-170, July 2017, Massachusetts Institute of TechnologyCambridge, MA
2. Electric Power Monthly with Data for June 2017, U.S. Energy Information Administration, August 2017
3. https://www.siemens.com/global/en/home/company/topic-areas/sustainable-energy.html#Power-generations
4. http://www.iowapolicyproject.org/
5. http://www.engineeringnews.co.za/article/electricity-consumption-to-increase-to-over-30-116-b-kwh-globally-in-2030-2009-04-17
6. B. Zohuri, *Plasma Physics and Controlled Thermonuclear Reactions Driven Fusion Energy*, 1st edn. (Springer, 2017)
7. B. Zohuri, *Inertial Confinement Fusion Driven Thermonuclear Energy*, 1st edn. (Springer, 2017)
8. B. Zohuri, *Magnetic Confinement Fusion Drive Thermonuclear Energy*, 1st edn. (Springer, 2017)
9. Center for Advanced Nuclear Energy Systems, MIT-ANP-TR-170, July 2017, Massachusetts Institute of Technology, Cambridge, MA
10. B. Zohuri, *Combined Cycle Driven Efficiency for Next Generation Nuclear Power Plants: An Innovative Design Approach*, 1st edn. (Springer, 2015)
11. K. Johnson, Is nuclear power renewable energy. Wall Str. J. 21 May 2009
12. B.L. Cohen, Breeder reactors: A renewable energy source. Am. J. Phys. **51**, 75 (1983)
13. J. Kanter, Is nuclear power renewable. New York Times, 3 Aug 2009
14. D. Chowdhury, Is nuclear energy renewable energy. Stanford Physics Department, 22 Mar 2012
15. http://www.wisconsinpublicservice.com/environment/crane_creek.aspx

Chapter 8
Energy Storage Technologies and Their Role in Renewable Integration

Today's world is at a turning point. Resources are running low, pollution is increasing, and the climate is changing. As we are about to run out of fossil fuels in the next few decades, we are keen to find substitutes that will guarantee our acquired wealth and further growth on a long-term basis. Modern technology is already providing us with such alternatives like wind turbines, photovoltaic cells, biomass plants, and more. But these technologies have flaws. Compared to traditional power plants, they produce much smaller amounts of electricity, and even more problematic is the inconsistency of the production. The global demand for electricity is huge, and it's growing by approximately 3.6% annually, but the sun isn't always shining nor is the wind always blowing. For technical reasons, however, the amount of electricity fed into the power grid must always remain on the same level as demanded by the consumers to prevent blackouts and damage to the grid. It leads to situations where the production is higher than the consumption or vice versa. This is where storage technologies come into play—they are the key element to balance out these flaws [1].

8.1 Introduction

With the growing importance of renewable energy sources, scientists and engineers are anxious to enhance efficiencies and to lower the costs of these technologies. Yet, there seems to be only a handful of technologies available that are efficient enough and also economical. Storing energy isn't an easy task, as most of us know. Our smartphone battery only lasts for about a day and laptops only a few hours; the range for electric cars is limited to only little more than a 100 km; and these are only examples for comparatively small devices. Now imagine the problem of storing energy at the level of hundreds to thousands of wind turbines and photovoltaic cells.

B. Zohuri, *Hybrid Energy Systems*,
https://doi.org/10.1007/978-3-319-70721-1_8

The way we handle the fluctuating energy demand today works fine—for now. But, as we approach the point of peak oil faster and faster, and as we are trying hard to replace these conventional plants with regenerative energy sources, the grid changes, whereas the demand will remain about the same. With renewable energy, the production is fluctuating in a way that is hardly predictable. We may be able to predict the weather for the next few days, but as we all know, the weather forecast isn't always right, and even then, a few days isn't enough to calculate in the context of a national or even transnational power grid to guarantee a secure energy supply. Also, when the wind stops, it stops, and foreseeing it won't change the fact that wind turbines won't produce the energy we need. So, there is a need to find ways to compensate for this fluctuation, to save the energy in times of sunny and windy days and use it for cloudy and windless days. Technology to do so exists, and we even use them today, but its capacity is not enough by a long shot—not if we're planning to go green and sustainable. The problem emerging is that we can't just simply build more of the existing storage technologies as each technology has its own flaws. For example, pumped hydro storage, the most reliable and so far only economical storage technology available, is extremely limited by few potential sites and strict laws on nature conservation.

In the following chapters, we will be introducing some basic knowledge of power grids, the most important storage technologies so far, as well as a critical observation of their benefits, problems, and possible impacts in the future, and a small glance at promising technologies still in their development and pilot phases.

As we have stated in past chapters, energy storage is the capture of energy produced at one time for use at a later time. A device that stores energy is sometimes called an accumulator or battery as it is illustrated in Fig. 8.1.

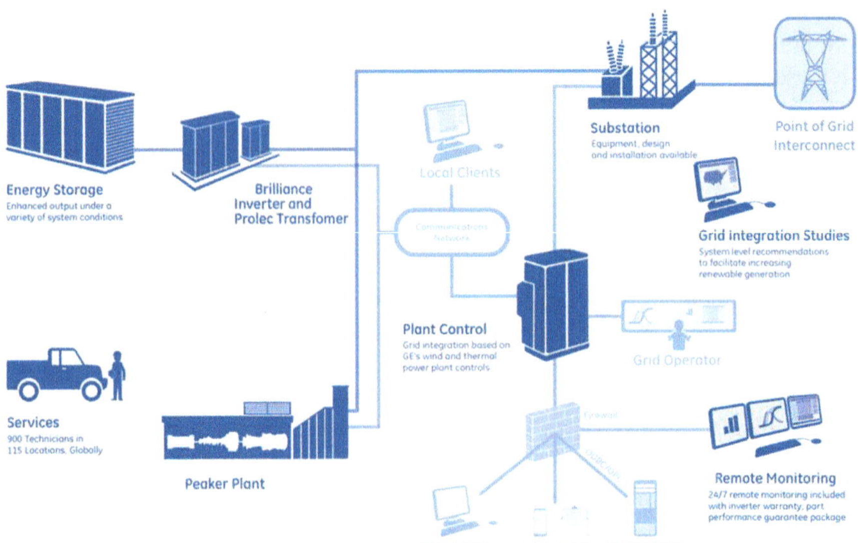

Fig. 8.1 Energy storage stages

Fig. 8.2 Chain of bees

To wrap up this introduction, it is notable that "when energy storage joins the block-chain the entire energy community benefits" as stated by Tim Larrison CFO Green Charge. Bees are an excellent example of joint community as it can be seen in Fig. 8.2.

Answering the call for increasing energy self-reliance, a grassroots electricity-sharing model is emerging. "Community microgrids," comprising community-owned or subscribed solar PV and other renewable energy sources, offer participants and surrounding consumers the security of energy resilience in times of grid failure and protection from energy price increases driven by volatile energy markets. They also give energy producers/consumers (aka "prosumers") more control over the renewable energy they generate. Figure 8.3 is a presentation of blockchain-based transaction, where two buildings with individual loads agree to a transaction of energy credits.

One of the most intriguing of such projects in the United States is LO3's Brooklyn Microgrid (BMG), where residents with rooftop solar PV sell their excess energy to their neighbors, instead of relying on net metering to sell it back to their utility, ConEdison. Although the law prohibits energy consumers from selling energy to one another, BMG members are able to execute these peer-to-peer energy transactions using credits in a blockchain ledger. Every BMG prosumer connects to the microgrid through a dual-purpose meter called a TransActive Grid Element (TAG-e). The TAG-e both measures the participant's energy production and consumption and communicates with other TAG-e devices to record transactions in the blockchain.

It isn't too much of a stretch to imagine future energy storage systems, incorporating open-source, microgrid-sensing software that could participate in blockchain-based transactions (see figure below). If, on a sunny day, there were community members with a glut of excess solar energy, they could use energy storage to keep the excess for later use or sale depending on the going rate for energy.

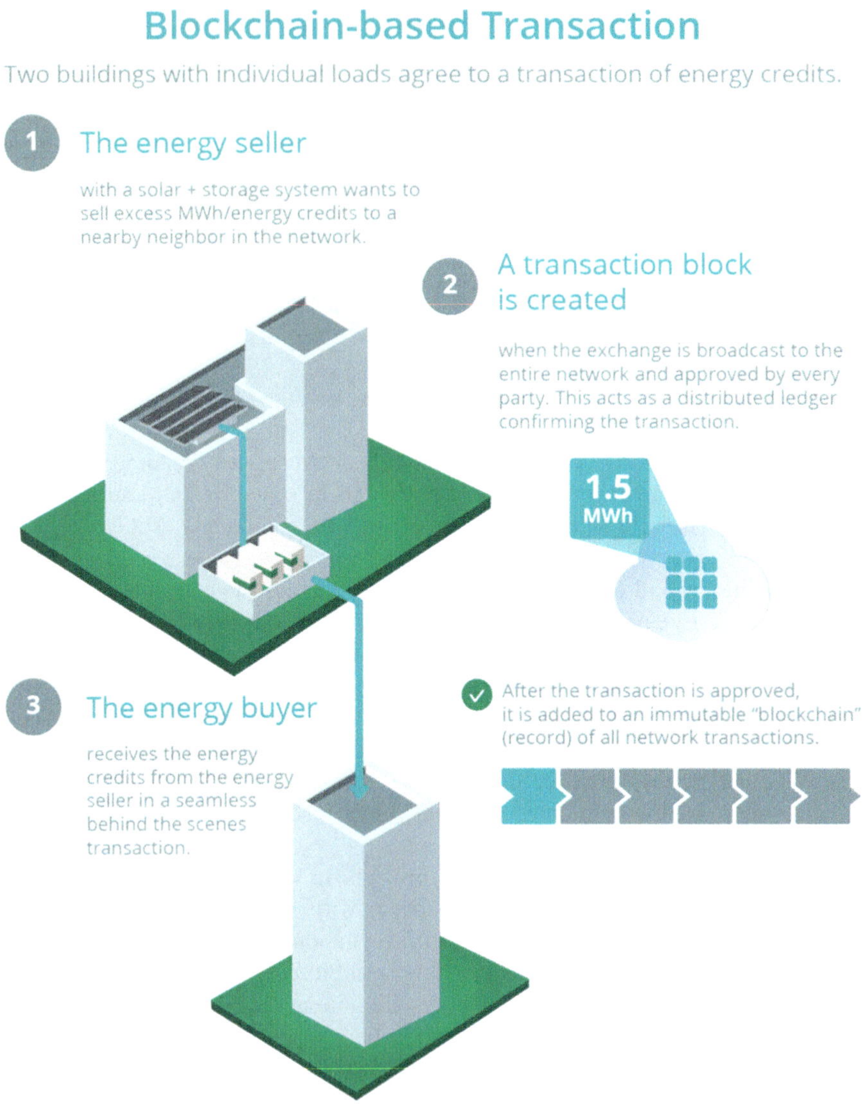

Fig. 8.3 Blockchain-based transaction illustration (Courtesy of Green Charge)

8.2 The Electric Grid

An electrical grid is an interconnected network for delivering electricity from producers to consumers. It consists of generating stations that produce electrical power, high-voltage transmission lines that carry power from distant sources to demand centers, and distribution lines that connect individual customers.

Power stations may be located near a fuel source or at a dam site to take advantage of renewable energy sources and are often located away from heavily populated areas. They are usually quite large to take advantage of economies of scale. The electric power which is generated is stepped up to a higher voltage at which it connects to the electric power transmission network. See Fig. 8.4.

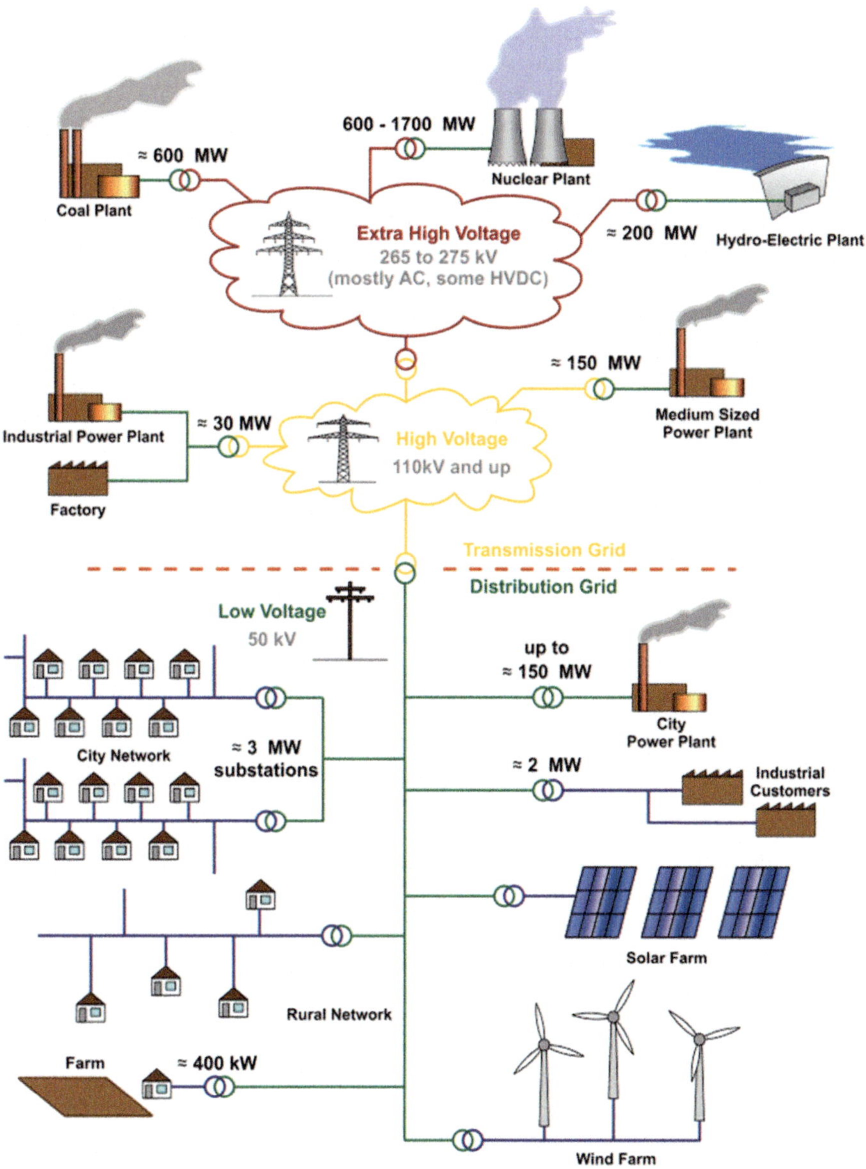

Fig. 8.4 General layout of electricity networks (Courtesy of Wikipedia)

Note that voltages and depictions of electrical lines are typical for Germany and other European systems.

The bulk power transmission network will move the power long distances, sometimes across international boundaries, until it reaches its wholesale customer (usually the company that owns the local electric power distribution network).

On arrival at a substation, the power will be stepped down from a transmission level voltage to a distribution level voltage. As it exits the substation, it enters the distribution wiring. Finally, upon arrival at the service location, the power is stepped down again from the distribution voltage to the required service voltage(s).

Electrical grids vary in size from covering a single building through national grids which cover whole countries to transnational grids which can cross continents.

Grids are designed to supply voltages at largely constant amplitudes. This has to be achieved with varying demand, variable reactive loads, and even nonlinear loads, with electricity provided by generators and distribution and transmission equipment that are not perfectly reliable. See Fig. 8.5 for the wide area synchronous grid of Europe, where most members are of the European Transmission System Operation (ETSO) association, and Fig. 8.6 is a presentation of the continental US power transmission grid which consists of about 300,000 km (186,411 mi) of lines operated by approximately 500 companies. The North American Electric Reliability Corporation (NERC) oversees all of them.

An entire grid runs at the same frequency. Where interconnection to a neighboring grid, operating at a different frequency, is required, a frequency converter is required. High-voltage direct current links can connect two grids that operate at different frequencies or that are not maintaining synchronism.

In a synchronous grid, all the generators must run at the same frequency and must stay very nearly in phase with each other and the grid. For rotating generators, a local governor regulates the driving torque, maintaining constant speed as loading changes. Droop speed control ensures that multiple parallel generators share load changes in proportion to their rating. Generation and consumption must be balanced across the entire grid, because energy is consumed as it is produced. Energy is stored in the immediate short term by the rotational kinetic energy of the generators.

Small deviations from the nominal system frequency are very important in regulating individual generators and assessing the equilibrium of the grid as a whole. When the grid is heavily loaded, the frequency slows, and governors adjust their generators so that more power is output (droop speed control). When the grid is lightly loaded, the grid frequency runs above the nominal frequency, and this is taken as an indication by automatic generation control (AGC) systems across the network that generators should reduce their output.

In addition, there is often central control, which can change the parameters of the AGC systems over timescales of a minute or longer to further adjust the regional network flows and the operating frequency of the grid. For timekeeping purposes, over the course of a day, the nominal frequency will be allowed to vary so as to balance out momentary deviations and to prevent line-operated clocks from gaining or losing significant time.

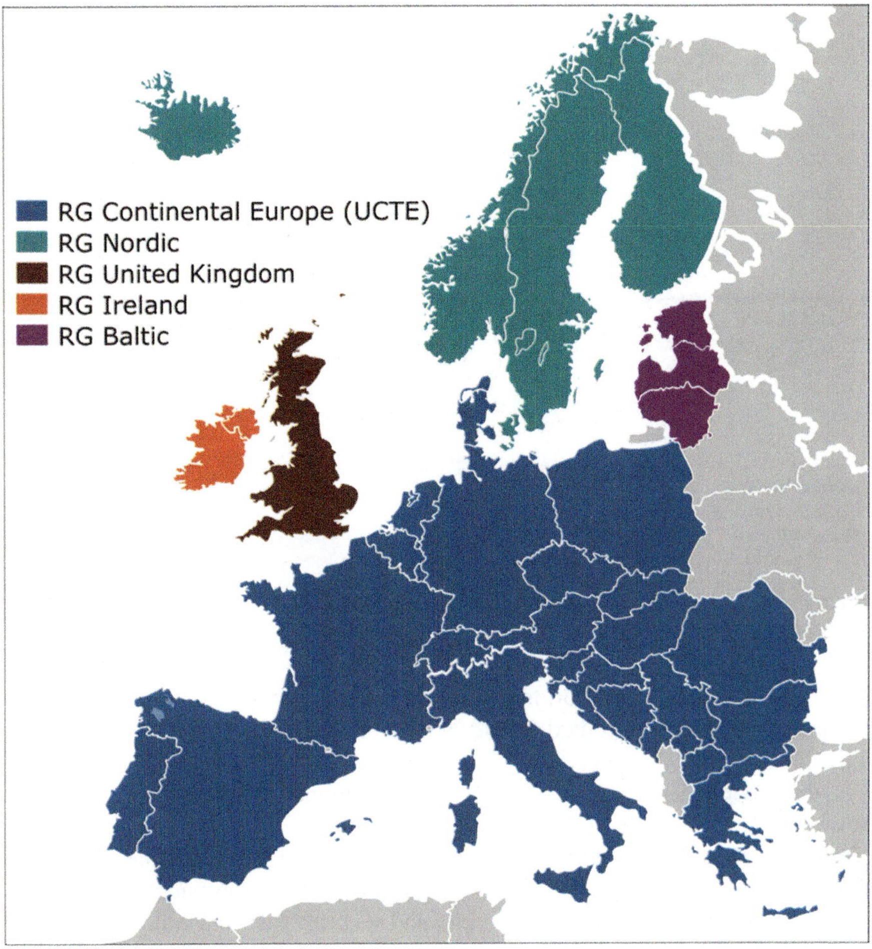

Fig. 8.5 Members of the European Transmission System Operators Association (Courtesy of Wikipedia)

Transmission networks are complex with redundant pathways. For example, see the map of the United States' (right) high-voltage transmission network.

The structure or "topology" of a grid can vary depending on the constraints of budget, requirements for system reliability, and the load and generation characteristics. The physical layout is often forced by what land is available and its geology. Distribution networks are divided into two types, radial or network [2].

The simplest topology for a distribution or transmission grid is a radial structure. This is a tree shape where power from a large supply radiates out into progressively lower-voltage lines until the destination homes and businesses are reached. However, single failures can take out entire branches of the tree.

Most transmission grids offer the reliability that more complex mesh networks provide. The expense of mesh topologies restricts their application to transmission

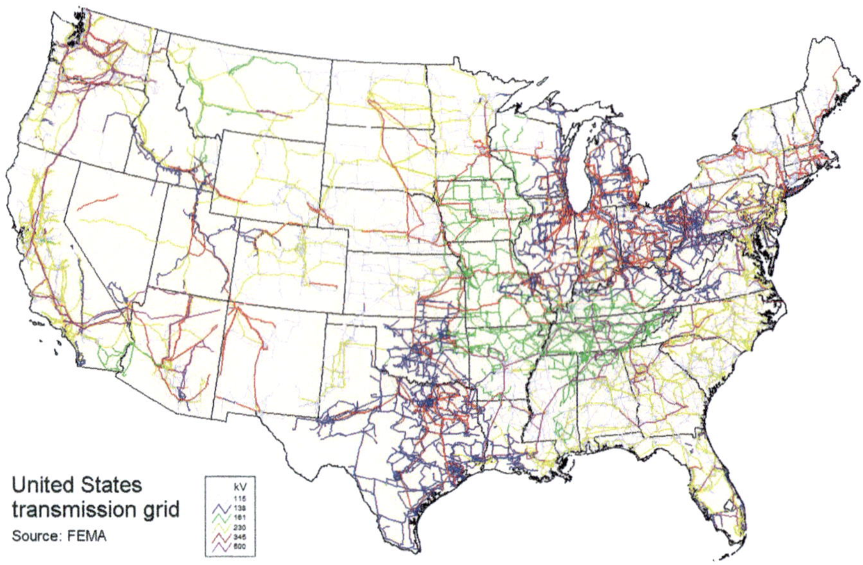

Fig. 8.6 The continental US power transmission grid (Courtesy of the North American Electric Reliability Corporation (NERC))

and medium-voltage distribution grids. Redundancy allows line failures to occur and power is simply rerouted, while workmen repair the damaged and deactivated line.

Other topologies used are looped systems found in Europe and tied ring networks.

In cities and towns of North America, the grid tends to follow the classic radially fed design. A substation receives its power from the transmission network; the power is stepped down with a transformer and sent to a bus from which feeders fan out in all directions across the countryside. These feeders carry three-phase power and tend to follow the major streets near the substation. As the distance from the substation grows, the fan-out continues as smaller laterals spread out to cover areas missed by the feeders. This treelike structure grows outward from the substation, but, for reliability reasons, usually contains at least one unused backup connection to a nearby substation. This connection can be enabled in case of an emergency, so that a portion of a substation's service territory can be alternatively fed by another substation.

A wide area synchronous grid or "interconnection" is a group of distribution areas all operating with alternating current (AC) frequencies synchronized (so that peaks occur at the same time). This allows transmission of AC power throughout the area, connecting a large number of electricity generators and consumers and potentially enabling more efficient electricity markets and redundant generation. Interconnection maps are shown of Europe (see Fig. 8.5, bottom left) and North America (see Fig. 8.6, right).

A large failure in one part of the grid—unless quickly compensated for—can cause current to reroute itself to flow from the remaining generators to consumers over transmission lines of insufficient capacity, causing further failures. One downside to a widely connected grid is thus the possibility of cascading failure and widespread power outage. A central authority is usually designated to facilitate communication and develop protocols to maintain a stable grid. For example, the North American Electric Reliability Corporation gained binding powers in the United States in 2006 and has advisory powers in the applicable parts of Canada and Mexico. The US government has also designated National Interest Electric Transmission Corridors, where it believes transmission bottlenecks have developed.

Some areas, for example, rural communities in Alaska, do not operate on a large grid, relying instead on local diesel generators [3].

High-voltage direct current (HVDC) lines or variable-frequency transformers can be used to connect two alternating current interconnection networks which are not necessarily synchronized with each other. This provides the benefit of interconnection without the need to synchronize an even wider area. For example, compare the wide area synchronous grid map of Europe (see Fig. 8.7) with the map of HVDC lines.

Electric utilities across regions are many times interconnected for improved economy and reliability. Interconnections allow for economies of scale, allowing energy to be purchased from large, efficient sources. Utilities can draw power from generator reserves from a different region in order to ensure continuing, reliable power and diversify their loads. Interconnection also allows regions to have access to cheap bulk energy by receiving power from different sources. For example, one region may be producing cheap hydropower during high water seasons, but in low water seasons, another area may be producing cheaper power through wind, allowing both regions to access cheaper energy sources from one another during different times of the year. Neighboring utilities also help others to maintain the overall system frequency and also help manage tie transfers between utility regions [4].

There are two types of grids that we can take under consideration and these are:

1. Super Grids

   Various planned and proposed systems to dramatically increase transmission capacity are known as super or mega grids. The promised benefits include enabling the renewable energy industry to sell electricity to distant markets, the ability to increase usage of intermittent energy sources by balancing them across vast geological regions, and the removal of congestion that prevents electricity markets from flourishing. Local opposition to sitting new lines and the significant cost of these projects are major obstacles to super grids. One study for a European super grid estimates that as much as 750 GW of extra transmission capacity would be required—capacity that would be accommodated in increments of 5 GW HVDC lines. A recent proposal by TransCanada priced a 1600 km, 3 GW HVDC line at $3 billion USD and would require a

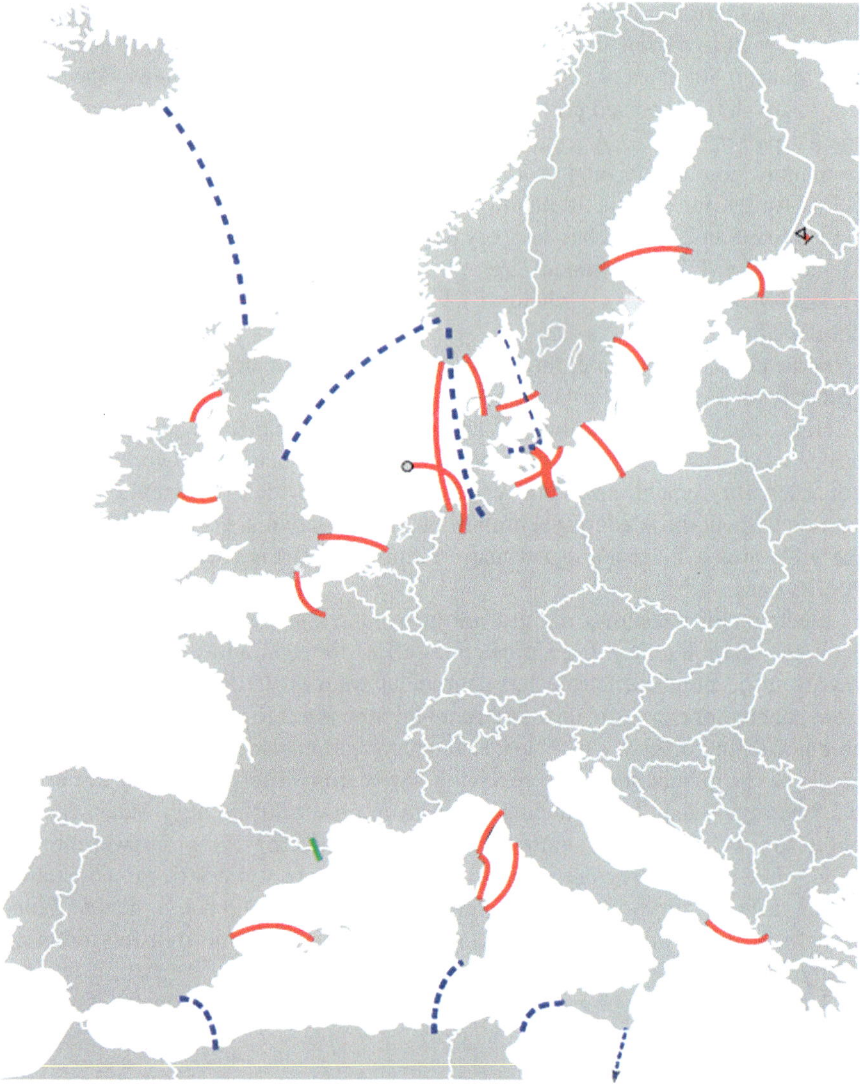

(Red colors are existing line and, green are under construction, and blue are proposed) (Courtesy of Wikipedia)

Fig. 8.7 High-voltage direct current interconnection in Western Europe (Red colors are existing line, and green are under construction, and blue are proposed; Courtesy of Wikipedia)

corridor wide. In India, a recent 6 GW, 1850 km proposal was priced at $790 million and would require a wide right of way. With 750 GW of new HVDC transmission capacity required for a European super grid, the land and money needed for new transmission lines would be considerable.

2. Smart Grids

The smart grid would be an enhancement of the twentieth-century electrical grid, using two-way communications and distributed so-called "intelligent" devices. Two-way electricity and information could improve the delivery network. Research is mainly focused on three systems of a smart grid—the infrastructure system, the management system, and the protection system [5].

The infrastructure system is the energy, information, and communication infrastructure underlying of the smart grid that supports:

- Advanced electricity generation, delivery, and consumption
- Advanced information metering, monitoring, and management
- Advanced communication technologies

A smart grid would allow the power industry to observe and control parts of the system at higher resolution in time and space [6]. One of the purposes of the smart grid is real-time information exchange to make operation as efficient as possible. It would allow management of the grid on all timescales from high-frequency switching devices on a microsecond scale, to wind and solar output variations on a minute scale, to the future effects of the carbon emissions generated by power production on a decade scale.

8.3 Power Generation

Electrical power usually starts at power plants. Although it may be coal, gas, or even nuclear power, almost every conventional plant produces electrical energy through steam-powered turbines. The fossil fuels are burned in order to make water boil and turn into steam which then enters the turbine and pushes against blades to turn the generator shaft to create electric current. See Fig. 8.8.

Right after the turbine, the steam is usually cooled down and turned into liquid form again in order to increase efficiency.

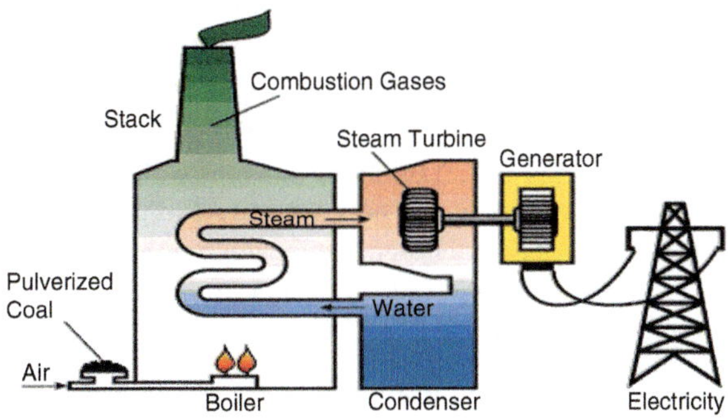

Fig. 8.8 Steam-powered power plant [7]

8.4 Transmission and Distribution

Power plants aren't located right next to your house; they tend to have sites where noise and emissions aren't disturbing issues for the population and near rivers for cooling purposes. Thus, in order to transport the electricity from the plants to the demanding locations, an electrical grid is needed. To minimize dissipation over long distance and to guarantee safety and functionality, different transmission grid types exist using different voltages. The closer the power gets to the consumer, the more the voltage decreases in the following order:

- *High-Voltage Grid*: Transformer stations transform the power to lower voltages, usually between 60 kV and 220 kV. These lines are supposed to carry the electricity into different regional areas with high population density or bigger industrial areas.
- *Distribution Grid*: Voltages rank between 6 kV and 60 kV. Its main task is to provide major institutions like facilities, schools or hospitals, and the transformer stations responsible for low-voltage grids delivering to private households.
- *Low-Voltage Grid*: Often referred to as the "last mile." Voltages between 110 V and 400 V are common. The low-voltage grid is the last station of the transmission and provides private households with power to use for everyday electronic devices.

8.5 Load Management

As mentioned before, it is necessary to produce the almost exact amount of electricity that is demanded by customers. In order to maintain grid stability, a frequency of 50 or 60 hertz (Hz) (depending on the country's standard) must be generated. Higher deviations (±2.5 Hz) will result in causing damage to the generators.

The curve progression in Fig. 8.9 varies over the day; but behavior is usually steady and pretty well known for each day and is used as a roadmap for production along general lines. For example, the need for electricity during the night is low, whereas at noon, when everybody starts cooking, it is at its peak.

Analyzing Fig. 8.9 indicates the load curve diagram shows that much of the electricity demand is in fact for continuous 24/7 supply (baseload), while some is for a lesser amount of predictable supply for about three quarters of the day and less still for variable peak demand up to half of the time; some of the overnight demand is for domestic hot water systems on cheap tariff. With overnight charging of electric vehicles, it is easy to see how the baseload proportion would grow, increasing the scope for nuclear and other plants which produce it.

Figure 8.9 also shows the concept of a load management which is separated into three different types:

- *Baseload*: This is the amount of electricity that is demanded and produced at any time. Nuclear, hydroelectric power or brown coal plants are known and common

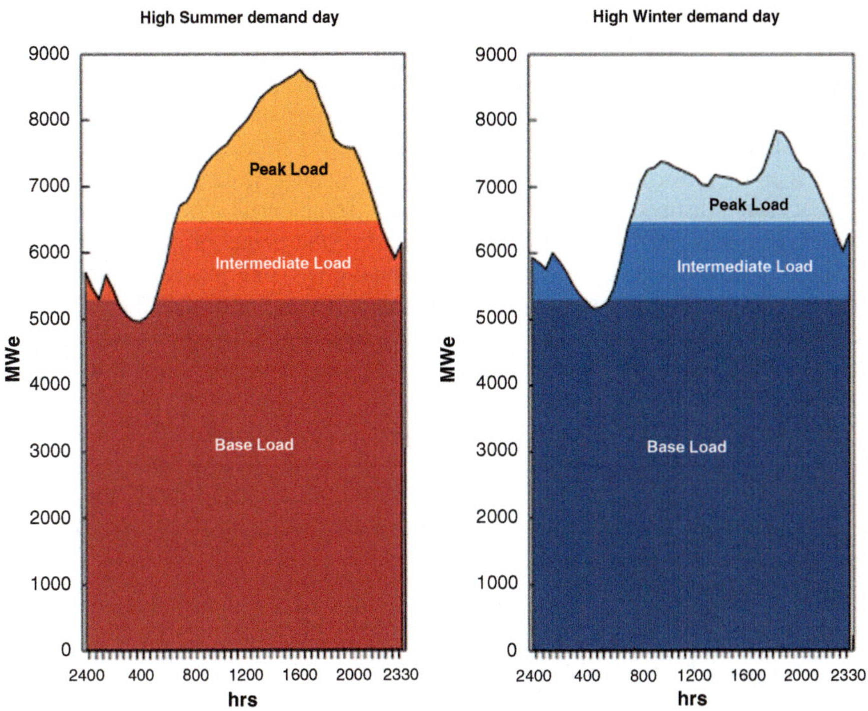

Fig. 8.9 Load curves for typical electric grid [8] (Courtesy of Vencorp)

to use as baseload plants due to the long start-up time and/or the low operating and fuel costs.

- *Intermediate Load*: Power plants that are easier and faster to regulate are used for the task of middle load. These plants are capable of working within minutes to an hour and have moderate operating costs. Black coal or wind plants are typical of middle loads.
- *Peak Load*: Peak load is the power demand outside of the daily "roadmap." Different events like unexpected hot and sunny days can lead to an extended use of air conditioners and therefore a higher electricity demand. Peak load plants have a fast response time, which means they're operational within seconds to a few minutes. A typical example would be gas turbine power plants or pumped-storage hydroelectricity.

Hydroelectric power plants are technically qualified for peak load but are used for baseload instead because not using the already flowing water would be a waste.

Middle load plants can be and are also used for this task; when not operated under full load, they bear reserves. In some countries like Germany, it is statutory that a certain amount of power plants must have these reserves for supply security reasons. Operating power plants on lower degrees however is to be avoided if possible as the efficiency of the turbines decreases.

Most electricity demand is for continuous, reliable supply that has traditionally been provided by baseload electricity generation. Some is for shorter-term (e.g., peak load) requirements on a broadly predictable basis. Hence if renewable sources are linked to a grid, the question of backup capacity arises, for stand-alone system energy storage is the main issue. Apart from pumped-storage hydro systems (see later section), no such means exist at present on any large scale [8].

However, a distinct advantage of solar and to some extent other renewable systems is that they are distributed and may be near the points of demand, thereby reducing power transmission losses if traditional generating plants are distant. Of course, this same feature sometimes counts against wind in that the best sites for harnessing it are sometimes remote from population and the main backup for lack of wind in one place is wind blowing hard in another, hence requiring a wide network with flexible operation [8].

In Europe, at the end of 2014, the five largest electricity markets (UK, Germany, France, Italy, and Spain) had 97.5 GWe of installed wind capacity and 77.6 GWe of installed solar capacity. In the first half of 2015, this produced 107.6 TWh from wind and 60.1 TWh from solar, giving (based on these figures) 25% capacity factor for wind and 17.7% for solar (20.8% in July).

8.6 Types of Storage Technology

Energy storage is the capture of energy produced at one time for use at a later time. A device that stores energy is sometimes called an accumulator or battery. Energy comes in multiple forms including radiation, chemical, gravitational potential, electrical potential, electricity, elevated temperature, latent heat, and kinetic. Energy storage involves converting energy from forms that are difficult to store to more conveniently or economically storable forms. Bulk energy storage is currently dominated by hydroelectric dams, both conventional and pumped. See Fig. 8.10, which is a depiction of the Llyn Stwlan dam of the Ffestiniog Pumped Storage Scheme in Wales. The lower power station has four water turbines which can generate a total of 360 MW of electricity for several hours, an example of artificial energy storage and conversion.

In 2015 hydro supplied about 3988 TWh from 1245 GWe (37% capacity factor), underlining its generally peak use.

Hydroelectric power, using the potential energy of rivers, is by far the best-established means of electricity generation from renewable sources. It may also be large scale—nine of the ten largest power plants in the world are hydro, using dams on rivers. China's Three Gorges leads with 22.5 GWe, then Itaipu in Brazil with 14 GWe, and Xiluodu in China, 13.9 GWe. In contrast to wind and solar generation, hydro plants have considerable mechanical inertia and are synchronous, helping with grid stability.

Hydropower using large storage reservoirs on rivers is not a major option for the future in the developed countries because most major sites in these countries having

Fig. 8.10 Llyn Stwlan dam of the Ffestiniog pumped storage scheme in Wales

potential for harnessing gravity in this way are either being exploited already or are unavailable for other reasons such as environmental considerations. Growth to 2030 is expected mostly in China and Latin America. China has commissioned the $26 billion Three Gorges dam, which produces 22.5 GWe and has a major role in flood control, but it has displaced over 1.2 million people. Brazil is planning to have 25 GWe of new hydro capacity by 2025, involving considerable environmental impact.

The chief advantage of hydro systems is their capacity to handle seasonal (as well as daily) high peak loads. In practice the utilization of stored water is sometimes complicated by demands for irrigation which may occur out of phase with peak electrical demands.

As part of pumped-storage hydroelectricity (PSH), pumped hydro plants, so far, are considered to be the only possible way to store energy in a huge amount while maintaining a high efficiency and being economical as well and have about 98% share of total global storage predominant in today's grid. The first plants of this type were built in Switzerland and Italy in the 1890s, making the concept over a hundred years old.

Conceptually, when you lift an object of a certain mass, you overcome gravity. In order to do so, you must supply a force over a height. The force required to lift is defined by the physical law (*m* for mass and *a* for acceleration), but in this case *a* is replaced by for the gravitational acceleration *g* (9.81 meters per square second [m/s^2]). The work, meaning the energy supplied and therefore stored in the object, is defined by $W = F * d$ (in this example the term *d* for distance can be replaced by for height). This results in $W = m * g * h$, meaning the energy stored equals the mass multiplied by the gravity and the height (Fig. 8.11).

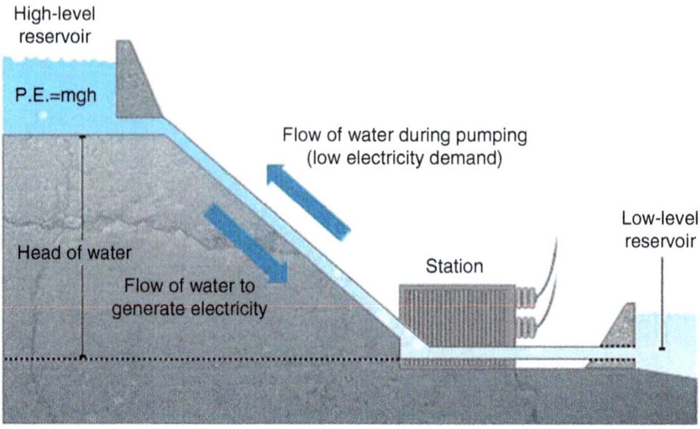

Fig. 8.11 Schematic of a pumped-storage plant

A PSH plant puts this math into practice. Basically, the system contains two water reservoirs at different elevations. In times of low electricity demand and high production, water is pumped from the lower reservoir into the higher, storing the electricity in the water in the form of potential energy. When needed, for example, on peak demand, the water can be released, flowing down the pipes again and back through the turbine which then generates the electricity. The general formula for the power output is $P = Q * h * \eta * g * \rho$, including the factors of volume flow rate passing the turbines (Q) and the hydraulic efficiency of the turbine (and the density of the water ρ).

Depending on the height difference, Pelton wheels and Kaplan or Francis turbines are used to maximize efficiency, each reaching roughly about 90%. These turbines are reversible and, therefore, capable of handling both the pumping and generating process. Capacities for PSHs are depending on the location and scale of the reservoirs as well as the altitude difference and can reach from a few MWh to several GWh [7].

Pros and cons are listed here with a sign of plus (pros) and minus (cons) in front of them as:

Pros

+ Mature technology, capable of storing huge amounts of energy
+ High overall efficiency (around 70–80%)
+ Fast response time
+ Inexpensive way to store energy

Cons

– Few potential sites
– Huge environmental impacts
– Requires a significant huge water source

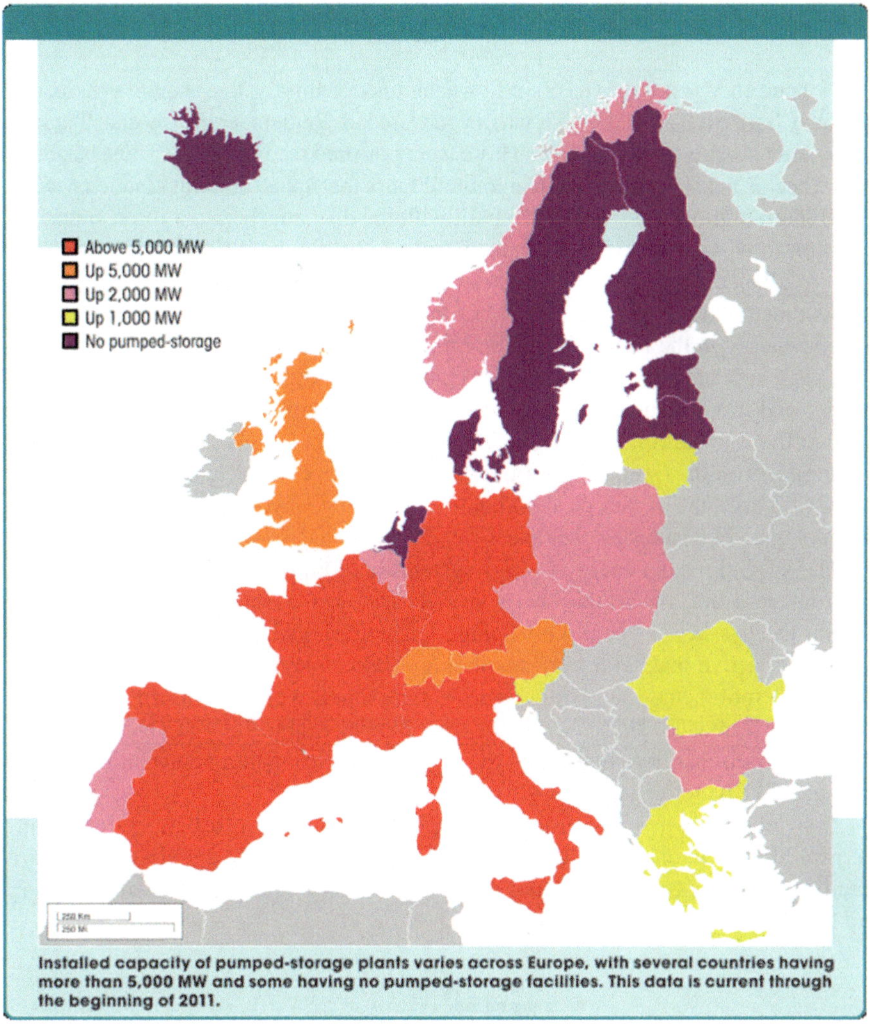

Fig. 8.12 Map of pumped-storage capacities in Europe

Figure 8.12 is a map of pumped-storage capacities in Europe.

Some technologies provide short-term energy storage, while others can endure for much longer.

A wind-up clock stores potential energy (in this case mechanical, in the spring tension), a rechargeable battery stores readily convertible chemical energy to operate a mobile phone, and a hydroelectric dam stores energy in a reservoir as gravitational potential energy. Fossil fuels such as coal and gasoline store ancient energy derived from sunlight by organisms that later died, became buried, and over time were then converted into these fuels. Food (which is made by the same process as fossil fuels) is a form of energy stored in chemical form.

8.6.1 Kinetic Energy Storage or Flywheel Concept

The kinetic energy storage (KES) or flywheel functionality system is quite simple, and you may have even played with it when you were kid. Remember the toy cars that kept going after spinning their wheels? Those were powered by a flywheel. Thus, basically a flywheel is a disk with a certain amount of mass that spins, holding kinetic energy.

Modern high-tech flywheels are built with the disk attached to a rotor in upright position to prevent gravity influence. They are charged by a simple electric motor that simultaneously acts as a generator in the process of discharging.

As illustrated in Fig. 8.13, The VYCON Direct Connect (VDC) ® system stores kinetic energy in the form of a rotating mass and is designed for high-power, short discharge applications. The patented technology within the VDC system includes a high-speed motor generator, active magnetic bearings that are used to levitate and sustain the rotor during operation, and a superior control system that can provide information on the system performance. These innovative technologies enable the VDC to charge and discharge at high rates for countless cycles making conventional technologies like batteries obsolete.

The VDC flywheel energy storage systems hold kinetic energy in the form of a rotating mass and convert this energy to electric power through patented technology within the flywheel system. Our unique technology includes a high-speed motor generator, active magnetic bearings that are used to levitate and sustain the rotor during operation, and a superior control system that can provide information on the system performance. These innovative technologies enable the VYCON flywheel to charge and discharge at high rates for countless cycles making conventional technologies like batteries obsolete.

This technology is ideal for a variety of applications as a battery-free uninterruptible power supply (UPS) solution, including data centers, healthcare facilities, broadcast stations, and other mission-critical operations that require up to 30–40 s of backup power. See Fig. 8.14.

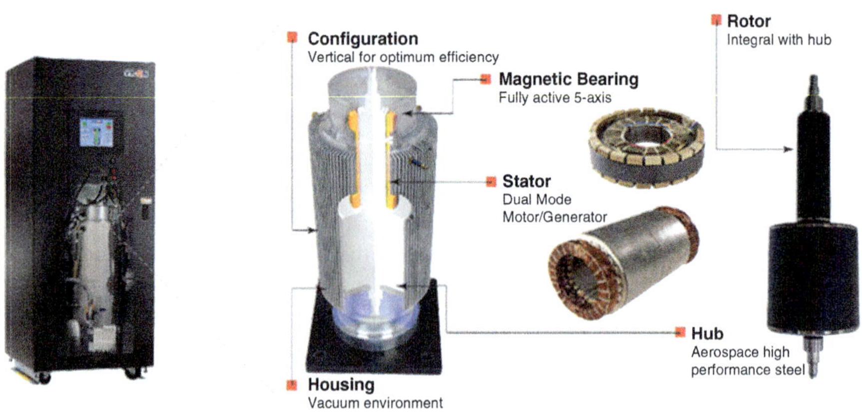

Fig. 8.13 Kinetic energy storage systems (Courtesy of VYCON)

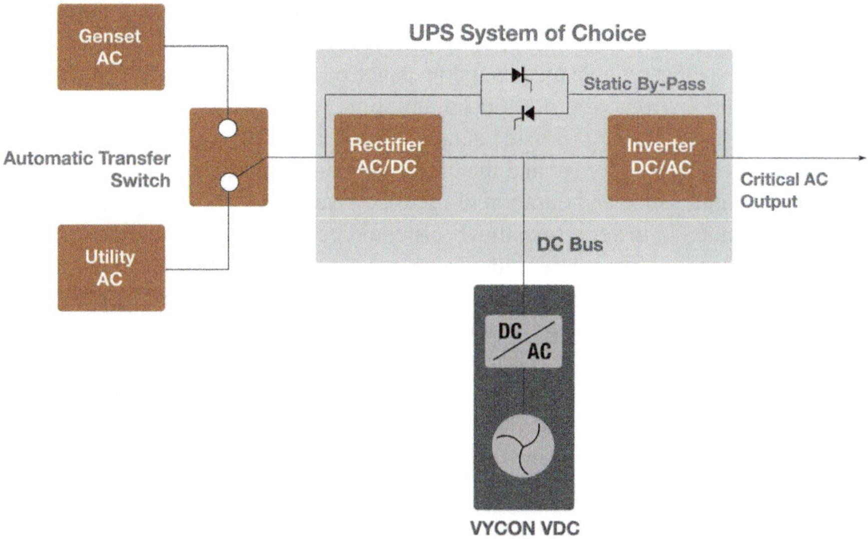

Fig. 8.14 Layout of uninterruptible power supply configuration (Courtesy of VYCON)

When dealing with efficiency however it gets more complicated, as stated by the rules of physics, they will eventually have to deal with friction during operation. Therefore, the challenge to increase that efficiency is to minimize friction. This is mainly accomplished by two measures: the first one is to let the disk spin in a vacuum, so there will be no air friction; and the second one is to bear the spinning rotor on permanent and electromagnetic bearings so it basically floats. The spinning speed for a modern single flywheel reaches up to 16.000 rpm and offers a capacity up to 25 kilowatt hours (kWh), which can be absorbed and injected almost instantly.

Pros and cons are listed here with a sign of plus (pros) and minus (cons) in front of them as:

Pros

+ Low maintenance and long life span: up to 20 years
+ Almost no carbon emissions
+ Fast response time
+ No toxic components

Cons

– High acquisition cost
– Low storage capacity
– High self-discharge (3–20% per hour)

Future prospects of kinetic energy storage systems are stated below here as:

Stephentown, New York, is successfully operating the largest and latest flywheel energy storage system since July, 2011. The facility is capable of storing up to

5 megawatt hours (MWh) with its 200 flywheels for several hours and required a budget of roughly $60 million2. This storage system has several advantages compared to others, most notable the low maintenance costs, the fast access to the stored energy, and the fact that you don't need any toxic resources as well as almost no carbon emissions. On the downside stands the low capacity compared to systems like the pumped hydro storage and the high acquisition costs, though compensated by the low maintenance and duration of up to 20 years [9].

On a side note, in lower terms, flywheels could be used in the transport sector to make vehicles more efficient by using their kinetic energy to charge them and therefore lower the need for energy through fuel. The Formula 1 is currently successfully making use of this technology kinetic energy recovery system (KERS) but to improve power, not efficiency. The project in Stephentown will show if flywheels are as good and economical as promised. Due to the high costs and low capacity however, it is likely that the flywheel technology will remain a niche market that requires fast response times as the high storage needs cannot be met by them.

8.6.2 Superconducting Magnetic Energy Storage

Superconducting magnetic energy storage (SMES) is a novel technology that stores electricity from the grid within the magnetic field of a coil comprised of superconducting wire with near-zero loss of energy.

SMES is a grid-enabling device that stores and discharges large quantities of power almost instantaneously. The system is capable of releasing high levels of power within a fraction of a cycle to replace a sudden loss or dip in line power. Strategic injection of brief bursts of power can play a crucial role in maintaining grid reliability especially with today's increasingly congested power lines and the high penetration of renewable energy sources, such as wind and solar. See Fig. 8.15 for illustration of top level depiction of SMES.

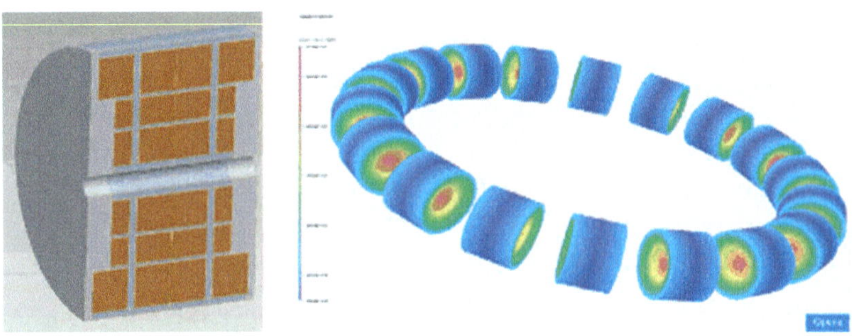

Fig. 8.15 Artistic depiction of superconducting magnetic energy storage (Courtesy of Brookhaven National Laboratory)

The concept of SMES is described in this section as well. The system consists of three major components: the coil, the power conditioning system (PCS), and a cooling system. The idea is to store energy in the form of an electromagnetic field surrounding the coil, which is made of a superconductor. At very low temperatures, some materials lose every electric resistance and thus become superconducting. The superconducting magnetic storage system (SMES) makes use of this phenomenon and—in theory—stores energy without almost any energy loss (practically 90–95% efficiency). See Fig. 8.16 for more details of the components of this system.

However, since relevant superconducting materials are only known to work below −253 °centigrade (C) (20 °kelvin[K]) (e.g., niobium-titanium −264 °C [9 K], niobium-tin −255 °C [18 K]), a system to cool the components down to those temperatures is required. This can be accomplished by liquefying helium; but, it is very expensive and the process lowers the efficiency, especially in standby mode. New high-temperature superconductors have been in development since 1986 reaching the state of superconductivity already at −163 °C (110 K), allowing them to be cooled by liquid nitrogen and thereby lowering the costs by a factor of 10–20. Known production methods for these materials however make them very brittle and difficult and expensive to process.

The PCS is the interface between the SMES coil and the power system. Its task is to convert alternating current (AC) into direct current (DC) and vice versa since the coil is only capable of storing and releasing the energy in the form of DC.

Figure 8.17 is an illustration of conceptual design of a superconducting coil, which is a toroidal magnet with an outside diameter for the cryostat of ~12 m [10]. A half-size prototype coil was constructed by Toshiba and had been recently tested at the time of the WTEC trip to Japan. The test coil used a forced-flow Nb-Ti cable-in-conduit conductor and demonstrated 20 kA at 2.8 T, which is the rated current for the basic design. The initial testing was conducted at the Japan Atomic Energy

Fig. 8.16 Components of a SMES system

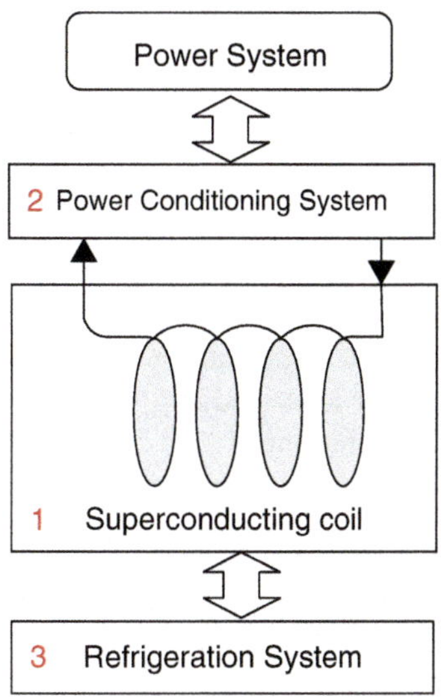

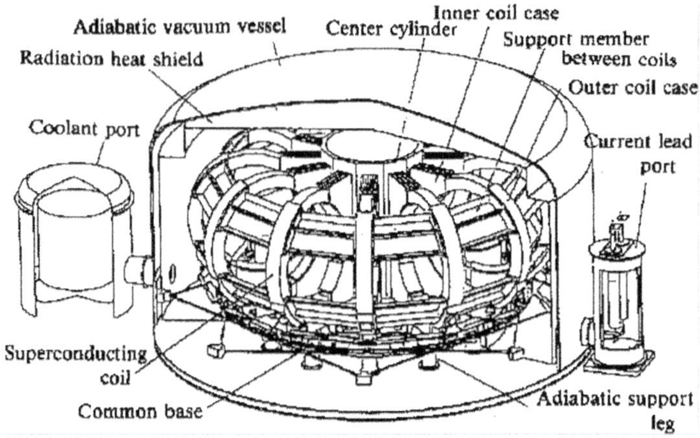

Fig. 8.17 Conceptual design of the ISTEC superconducting coil for the 100 kWh small-scale SMES

Research Institute (JAERI), with further tests planned at Lawrence Livermore National Labs (LLNL) in the United States. Additional information on this SMES program is presented in the ISTEC, Chubu, and Toshiba site reports.

In the United States of America, the most significant program on energy storage in the world is being carried out by Babcock and Wilcox (B&W). This ~$50 million program is cost shared by industry (70%) and the federal government (30%) through DARPA. B&W will construct and install a 500 kWh SMES primarily to provide spinning reserve to the Anchorage Municipal Light and Power (AML&P) utility. The Anchorage utility is part of the "Alaskan Railbelt System," one of the most isolated utility networks in the United States. The Anchorage area served by AML&P uses almost half of the railbelt system's peak load, which reaches approximately 600 MW during winter. AML&P is required as part of the railbelt interconnection agreement to designate ~30% of its generating capacity for spinning reserve; part of this reserve is provided by a hydroelectric facility at Bradley Lake. Physical restrictions at Bradley Lake result in dispatch time for this hydro capacity of approximately one minute or more. This is too long a time to prevent additional load shedding during an event such as a generator outage, which would lead to frequency instability on the system, resulting in programmed load shedding. The planned SMES system will instantly dispatch ~30 MW over a period of ~1 min, which will provide sufficient time to ramp up the hydro capacity and put it on line to prevent further load shedding. As designed, the SMES will store 1800 MJ in a low-aspect solenoid almost 7 m in diameter using an aluminum-stabilized Nb-Ti conductor operating in a "cryostable cooled" mode [11].

Several US companies are producing small SMES systems, called micro-SMES, primarily to provide power quality improvements to selected customers rather than as grid or network solutions. These ~1 MW units with a few MJ stored energy are commercially produced by Superconducting, Inc. (SI), of Madison, Wisconsin, and by IGC of Latham, New York. SI and IGC have supplied micro-SMES systems to the United States Air Force (USAF) as part of a program to provide uninterruptable power capability and power conditioning, primarily for voltage stabilization, to selected USAF control centers. At present this "power quality" market is also served by battery storage or flywheel systems, especially in Japan and Germany. Outside of the United States, there is no comparable activity for micro-SMES commercialization.

The most active effort on SMES in Germany is carried out at the Forschungszentrum Karlsruhe (FZK) laboratory. FZK is constructing a 250 kJ SMES with a toroidal field design to address a power quality problem due to the frequent start-up of large motors at a sawmill. Analysis indicates that a SMES system would be ideal for reducing this flicker problem, due to its fast response capability. FZK is also investigating at the Deutsches Elektronen-Synchrotron (DESY) laboratory the use of SMES as a pulse power source for providing ~10 GW pulses with a 1.0 ms duration at a 10 Hz rate in order to power RF klystrons [12].

At Siemens, the interest in SMES over the last few years has been mostly restricted to design and evaluation studies, with no current plans for development or construction. Siemens, RWE (the largest German utility), and PreussenElektra have completed an evaluation and conceptual design of 2 MWh/50 MW SMES for use in providing frequency stabilization to the electric system. SMES continues to be of interest, but recent economic studies by Siemens indicate it may be too expensive compared to other storage technologies [13].

It is thought that using HTS current leads, which are now a commercial reality and can be purchased from a number of manufacturers including Hoechst and ASC, would reduce the heat leak in a SMES system. Use of HTS conductors to fabricate a high field, high density SMES is also projected as an interesting future application. An HTS SMES operating at 10–20 T would require a "composite conductor" capable of carrying thousands of amperes with low AC losses to minimize heating for multiple charge and discharge cycles.

The Technical University of Munich has also been conducting research on SMES and is constructing 1.4 MJ toroidal field system using LTS conductors [14]. At ABB in Switzerland, energy storage is considered an important area; it has been strongly followed in the past, and an LTS SMES has been constructed for experimental evaluation. At the time of the WTEC visit, a major SMES system using LTS conductor under development for the Swiss railroad had been terminated due to realization of an alternative, less costly, solution. ABB had no plans for SC storage but did plan to continue to evaluate the technology.

In summary, the Japanese Super-GM program to develop SC generators represents the major activity worldwide directed at the commercialization of superconductivity in an electric power application. This conclusion is based on the duration of the program, the total money invested by industry and the Japanese government, and the number of institutions and people involved. The future for this program, however, is highly dependent on the "complete" success of the planned testing through 1998 on the three rotor configurations for the generator. The follow-on program to construct a 200 MW-class pilot machine is also highly dependent on a significant improvement in the market forecast for this generator. Without an increase in demand, it is quite likely that even with unqualified success for the generator tests that Japanese industry may not be willing to commit the necessary cost share to go forward with the program. Interest in using HTS conductors in this future program will depend on progress in achieving higher transport properties at higher fields.

The current US DOE SPI program on the development of an HTS synchronous motor is unique. The B&W "spinning reserve" program with Anchorage Municipal Light and Power is the world's largest SMES program and provides an acceptable performance and cost-effective solution to a utility problem.

The benefits of superconducting magnetic energy storage are listed below as:

- Improves power quality for critical loads and provides carryover energy during momentary voltage sags and power outages.
- Improves load leveling between renewable energy sources (wind, solar) and the transmission and distribution network.
- Environmentally beneficial as compared to batteries; superconductivity does not rely on a chemical reaction and no toxins are produced in the process.
- Enhances transmission line capacity and performance—SMES features a high dynamic range, an almost infinite cycling capability, and an energy recovery rate close to 100%.
- Ultrahigh field operation enables long-term storage SMES systems in a compact device with cost advantages in material and system costs.

Pros and cons are listed here with a sign of plus (pros) and minus (cons) in front of them as:

Pros

+ Capable of partial and deep discharge
+ Fast response time
+ No environmental hazard

Cons

– High energy losses (~12% per day)
– Very expensive in production and maintenance
– Reduced efficiency due to the required cooling process

Future prospects of superconducting magnetic energy storage is stated below here as:

Future prospects are difficult to determine because they depend on further development in superconducting materials. The discovery of a suitable material with these properties on room temperature would change nearly anything (hence, the consideration for being the holy grail of physics) and would make energy storage and transmission easy, safe, and cheap. However, it is uncertain if such a material even exists.

Right now, SMES systems are pretty much like flywheels, considered a niche market, requiring fast response times. Because of the difficult and expensive procedure to process high-temperature superconductors, it is expected that low-temperature materials will come to action in short and medium terms. Right now, the development focus lies on micro-SMES systems with capacities up to 10 kWh, applied mainly for power quality and uninterrupted power sources (UPS) and therefore of no relevant significance for renewable energies right now. Further technological improvements and achievements in processing high-temperature superconductors could change the course and make SMES systems more economical and relevant for energy storage in the future though.

8.6.3 Batteries

Energy storage systems are comprised of three main modules:

- The direct current (DC) battery where energy is stored
- The alternate current (AC) power conversion where the energy is converted from AC to DC
- The control system that manages the operation of any energy storage system

As a battery manufacturer, the researches within the company should have a deep technical understanding of cells, thermal management, and other key battery-related elements that helps us to configure the right DC battery technology to meet our customers' needs.

A battery is a device that produces electrical energy from chemical reactions. There are different kinds of batteries with different chemicals. The idea behind them is that the two different chemicals within a battery cell have different loads and are connected with a negative (cathode) and the other with a positive electrode (anode). When connected to an appliance, the negative electrode supplies a current of electrons that flow through the appliance and are accepted by the positive electrode. For the use of storing energy produced by renewable energy sources, only rechargeable batteries are relevant and will be considered. Figure 8.18 is cross-sectional view of the inside of a lead-acid battery.

Figure 8.18a shows the battery cross section consists of six 2 v cells connected in series, while Fig. 8.18b is an illustration of each component that is composed of several negative (cathode) and positive (anode) electrodes made of pure spongy lead and lead oxide, respectively. The electrodes, connected in parallel, are immersed in a dilute solution of sulfuric acid.

8.6.3.1 Lead-Acid Batteries

A lead-acid battery is an electrochemical battery that uses lead and lead oxide for electrodes and sulfuric acid for the electrolyte. Lead-acid batteries are the most commonly used in photovoltaic (PV) and other alternative energy systems because their initial cost is lower and because they are readily available nearly everywhere in the world. There are many different sizes and designs of lead-acid batteries, but

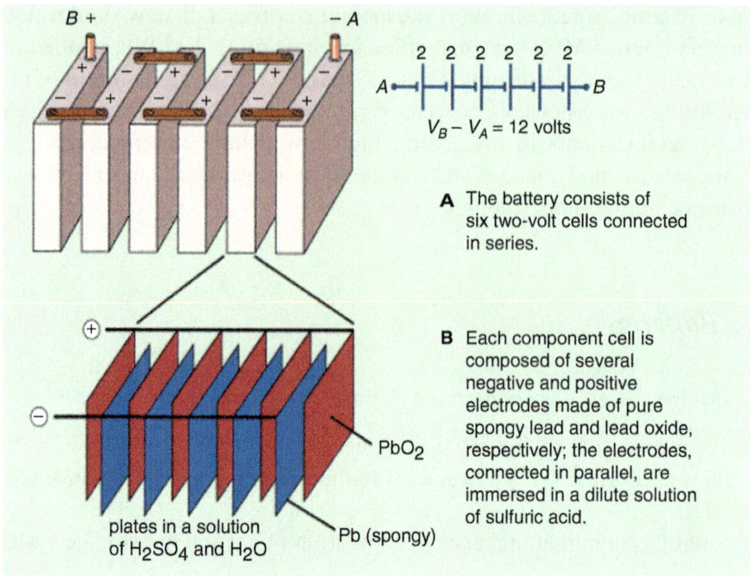

Fig. 8.18 The inside of a lead-acid battery

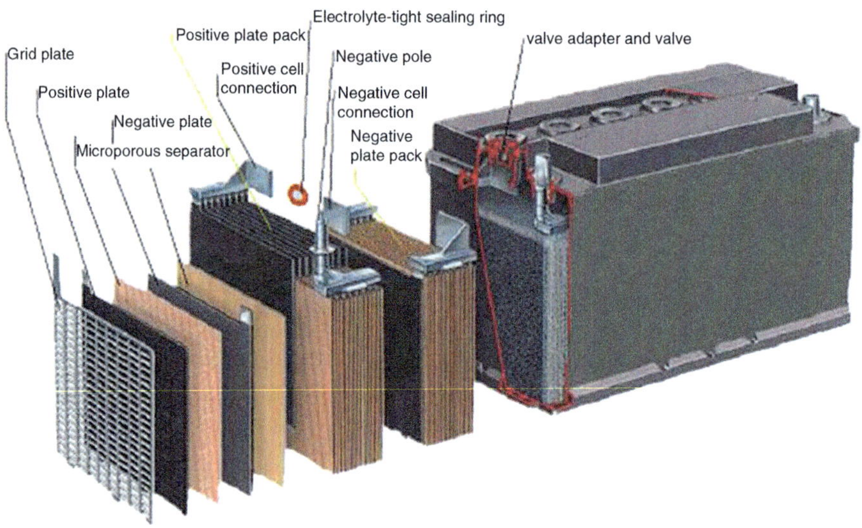

Fig. 8.19 Lead-acid battery

the most important designation is whether they are deep cycle batteries or shallow cycle batteries. See Fig. 8.19 for a diagram of lead-acid battery with its internal components.

Lead-acid battery is the best solar deal available now—up to \$4000 in maximum savings in today's market—and they can be found in three different types of design:

1. *Shallow Cycle Batteries*: like the type used as starting batteries in automobiles, they are designed to supply a large amount of current for a short time and stand mild overcharge without losing electrolyte. Unfortunately, they cannot tolerate being deeply discharged. If they are repeatedly discharged more than 20%, their life will be very short. These batteries are not a good choice for a photovoltaic (PV) system.
2. *Deep Cycle Batteries*: These types of batteries are designed to be repeatedly discharged by as much as 80% of their capacity, so they are a good choice for power systems. Even though they are designed to withstand deep cycling, these batteries will have a longer life if the cycles are shallower. All lead-acid batteries will fail prematurely if they are not recharged completely after each cycle. Letting a lead-acid battery stay in a discharged condition for many days at a time will cause sulfating of the positive plate and a permanent loss of capacity.
3. *Sealed Deep Cycle Lead-Acid Batteries*: These batteries are maintenance-free. They never need watering or an equalization charge. They cannot freeze or spill, so they can be mounted in any position. Sealed batteries require very accurate regulation to prevent overcharge and overdischarge. Either of these conditions will drastically shorten their lives. Sealed batteries are well-suited for remote, unattended power systems.

Conceptually, a lead-acid battery usually has several in-series connected cells, each delivering 2 volts (V) and each consisting several spongy pure lead cathodes, positive loaded lead oxide anodes, and a 20–40% solution of sulfuric acid that acts as an electrolyte. When discharged, both the anode and the cathode undergo a chemical reaction with the electrolyte that progressively changes them into lead sulfate that releases electrical energy in the process. This reaction can be almost completely reversed by supplying the electrodes with electricity, which is the reason a lead-acid battery can be recharged.

The life cycle and the ability to tolerate deep discharges depend on the type. Starting-lighting-ignition (SLI) batteries used in cars are not designed to be discharged to more than 50% as they have thinner lead plates. Doing so on a regular base will damage them and shorten their life cycle dramatically, whereas deep cycle batteries with thicker plates can handle this much better but are as a result heavier and bulkier.

Pros and cons are listed here with a sign of plus (pros) and minus (cons) in front of them as:

Pros

+ Easy and therefore cheap to produce
+ Mature technology, more than 150 years of experience and development
+ Very high surge-to-weight-ratio, capable of delivering a high jolt of electricity at once, which is why they are so suitable as car starters
+ Easily recyclable

Cons

– Very heavy and bulky.
– Rather short lived.

– Environmental concerns; although pretty safe, leads are a very toxic element and exposure can cause severe damage to people and animals.
– Corrosion caused by the chemical reactions.

Future prospects of superconducting magnetic energy storage is stated below here as:

Lead-acid batteries have pretty much reached the end of the rope in terms of development. It is clear that no significant improvements can be made in capacity, density, or weight. Therefore, resources on future development should concentrate on other battery technologies with higher potentials.

Nonetheless because of the cost-effectiveness, lead-acid batteries are an important part of today's technology systems that can't be denied. Until other battery technologies emerge, they will remain first choice for many applications; however, grid storage is unlikely to be one of them simply because these batteries are not capable of storing huge amount of energy compared to other systems like a pumped-storage hydroelectricity (PSH) plant while staying cost-effective as the energy density is just too low. It is possible to integrate battery banks for few smaller decentralized systems like photovoltaic (PV) systems on rooftops; but, it cannot be used as a definite solution, just for the simple reason that the amount of resources is not available for the required capacity scale. Also, these batteries have a limited life cycle of a few years and, therefore, have to be replaced by new ones.

Future prospects and outlooks for other battery technologies and development indicate that lead-acid-based batteries will probably play the role of an inexpensive transitory technology.

8.6.3.2 Lithium-Ion Batteries

Lithium-ion batteries are incredibly popular these days. You can find them in laptops, personal digital assistants (PDAs), cell phones, and iPods. They are so common because, pound for pound, they are some of the most energetic rechargeable batteries available. The technology of the lithium battery has been slowly improving to create much more stable products. Learn about plug-in hybrid electric vehicle (PHEV) and lithium battery technology. See Fig. 8.20.

Lithium-ion batteries have also been in the news lately. That is because these batteries have the ability to burst into flames occasionally. It is not very common—just two or three battery packs per million have a problem—but when it happens, it's extreme. In some situations, the failure rate can rise, and when that happens, you end up with a worldwide battery recall that can cost manufacturers millions of dollars.

Conceptually, lithium is the lightest metal with the highest potential due to its very reactive behavior, which, in theory, makes it very fitting as a compound for batteries. Just as the lead-acid and most other batteries, the lithium-ion battery by definition uses chemical reactions to release electricity. Although all are called lithium-ion batteries, there's a variety of types with slightly different chemical

Fig. 8.20 A portable lithium-ion battery

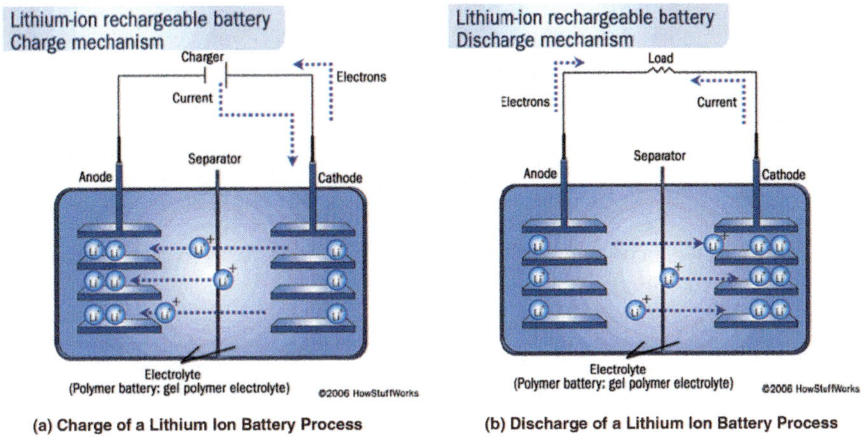

Fig. 8.21 Lithium-ion battery charge and discharge processes

compounds. The construction looks somehow similar to a capacitor, using three different layers curled up in order to minimize space. The first layer acts as the anode and is made of a lithium compound; the second one is the cathode and is usually made of graphite. Between anode and cathode is the third layer—the separator that, as suggested by the name, separates them while allowing lithium ions to pass through. The separator can be made of various compounds allowing different characteristics and, with that, different benefits and flaws. In addition, the three layers are submerged in an organic solvent, the electrolyte, allowing the ions to move between the anode and the cathode. See Fig. 8.21 for a layout of lithium-ion battery for rechargeable battery charge and rechargeable battery discharge mechanism.

In the charging process, the lithium ions pass through the microporous separator into spaces between the graphite (though not compounded), receiving an electron from the external power source.

During the discharge process, the lithium atoms located between the graphite release its electrons again that migrates over the external circuit to the anode providing a current. The lithium ions move back to the anode as well, parallel to their released electrons.

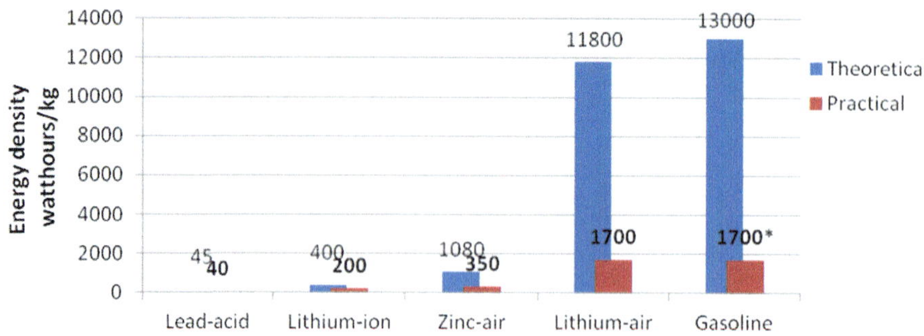

Fig. 8.22 Comparison of energy densities

Because lithium is a very reactive compound and can burst into flames, safety measures have to be included, such as onboard control chips to manage the temperature and prevent a complete discharge.

Pros and cons are listed here with a sign of plus (pros) and minus (cons) in front of them as:

Pros

+ Highest energy density in commercial available batteries with huge potential.
+ Provides higher voltage per cell (3.7 V compared to 2.0 V for lead-acid).
+ Low energy loss; only about 5% per month.
+ Lithium and graphite as resources are available in large amount.

Cons

– Very expensive.
– Complete discharge destroys the cells.
– Deteriorates even if unused (life cycle of about 5 years).
– Lithium is flammable in contact with atmospheric moisture.

Future prospects of superconducting magnetic energy storage are stated below here as:

Lithium-ion batteries would be suitable for storing large amounts of energy if it were not for the costs. The rather expensive processing and the safety measures make them too expensive for commercial use besides small electronic devices like smartphones and laptops. Even for small decentralized systems, competitors like lead-acid batteries are more cost-effective right now, although that will change as they become cheaper.

However, lithium-based batteries have an incredibly huge potential. IBM is currently working on a project called Battery 500. This project's goal is to develop a battery using lithium and the air of the atmosphere as components (both the two lightest elements suitable for this purpose), capable of storing enough energy to power an electric car for 500 miles (~804 km) [15].

Commercial use is targeted somewhere between 2020 and 2030 as there are still a lot of obstacles to overcome. Figure 8.22 is a chart of comparison of energy densities among different batteries that commercially are in existence for combustion engine and companies like IBM are involved with this matter.

Besides the ambitious IBM project, several other companies worldwide are working and experimenting on new suitable compounds for lithium-based batteries, and it is very possible that this technology will reach a point where it becomes cost-effective for storing grid energy.

Despite the ongoing rumor, lithium is not short in supply, and even though in small amounts (0.1–0.2 parts per million (PPM)), it is available in saltwater and can be extracted through technical methods, making the supply almost infinite.

8.6.4 Other and Future Batteries in Development

While smartphones, smart homes, and even smart wearables are growing ever more advanced, they are still limited by power. The battery has not advanced in decades. But we are on the verge of a power revolution.

Big technology and car companies tare all too aware of the limitations of lithium-ion batteries. While chips and operating systems are becoming more efficient to save power, we are still only looking at a day or two of use on a smartphone before having to recharge. Thankfully, universities are getting involved.

While it may be some time before we get a week's life out of our phones, development is progressing well. We've collected all the best battery discoveries that could be with us soon, from over the air charging to super-fast 30-second recharging. Hopefully you'll be seeing this tech in your gadgets soon.

We rarely cover claims of "battery breakthroughs" because it's extremely difficult to distinguish what has real potential and what is simply public relation (PR) fluff. We hint at new battery breakthrough on the horizon of this technology, and we are walking into new dawn of smart batteries for smart applications in our day-to-day life as necessity.

No one has been harder on those claims than Elon Musk, but now Tesla's CEO himself tentatively hints a promising upcoming battery breakthrough on the horizon. As we know, Tesla is a company pioneering the plug-in hybrid electric vehicle (PHEV) and lithium batteries technology.

As the world biggest consumers of lithium-ion batteries, Tesla is in a unique position to have access to the latest technology that battery scientists are trying to bring to production.

Tesla previously disclosed that they often receive battery samples and track hundreds of battery research programs to which they assign points based on potential. Previously, CTO JB Straubel, who leads battery development at Tesla, said that very few programs have substantial potential.

One of the most recent battery breakthrough claims that received a lot of attention was a new solid-state battery technology developed by John Goodenough, who is credited as the coinventor of the lithium-ion (Li-ion) battery cell. See Fig. 8.23, which is an illustration of lithium-ion battery versus solid-state battery layout.

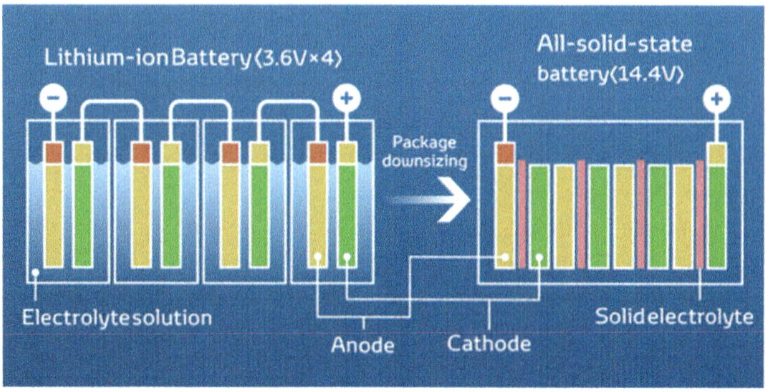

Fig. 8.23 Lithium-ion battery vs. solid-state battery (Courtesy of Toyota Corporation)

While battery breakthroughs are alluring, they are not actually needed to achieve the battery cost and capacity for all cars to go electric. The incremental improvements to Li-ion batteries that we have seen over the years are believed to be enough to soon be competitive with internal combustion engines before the cost of operation or incentives.

Nonetheless, breakthroughs or other battery cell technologies could be necessary to electrify other modes of transport, like air transport.

Lithium-ion batteries represent a landmark technology that has made the current generation of electric vehicles possible. However, the day of their demise, while it still lies years in the future, is within view. Lithium-ion chemistries have a certain maximum energy density, dictated by those pesky laws of physics, and today's batteries are not so far from that theoretical maximum. If drivers keep demanding longer ranges and faster charging times, then a better technology will have to be found.

With progress in development of new generation of Li-ion, safety issues are also growing and have to be taken seriously under consideration. The spectacular explosions and fireballs that some documentary makers revel in are not the norm (when was the last time your phone or computer caught fire?), but Li-ion batteries do have to be handled carefully, and necessary safety features add complexity and cost to battery packs. A new chemistry that is safer could also prove to be cheaper.

Researchers around the world are working on "beyond lithium" projects, and the past year has seen several significant breakthroughs. Of course, advances in the lab take years to make their way to the marketplace, but if and when one of these promising technologies can be commercialized, we could see game-changing improvements in the performance and cost of electric vehicles (EVs).

However, there has been some breakthrough in other type of batteries such is the one in magnesium batteries that was announced by the University of Houston researchers in August 24, 2017, where they have claimed that magnesium batteries are safe and unlike their counterpart lithium-ion batteries they are not flammable or subject to exploding—but until now their ability to store energy has been limited. Researchers have reported a new design for the battery cathode, drastically increasing the storage capacity and upending conventional wisdom that the magnesium-chloride bond must be broken before inserting magnesium into the host.

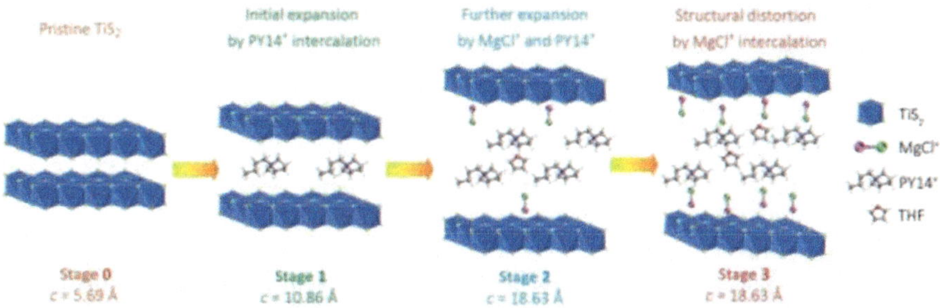

Fig. 8.24 Artistic schematic structure of the magnesium batteries (Courtesy of University of Houston)

Figure 8.24 is the artistic schematic of the structural evolution of titanium disulfide at different stages of intercalation. Interlayers are expanded or distorted as different amounts of pillaring molecules, complex cautions, and solvents are intercalated into the van der Waals gap of a host material at each stage.

Magnesium batteries offer a promise for safely powering modern life—unlike traditional lithium-ion batteries, they are not flammable or subject to exploding—but their ability to store energy has been limited.

The work on this type of battery was first conceived by Yao and postdoctoral fellow Hyun Deog Yoo in 2014; the project spanned several years and involved scientists from three universities and three national laboratories, working both experimentally and theoretically.

"Magnesium ion is known to be hard to insert into a host," said Yoo, first author on the paper. "First of all, it is very difficult to break magnesium-chloride bonds. More than that, magnesium ions produced in that way move extremely slowly in the host. That altogether lowers the battery's efficiency."

The new battery stores energy by inserting magnesium monochloride into a host, such as titanium disulfide. By retaining the magnesium-chloride bond, Yao said, the cathode demonstrated much faster diffusion than traditional magnesium versions.

The researchers report the new battery has storage capacity of 400 mAh/g, compared with 100 mAh/g, for earlier magnesium batteries. Commercial lithium-ion batteries have a cathode capacity of about 200 mAh/g, said Yao, who is also a principal investigator with the Texas Center for Superconductivity at UH.

Voltage of the new battery remains low at about 1 volt. That compares to 3–4 v for lithium batteries.

The high voltage, coupled with their high energy density, has made lithium-ion batteries the standard. But lithium is expensive and can develop breaches in its internal structure, a condition known as dendrite growths, which can cause the batteries to catch fire. As an earth-abundant resource, magnesium is cheaper and does not form dendrites. Until now, however, it has been held back by the need for a better cathode—the electrode from which the current flows—and more efficient electrolytes, the medium through which the ionic charge flows between cathode and anode.

The key, Yoo said, is to expand the titanium disulfide to allow magnesium chloride to be inserted—a four-step process called intercalation—rather than breaking the magnesium-chloride bonds and inserting the magnesium alone. Retaining the magnesium-chloride bond doubled the charge the cathode could store.

The magnesium monochloride molecules are too large to be inserted into the titanium disulfide using conventional methods. Building upon their earlier work, the researchers created an open nanostructure by expanding the gaps in the titanium disulfide by 300%, using organic "pillars."

The opening still was small, increased from 0.57 nanometers to 1.8 nanometers, Yao said, that allowed for the magnesium chloride to be inserted.

"Combined theoretical modeling, spectroscopic analysis, and electrochemical study reveal fast diffusion kinetics of magnesium monochloride cations without scission of magnesium chloride bond," the researchers wrote. The large capacity accompanies excellent rate and cycling performances even at room temperature, opening up possibilities for a variety of effective intercalation hosts for multivalent-ion batteries.

One technology that has been getting a tremendous amount of attention from researchers is the solid-state battery, which uses a solid electrolyte instead of the liquid electrolyte used today. Solid-state batteries could theoretically have doubled the energy density of current batteries and last several times longer. They also use a nonflammable electrolyte—usually glass, polymer, or a combination—so they would eliminate the safety issues that plague Li-ion cells.

Lithium-air batteries likewise could offer far greater energy density—maybe as much as ten times more—but they suffer from poor life cycle. In 2015, Cambridge scientists wowed the battery world with an announcement that they had demonstrated a highly efficient and long-lasting lithium-oxygen battery. Alas, researchers from several universities and national labs have since been unable to duplicate the original results. Figure 8.25 is an artistic layout of a lithium-air battery and the process of how it works.

Other promising battery chemistries use other elements in place of lithium. Sodium batteries powered Jules Verne's futuristic submarine in *20,000 Leagues Under the Sea*. More recently, in 2015, researchers created a prototype sodium-ion battery in the industry-standard 18,650 cylindrical format.

According to a recent article in the *Nikkei Asian Review*, battery research has seen a big shift in recent years. At one time, nearly half of the presentations at the Battery Symposium in Japan were about fuel cells and Li-ion battery cathode materials. But since 2012, these topics have been supplanted by presentations about solid-state, lithium-air, and non-lithium batteries.

Toyota has been focusing on solid-state and Li-air batteries. At the latest Battery Symposium, battery researcher Shinji Nakanishi discussed a scenario for transitioning from Li-ion batteries to solid-state and then Li-air batteries. "We want our electric cars to go 500 km" on a single charge, he said. "And for this, we want rechargeable batteries that can generate 800 to 1,000 watt-hours per liter." That would be two to three times the energy density of today's best Li-ion batteries.

Panasonic, Tesla's battery supplier, is also taking a hard look at solid-state technology. "We think the existing technology can still extend the energy density of Li-ion batteries by 20% to 30%," President Kazuhiro Tsuga told Nikkei. "But there is a trade-off between energy density and safety. So if you look for even more

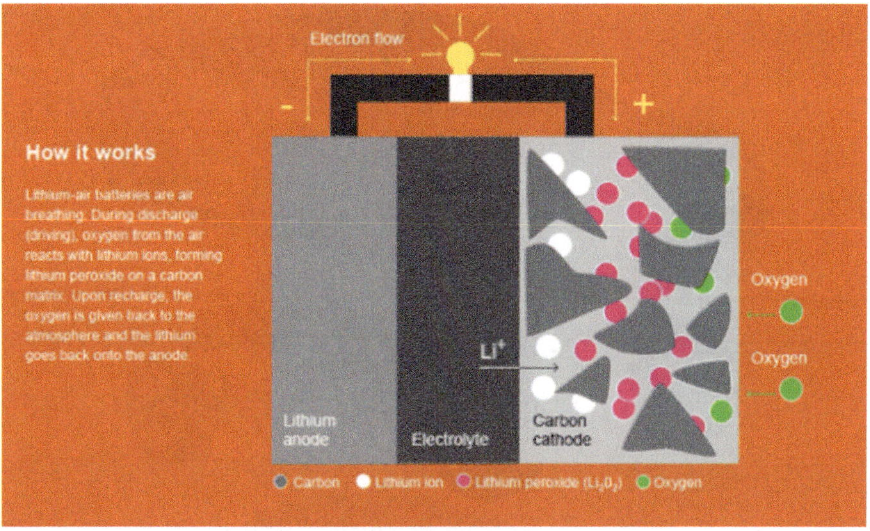

Fig. 8.25 How lithium-air battery works (Courtesy of Money Incorporation)

density, you have to think about additional safety technology as well. Solid-state batteries are one answer."

Engineers have been pushing the limits of Li-ion technology for decades. Today's best Li-ion cells can reach an energy density of about 300 W/kg, which is getting close to the theoretical maximum. "Existing Li-ion batteries still have room to improve their energy density because you can raise the density by introducing a nickel-based cathode material, so you can expect the batteries will still be used in the next few years," said battery expert Naoaki Yabuuchi of Tokyo Denki University. He expects lithium-ion technology to reach its limits around 2020.

"Tesla has one of the largest cell characterization laboratories in the world – we have just about every cell you can imagine on test," Tesla Product Planner Ted Merendino has stated, back in 2013. However, both Elon Musk and JB Straubel have said that so far, they have seen no viable replacement for lithium-ion and, believe me, they have been asked the question many times.

When Model 3 was announced, some EV watchers opined that in order to deliver the new vehicle at the desired price point, Tesla would need to make a major battery breakthrough. In the event, Tesla has developed a new battery for Model 3, but it looks more like an incremental improvement than a paradigm shift. The new 2170 cell, which is now being produced at the Gigafactory, is slightly larger than the trusty 18,650 and can store more energy. According to Elon Musk, it's "the highest energy density cell in the world and also the cheapest." Advances in the way the cells are assembled into modules and packs are also expected to yield a significant reduction in battery costs.

Figure 8.26 is Tesla's battery pack in the floor pan of the Model S.

Therefore, it appears that lithium will continue its reign for a few more years at least. However, the post-lithium holy grail is still out there, and as likely as not, the

Fig. 8.26 Tesla's battery pack in the floor pan of the model S (Courtesy of Tesla Corporation)

knights of Tesla's round table will be the ones to bring it home. Battery superstar Jeff Dahn and his colleagues aren't working for Tesla just to make speeches at conferences. It is entirely possible that, at some super-secret facility in California or Nevada, test mules are being powered by solid-state or lithium-air batteries even as we speak.

However we still need to see how this technology is progressing in its development and these types of batteries, so for us consumer to be able to see even a single example of a cell battery working at the laboratory level which is better than the one we have or the ones that we expect to come out with, said Elon Mask in 2014.

There are batteries under development and worth to mention here:

- *Redox Flow Battery*
  These batteries technically are similar to conventional batteries, except that the electrolytes (there are different forms, using one or two different fluids) can be exchanged, meaning that if the battery is discharged, the fluids are replaced with loaded ones. This concept could, in theory, become very handy for electric cars as you could charge your car simply by refueling just as you do now. However, the energy density is about 35 watt-hours per kilogram (Wh/kg) in the same region as lead-acid batteries right now and therefore considerately low, although the Fraunhofer Institute in Germany claims to have managed to increase density up to the level of lithium-ion batteries (200 Wh/kg). Other advantages are the long life span of roughly 40 years and the fact that capacity can be increased by simply increasing the tanks and adding more electrolytes.

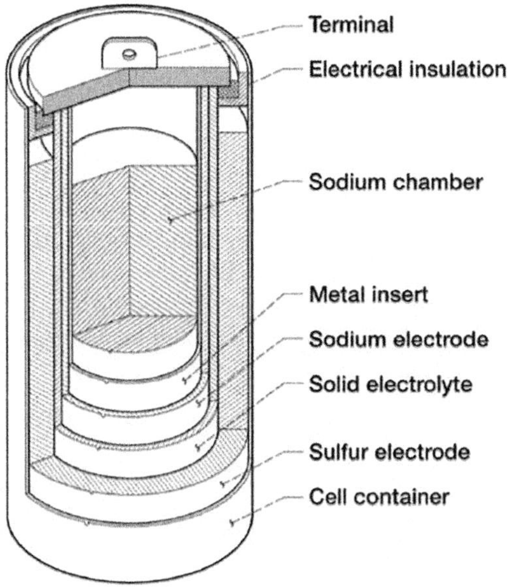

Fig. 8.27 Sodium battery layout

For the purpose of grid storage, there are commercial available plants; but the value is limited similar to flywheels, SMES, or other battery storage types due to the yet low energy density. Pilot projects are in operation, most recently in California for an agricultural processing facility with a capacity of 3.6 MWh [16].

- *Sodium Battery*
  The liquid sodium sulfur battery (See Fig. 8.27) is yet another type of battery in development, but already operational in some countries like Japan. About 250 megawatts (MW) of sodium battery power have been installed there [17]. Sodium batteries have the advantage of a relatively high density with up to 240 Wh/kg, a long life span of 10–15 years, and high efficiency (75–90%); but they need to be operated at high temperatures (350 °C/623 K) to get the sodium liquid, which not only makes it more difficult and expensive to operate but also more dangerous as the liquid sodium reacts easily with the water in the atmosphere. Since the Nippon Tokushu Togyo Kabushiki Kaisha Co. Ltd. (NGK) and the Tokyo Electric Power Togyo Kabushiki Kaisha Co. Ltd. (NGK) and the Tokyo Electric Power Co. Ltd. (TEPCO) began shipping out sodium batteries in 2002, three incidents resulting in fires have occurred, setting the development back.
- *Zinc-Air Battery*
  Just like the lithium-air battery, the zinc-air battery uses air as a second component. Zinc-air has been a focus in development for a while because of its safety aspects and potential in density but was dropped due to the low efficiency and short life cycles. Two independent companies claim to have solved these problems and plan to begin small-scale field tests this year, but have yet to present reliable data [18, 19].

Zinc-air, just like lithium-air, holds potential for grid storage due to its density and the fact that zinc is a commonly found metal, but as long as no data is presented and no field tests have been made, it remains as an idea with theoretical potential.

- *Flexible Battery*

  Flexible batteries are powered by IV cellular fluids. IV and cellular fluids power flexible batteries are recent attention of researchers and scientists in the field of new batteries generation development.

  Researchers have engineered bendable batteries that can run on body-inspired liquids such as normal IV saline solution and cell culture medium. The authors designed alternatives to lithium-ion batteries by focusing on the mechanical stress demands of wearable electronics such as smart-watches and the safety requirements of implantable electronics.

Figure 8.28 is an artistic rendering of fiber-shaped implantable batteries using biocompatible electrolytes.

Researchers in China have engineered bendable batteries that can run on body-inspired liquids such as normal IV saline solution and cell culture medium. In their work, published August 10 in the journal *Chem*, the authors designed alternatives to lithium-ion batteries by focusing on the mechanical stress demands of wearable electronics such as smart-watches and the safety requirements of implantable electronics.

"Current batteries like the lithium-ion ones used in medical implants generally come in rigid shapes," says co-senior author Yonggang Wang, a chemistry professor at Fudan University and the Collaborative Innovation Center of Chemistry for Energy Materials. "Additionally, most of the reported flexible batteries are based on flammable organic or corrosive electrolytes, which suffer from safety hazards and poor biocompatibility for wearable devices, let alone implantable ones."

Fig. 8.28 An artistic rendering of fiber-shaped implantable batteries [20]

Safety measures for wearable and implantable batteries have generally involved structural reinforcement to prevent hazardous chemicals from leaking out. Instead, the researchers, led by Wang and macromolecular science professor Huisheng Peng, swapped out those toxic and flammable liquids for cheap and environmentally conscious sodium-ion solutions. Among those solutions were two biocompatible ones suitable for implantable devices, given that they pose no harm to the surface or interior of the body. Although electrolyte leakage is still undesired, its danger is minimized by the use of either the normal saline solution pumped into the body in most IV treatments or a cell culture medium that contains amino acids, sugars, and vitamins in addition to sodium ions and thus mimics the fluid that surrounds human cells.

Freed from leakage concerns, which can require so much protective material that batteries become bulky and unbendable, the researchers designed two types of flexible batteries—a 2D "belt"-shaped battery for which they adhered thin electrode films to a net of steel strands and a 1D fiber-shaped battery for which they embedded nanoparticles of electrode material around a carbon nanotube backbone. Besides testing biocompatible fluids, the authors also tested ordinary sodium sulfate, a safe and fairly inert solution, as a liquid electrolyte suitable for use in external wearable devices.

With sodium sulfate solution as the electrolyte, both battery types outperformed most of the reported wearable lithium-ion batteries in terms of charge-holding capacity (an indicator of how long a battery can function without recharging) and power output for their size. That performance held up when the authors folded and bent the batteries to mimic the impact of wrapping a sensor, watch, or similar device around one's arm. Charge-holding capacity was only marginally reduced for the saline- and cell culture-based batteries, most likely because they had slightly lower sodium-ion content than the sodium sulfate solution.

An undesired side reaction involving their fiber-shaped battery is even pointing the researchers toward possible biomedical applications. The same carbon nanotubes that make up the skeleton of the 1D battery can also accelerate the conversion of dissolved oxygen into hydroxide ions, a process that harms battery effectiveness if left uncontrolled but as a stand-alone process boasts therapeutic potential for treating cancer and bacterial infections.

"We can implant these fiber-shaped electrodes into the human body to consume essential oxygen, especially for areas that are difficult for injectable drugs to reach," says Wang. "Deoxygenation might even wipe out cancerous cells or pathogenic bacteria since they are very sensitive to changes in living environment pH. Of course, this is hypothetical right now, but we hope to investigate further with biologists and medical scientists."

8.7 A Battery-Inspired Strategy for Carbon Fixation

Scientists working toward the elusive lithium-air battery discovered an unexpected approach to capturing and storing carbon dioxide away from the atmosphere. Using a design intended for a lithium-CO_2 battery, researchers have developed a way to isolate solid carbon dust from gaseous carbon dioxide, with the potential to also separate out oxygen gas through the same method.

Scientists working toward the elusive lithium-air battery discovered an unexpected approach to capturing and storing carbon dioxide away from the atmosphere. Using a design intended for a lithium-CO_2 battery, researchers in Japan and China have developed a way to isolate solid carbon dust from gaseous carbon dioxide, with the potential to also separate out oxygen gas through the same method. Their work appears August 9 in *Joule*, a new interdisciplinary energy journal from *Cell Press*.

Converting carbon dioxide emissions into other carbon-containing compounds is desirable due to carbon dioxide's contribution to the greenhouse effect and global warming. Examples range from natural processes, such as plants turning CO_2 into oxygen and sugars, to human-made ones, such as injecting carbon dioxide into rock formations to be trapped as carbonate minerals.

"The problem with most physical and chemical pathways for CO_2 fixation is that their products are gases and liquids that need to be further liquefied or compressed, and that inevitably leads to additional energy consumption and even more CO_2 emissions," says senior author Haoshen Zhou of Japan's National Institute of Advanced Industrial Science and Technology and China's Nanjing University. "Instead, we are demonstrating an electrochemical strategy for CO_2 fixation that yields solid carbon products, as well as a lithium-CO_2 battery that can provide the energy necessary for that process."

The researchers encountered the carbon fixation strategy when they tried to recharge a lithium-CO_2 battery prototype. Instead of fully regenerating lithium ions and CO_2 from the lithium carbonate and carbon produced during battery discharge, as would have taken place with a reversible Li-CO_2 battery, the lithium carbonate decomposed, yielding additional carbon, as well as oxygen gas that was not isolated due to rapid reaction with the battery electrolyte. Typically, this kind of buildup causes physical degradation and reduced functional life span for a battery, but instead, the deposition of solid carbon boasts a separate advantage, pointing to a promising approach to fix carbon in a stable and easy-to-dispose-of form.

Figure 8.29 is a depiction of flowchart of energy storage and carbon fixation using Li-CO_2 technology.

"What is impressive about this work is the possibility to convert one-third of the CO_2 species to carbon with high theoretical energy efficiency above 70%," says *Joule* scientific editor Rahul Malik. "Battery architecture is an unforeseen but intriguing way to look at carbon fixation."

Since generating carbon solids both realizes carbon fixation decreases battery performance, the researchers were not able to simultaneously satisfy both goals within a single device. However, by incorporating a tiny amount of ruthenium metal into their design as a catalyst, they were able to avoid extensive carbon deposition and induce better reversibility, converting their carbon-fixing apparatus into a functioning Li-CO_2 battery.

A remaining challenge for both carbon fixation and battery performance is to move from pure CO_2 to ambient air, a jump that would potentially allow for treating atmospheric CO_2 in the first case and would advance toward the theoretically powerful but not-yet-stable lithium-air battery technology in the second case.

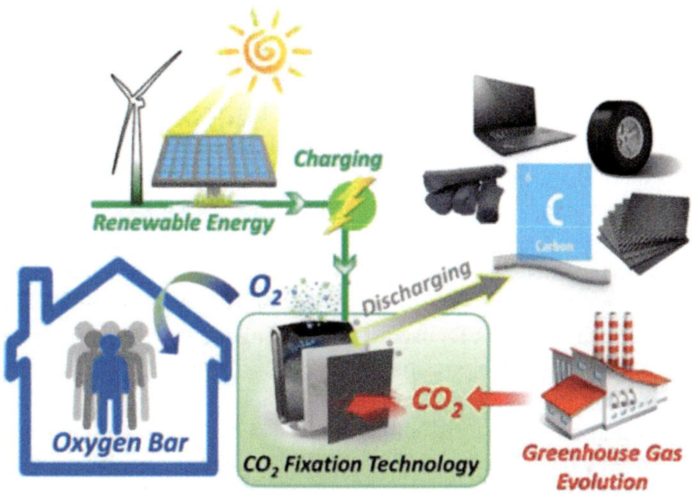

Fig. 8.29 Industrial utilization of Li-CO_2 [20]

The fixation technique might also be adapted to scrub other harmful or polluting gases such as carbon monoxide, sulfur dioxide, nitric oxide, and nitrogen dioxide from the atmosphere, Zhou says.

Looking ahead, the researchers are also excited by their system's potential to perhaps lead to a pathway for converting carbon dioxide into pure carbon and oxygen gas. "Attaining the release of oxygen gas upon charging, coupled with the accumulation of solid carbon, would realize an electrochemical carbon dioxide fixation strategy analogous to photosynthesis," says Zhou.

8.8 Saliva-Powered Battery

New battery is activated by your spit; saliva-powered battery could be helpful in extreme conditions.

Researchers have developed the next step in microbial fuel cells (MFCs)—a battery activated by spit that can be used in extreme conditions where normal batteries do not function.

For the last 5 years, Binghamton University Electrical and Computer Science Assistant Professor Seokheun Choi has focused on developing micro-power sources for the use in resource-limited regions to power point-of-care (POC) diagnostic biosensors; he has created several paper-based bacteria-powered batteries.

"On-demand micro-power generation is required especially for point-of-care diagnostic applications in developing countries," said Choi. "Typically, those applications require only several tens of microwatt-level power for several minutes, but commercial batteries or other energy harvesting technologies are too expensive and over-qualified. Also, they pose environmental pollution issues."

Choi, along with research assistant Maedeh Mohammadifar, created a high-performance, paper-based, bacteria-powered battery by building microbial fuel cells with inactive, freeze-dried exoelectrogenic cells which generates power within minutes of adding saliva. The proposed battery generated reliable power from with one drop of saliva, supplying onboard power that could be used by the next generation of disposable, paper-based POC diagnostic platforms.

"The proposed battery has competitive advantages over other conventional power solutions because the biological fluid for on-demand battery activation is readily available even in the most resource-constrained settings, and the freeze-drying technology enables long-term storage of cells without degradation or denaturation," wrote the researchers. Choi is focused on improving the battery's power density so that more applications can be powered.

"Now, our power density is about a few microwatts per centimeter square. Although 16 microbial fuel cells connected in a series on a single sheet of paper generated desired values of electrical current and voltage to power a light-emitting diode (LED), further power improvement is required for other electronic applications demanding hundreds of milliwatts of energy," said Choi.

8.9 Summary

Energy storage, a potential solution for integrating intermittent renewables and improving grid stability, again saw rapid growth this past year. A "transition year" for US energy storage, 2016, saw a more diverse market emerged "both in terms of the types of systems (market segments) deployed and the business models." These trends are expected to continue in the United States with combined residential, commercial, and industrial energy storage deployments predicted to surpass 2 GW by 2021.

Despite this rapid growth, there are those who claim that the adoption of energy storage has been slowed by a "web" of regulations at all levels. Some argue that regulations, which can vary from state to state, are in need of modernization in order to facilitate easier integration of energy storage into our nation's infrastructure. While not necessarily advocating national uniformity for these regulations, energy storage developers hope to avoid a framework as disjointed as that of solar, where they must "navigate 50 different markets with 50 different sets of regulations."

Recently, the Federal Energy Regulatory Commission (FERC) released a policy statement addressing cost recovery for energy storage, stating these facilities are permitted to earn both cost- and market-based revenue streams in light of the multiple services they provide. This decision serves as an example of the type of modernization that energy storage proponents hope for and serves to ease barriers to grid integration. While there is not necessarily opposition to expanding the nation's energy storage capacity, there will likely be challenges and objections to specific rules in the future. FERC's recent statement, for example, was met by some with concerns over equity, possible over-recovery at the expense of the rate payer, and negative impacts on competition within the industry.

References

1. http://www.engineeringnews.co.za/article/electricity-consumption-to-increase-to-over-30-116-b-kwh-globally-in-2030-2009-04-17
2. A.A. Sallam, O.P. Malik, *Electric Distribution Systems* (IEEE Computer Society Press, Hoboken 2011), p. 21. isbn:9780470276822
3. *Energy profile of Alaska, United States*, ed. by C.J. Cleveland. Last Updated: 30 July 2008—Encyclopedia of Earth
4. J.D. Glover, M.S. Sarma, T.J. Overbye, *Power System and Analysis*, 5th edn. (Cengage Learning, 2010), p. 10
5. *Smart Grid—The New and Improved Power Grid: A Survey*; IEEE Communications Surveys and Tutorials 2011, ed. by X. Fang, S. Misra, G. Xue, and D. Yang. doi:https://doi.org/10.1109/SURV.2011.101911.00087
6. A. Von Meier, Electrical engineer 137A: electric power systems. *Lecture 2: Introduction to Electric Power Systems*, Slide 33 (2013)
7. Bath County Pumped Storage Station, Virginia, USA is currently the largest PSH worldwide with a storage capacity of 30GWh and a 3GW power output
8. http://www.oncor.com/community/knowledgecollege/energy_library/generating/generating01.aspx
9. http://investors.beaconpower.com/releasedetail.cfm?ReleaseID=593208
10. K. Kamiyama, ISTEC J. **7**, 39 (1994)
11. X. Huang, S.F. Kral, G.A. Lehman, Y.M. Lvovsky, M. Xu, IEEE Trans. Appl. Supercond. **5**, 428 (1995)
12. K.-P. Jungst, *Supraleitung in der Energietechnik II* (VDI-Verlag, 1995), p. 133
13. K. Prescher, W. Nick, H.-E. Vollmar, M. Kleinmaier, U. Radtke, *Supraleitung in der Energietechnik II* (VDI-Verlag, 1995), p. 115
14. H.W. Lorenzen, V. Brammer, J. Meinke, F. Rosenbauer, J. Schaller, R.M. Schottler, *Supraleitung in der Energietechnik II* (VDI-Verlag, 1995), p. 149
15. http://www.ibm.com/smarterplanet/us/en/smart_grid/article/battery500.html
16. http://www.sustainableplant.com/2012/05/vanadium-flow-battery-to-provide-grid-level-storage-for-gills-onions/
17. http://www.greentechmedia.com/articles/read/is-sodium-the-future-formula-for-energy-storage/
18. http://phys.org/news/2012-01-eos-zinc-battery-recipe-energy.html
19. http://www.treehugger.com/clean-technology/very-promising-zinc-air-battery-could-hold-300-more-energy-than-lithium-ion.html
20. Z. Guo et al., Multi-functional flexible aqueous sodium-ion batteries with high safety. Chem (2017). https://doi.org/10.1016/j.chempr.2017.05.004
21. http://www.world-nuclear.org/info/inf10.html

Appendix A: Global Energy Interconnection

Global energy interconnection (GEI) represents the ultimate evolution of the trend toward greater interconnection of power systems. It embodies high-level integration of the flow of energy, flow of information, and flow of business as an intelligent, automated, and networked-based system for ensuring energy security on a universal scale.

Introduction

Fueled by global economic growth, world energy consumption rose from 5.4 billion tons of coal equivalent in 1965 to 18.5 billion tons in 2014. Fossil energy accounted for more than 85% of the total. The world's energy consumption will maintain a growing trend in the future, as it is difficult to reverse the long-established patterns of intensive energy consumption [1].

Seeking a solution to these trends, the implementation of global energy interconnection (GEI) would integrate a large-scale deployment of clean energy led by variable renewables with a smart grid incorporating high levels of interoperability and supported by an ultrahigh-voltage (UHV) grid backbone including extensive interconnections across countries, continents, technical domains, hierarchies, and equipment life cycle phases.

Though such levels of deployment are highly ambitious, the technologies themselves are largely available or are currently in the pipeline.

The technical difficulties for large-scale, transcontinental, or global energy interconnection, on the other hand, will come from the unprecedented degree of system integration that will be required. To help surmount this challenge, consensus-based international standards and specifications will form an indispensable basis on which to build concrete solutions. Standards, specifically those at the systems level, will facilitate procurement and national and international acceptance and will play a stabilizing role by pursuing research activities on which real market opportunities are built.

B. Zohuri, *Hybrid Energy Systems*,
https://doi.org/10.1007/978-3-319-70721-1

The readiness of potential markets for GEI identifies the technical and economic trends in related technologies and evaluates at a high level the impact on energy, environment, technologies, and policies. Taking the large-scale concepts connected with GEI to actual realization will require significant efforts in standardization, e.g., development of initiatives to enable multi-system interoperability. Thus this white paper aims to highlight the concept of GEI and begin laying the foundations for identifying and addressing the standardization needs for large-scale, transcontinental, and global energy interconnection.

Achieving a sustainable, secure, and affordable supply of energy has traditionally been the goal of both national and international energy policies. At the center of achieving sustainability in the energy system lies the challenge of climate change, a factor recently brought to the fore by the Paris accords. The rise of electricity as the key energy carrier due to its high quality and versatility has determined its current role as a central vehicle for decarbonizing the supply of energy. Dramatic cost reductions in renewable energy (RE), particularly wind and solar, have made extensive deployment of such energies attractive on a global scale, with emphasis being placed on how to integrate these resources widely. Reducing reliance on fossil fuels by substituting low-carbon electricity for the input of energy end uses that rely on them (e.g., through electric vehicles in transport or electrification of industrial processes) will only increase overall demand for electricity. At the same time, large portions of the global population remain without access to electricity. Following a logical progression in power systems, as generation and demand continue to evolve in response to these trends, power grids will become more and more interconnected at the transnational and regional levels. This is reflected in the recent introduction of the concept of global energy interconnection (GEI).

GEI would represent the ultimate stage in the evolution of power grids toward greater levels of interconnectivity: a global energy network of intercontinental and cross-border backbone networks of high and ultrahigh voltage (UHV), as well as smart power grids (transmission and distribution networks) in all interconnected countries at various voltage levels. A GEI could connect the power grids of all continents and take advantage of the diversity of different time zones and seasons, thus supporting a balanced coordination of power supply for all interconnected countries.

As one of the international organizations participating in the United Nations Sustainable Energy for All Initiative (SE4ALL), the International Electrotechnical Commission (IEC) and its international standards play a major role in meeting fundamental energy challenges. IEC's purpose in issuing this white paper is to highlight the concept of GEI and begin laying the foundations for identifying and addressing the standardization needs of large-scale, transcontinental, and global energy interconnection.

The main objectives are as follows:

- To provide a high-level assessment of the potential worldwide needs, benefits, and conditions of GEI
- To examine the readiness of potential markets for the technologies that would underpin GEI

- To identify technical and economic trends in related technologies
- To evaluate at a high level the impact on energy, environment, technologies, policies, and relevant standards
- To provide an outline of how standardization could be conducted from a high level and recommendations for different stakeholders to participate in the standardization work

Global Energy Challenges

Global primary energy supply has grown tenfold in the last 100 years and more than doubled in the last 40 years. But for the first time in 2006, developing countries (i.e., those not members of the Organization for Economic Co-operation and Development (OECD)) accounted for a larger share of energy use than developed economies; in 2013 the ratio was 61:39. The shifting of the traditional centers of energy demand to China, India, and South Asia is reflected in global trends: China has accounted for the largest increase in primary energy demand and CO_2 emission over the last decade, and yet the added pace of RE deployment and drastic improvements in energy intensity has reduced its annual growth in CO_2 emissions to levels not seen before 2004, with recent signs of decoupling. India alone has been responsible for almost 10% of the increase in global energy demand since 2000, while Indonesia has seen the largest growth in coal use globally. A combination of emission caps, a reduction in economic activity, and rapid growth in renewables has drastically altered the energy landscape in Europe. Developments in unconventional oil and gas technology and exploitation are dramatically changing energy prospects in the United States and its status as an energy importer. Moreover clean energy added more capacity for the first time in 2014 than all other power generation sources combined.

In spite of these developments, anthropogenic energy-related global CO_2 emissions reached a record 31.6 gigatons CO_2 ($GtCO_2$) in 2012, warming reached a 1 °C increase above preindustrial levels, and 1.3 billion people remain without access to clean energy around the world.

The challenges facing the world in terms of providing secure, affordable, and clean energy are greater than they have ever been, framed against an array of pressures at an unprecedented scale and a landscape of rapid technological change.

Energy Security

Fueled by global economic growth, world energy consumption rose from 5.4 billion tons of standard coal in 1965 to 18.5 billion tons in 2014. Fossil energy accounted for more than 85% of the total. The world's energy consumption will maintain a growing trend in the future, as it is difficult to reverse the long-established pattern of intensive energy consumption.

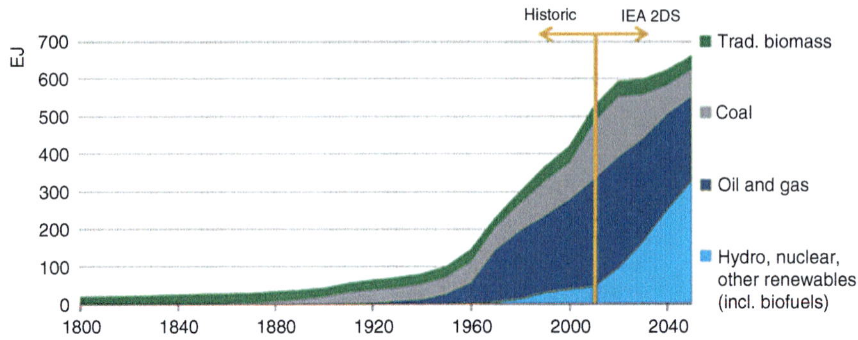

Fig. A.1 Primary energy supply by source, Historic 1800–2013, and IEA 2-Degree Scenario 2013–2050 (IEA) (Courtesy of International Energy Agency)

Energy security can be understood as "the uninterrupted availability of energy sources at an affordable price." Energy security has many dimensions: long-term energy security mainly involves timely investments to supply energy in line with economic developments and sustainable environmental needs. Short-term energy security focuses on the ability of the energy system to react promptly to sudden changes within the supply demand.

Concern related to physical unavailability of supply is more prevalent in energy markets in which networks and systems must be kept in constant balance, such as electricity and, to some extent, natural gas. This is particularly the case in instances where capacity constraints exist or where prices cannot function as an adjustment mechanism to balance supply and demand in the short term.

The world has plentiful energy resources, including those from fossil energies. In the long term, however, fossil-based energies are exhaustible and are heavily location-constrained. Geopolitics, the changeable economic outlook, the prevailing investment climate, and a rapidly shifting technology landscape mean that the circumstances surrounding exploitation of fossil fuels are ever changing. At the same time, the world's renewable energy (RE) resources are vast (see Fig. A.1) and, if the full technical potential of such resources is captured, would meet the world's energy needs many times over.

Climate Change

The Intergovernmental Panel on Climate Change (IPCC) has concluded that, in the absence of fully committed and urgent action, climate change will have severe and irreversible impacts across the world. Avoiding dangerous climate change and its environmental consequences will require sustained and important reductions in greenhouse gas (GHG) emissions.

Energy production and use account for two-thirds of global GHG emissions generated by human activity. Therefore, transforming the energy sector is essential for addressing the climate challenge. Despite efforts to decarbonize the world's energy

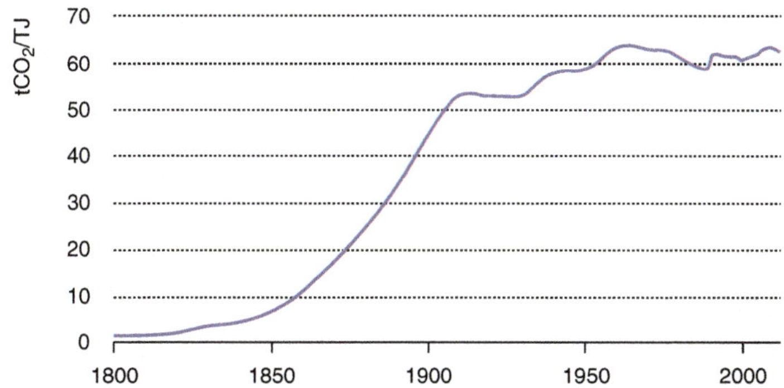

Fig. A.2 Carbon intensity of global energy supply, 1800–2013 (IEA) (Courtesy of International Energy Agency)

system, the past 30 years have seen little change in the share of fossil fuels in global energy supply, which totaled around 81% in 2013. Meanwhile, coal power accounted for the highest contribution to the energy mix in the past 40 years.

Since the Industrial Revolution, annual CO_2 emissions from fuel combustion have dramatically increased from nearly 0 to over 32 $GtCO_2$ in 2013, and yet since the early 1900s, the carbon intensity of the energy supply (the amount of emissions per unit of energy going into the system) has barely changed (see Fig. A.2). Between 1990 and 2013, a stable carbon intensity of supply combined with an increase in population (35%) and in per capita gross domestic product (GDP) (60%), leading to a dramatic increase in global CO_2 emissions of nearly 60%. The atmospheric concentration of these gases has increased steadily to 445 parts per million carbon dioxide equivalent (ppm CO_2-eq) in 2014.

The global energy system is thus at a crossroads in the race against climate change. The recent UN Climate Summit delivered an unprecedented response to this challenge. Countries from around the world gathered in Paris for the 21st Conference of the Parties (COP21) to negotiate an international agreement and set a direction for combating climate change within the next decade and beyond—one that aims to reach global peaking of GHG emissions as soon as possible, with an ambition to limit the global average temperature rise to well below 2 °C and pursue efforts to limit the temperature increase to 1.5 °C. The international commitment to keep the increase in long-term average temperatures to within this target of temperature rise relative to preindustrial levels will require substantial and sustained reductions in global emissions.

GHG remain in the atmosphere for many years—what matters for climate change is the concentration of GHG that accumulates over a period of time. The world had emitted an estimated 1970 $GtCO_2$ by 2014. The IPCC estimated that the cumulative amount of CO_2 emitted between 1991 and 2100 would have to remain below approximately 3000 Gt to maintain a 50% chance of keeping global warming below 2 °C. Taking into account estimated nonenergy-related emissions of GHG, around 880–1180 Gt could be emitted by the energy sector between now and 2100—around 60% of what was emitted during the last century.

Environmental Pollution

An estimated 6.5 million annual deaths are linked to air pollution, a number that will only increase unless the energy sector takes greater action to curb emissions. Premature deaths due to outdoor air pollution are projected to rise from 3 million today to 4.5 million by 2040, concentrated mainly in developing Asia. The IEA estimates that under a clean air scenario, premature deaths from outdoor pollution would decline by 1.7 million in 2040—a scenario that would only require a 7% increase in total energy investment.

In conclusion of all these issues around global energy interconnection (GEI), we can state that GEI is an ambitious concept, integrating a large-scale deployment of clean energy led by variable renewables: a smart grid with various levels of interoperability and with information technology (IT) and operational technology (OT) integration capabilities currently not deployed on an adequate scale, the large-scale deployment of advanced technologies for transmitting power over vast distances and forming high-capacity and high-voltage networks, and electrification of a large number of energy services involving new equipment.

Many of the building blocks of large-scale energy networks and an eventual GEI are available today. Large-scale renewable energy (RE) is a reality, and achievements in ultrahigh-voltage (UHV) power transmission are redrawing the boundaries of how much power can be cost-effectively transmitted and across what distances. While the required technologies themselves are largely available or are in the pipeline, the challenge for large-scale, transcontinental, or global energy interconnection is one of the implementations, in which barriers and challenges will need to be overcome as the process unfolds.

More details of this appendix subject can be found in Ref. [1] white paper by the International Electrotechnical Commission.

Reference

1. International Electrotechnical Commission white paper on Global Energy Interconnection

Appendix B: Grid Integration of Large Capacity

We briefly touch up in this Appendix a part of grid integration of large capacity. We will also see in this appendix the renewable energy sources and use of large-capacity electrical energy storage. The white paper that is produced by the International Electrotechnical Commission provides more details and shows the market strategy board, responsible for analyzing and understanding of this matter [1].

Introduction

The proportion of renewable energies (RE) is called upon to increase in all major electricity markets. The reasons for this are not examined closely here, since they have been fully treated elsewhere. This paper explores what is needed to integrate large quantities of renewables into existing electricity grids, given various characteristics and difficulties which necessarily accompany such a change.

Reference [1] examines these characteristics, describes the difficulties, and analyzes the consequent challenges for grid operators as well as for producers of electricity, both renewable and conventional.

We can also express today's method and responses to the challenges in state of the art in integrating large-capacity renewable energy (RE), which includes the form of RE, such as wind, photovoltaic, and concentrated solar power generation.

Although system-wide level RE power plants generate electricity just like any other power plants, RE power has quite distinctive characteristics in generation, transmission, and operation technology when compared to conventional generation. Understanding these distinctive characteristics and their interaction with the other parts of the power system is the basis for the integration of large-capacity RE power in the grid. The state of the art of the technologies and practices related to large-capacity RE integration was briefly described in Chap. 1 and consequent chapters as well, and further information can be found in Ref. [1] provided by the International Electrotechnical Commission in the white paper.

B. Zohuri, *Hybrid Energy Systems*,
https://doi.org/10.1007/978-3-319-70721-1

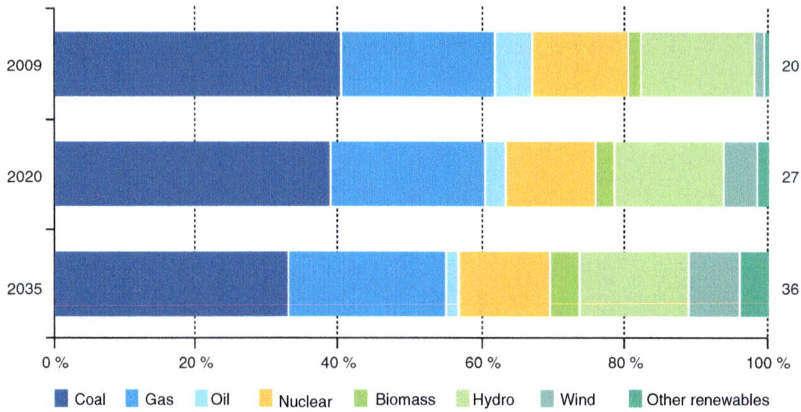

Fig. B.1 Share of world electricity generation by fuel in IEA's new policies scenario (Courtesy of International Energy Agency)

These technological descriptions and practices related to large-capacity RE integration facilitate the understanding of their interaction with the power grid, and it can be further divided into the renewable energy generation technology itself, namely, the *transmission* technology and the *operational* technology and *practices*.

As we have stated before under subject of energy security, renewable energy driven by the wind, the sun, and the waves has no fuel cost. This zero-fuel-cost aspect of RE manifests itself in two benefits. First, average energy costs tend to decline over time for renewable generation, as variable costs are limited to operations and maintenance and do not include fuel. Second, RE assets are insulated from fluctuations in fossil fuel prices, which are historically volatile and subject to geopolitical disruptions. Coal-, gas-, and oil-fired generation costs, in contrast, increase when the cost of the relevant fuel increases.

Figure B.1 depicts the International Energy Agency's (IEA) projection 1 for the share of world electricity generation by fuel up to 2035 and shows a displacement of coal- and oil-based generation's shares by wind, biomass, and other renewables as governments continue to promote RE.

Because fossil fuel supplies are both unevenly distributed and ultimately exhaustible, many countries have identified a long-term energy security proposition in gradually decreasing dependence on them in the production of electricity. In comparison to fossil resources, renewable resources are better distributed throughout the world and do not diminish as they are used. A country's investment in RE results in a zero-fuel-cost generation resource that is domestically located. Thus even countries with substantial fossil fuel resources, such as China, have set aggressive wind power targets. And despite a recent boom in natural gas production in the United States, states have made no indication of any intent to remove RE goals. RE can also prove useful for short-term energy security concerns. Many electric utilities have diversified their generation mixes with renewables so as to hedge against volatile fossil fuel prices on the oil, gas, and coal markets.

Renewable energy (RE) is a growing component of electricity grids around the world due to its contributions to:

1. Energy system decarbonization
2. Long-term energy security
3. Expansion of energy access to new energy consumers in the developing world

As stated in the market strategy board (MSB) of the International Electrotechnical Commission (ICE) Electrical Energy Efficiency (EEE) Report:

Note that the MSB encouraged the IEC to take the lead on the discussion on energy efficiency and smart grids. To allow proactive instead of reactive activities, the IEC was asked to put in place a technology watch and roadmapping solutions.

In short, the challenge is ensuring energy availability and preserving the environment. The key elements are the following:

1. Stabilizing climate impact from fossil fuel use
2. Meeting the energy demand of a growing population
3. Bringing electricity to the 1.6 B people without access
4. Ensuring stable and secure energy access for all nations
5. Transporting electricity long distances from where it is generated to where it is used [msb10]

Integration of RE is a poly-nodal problem involving multiple decision-makers at a variety of spatial and temporal scales and widely varying degrees of coordination. These decision-makers include operators of RE and energy storage resources, grid operators, energy market operators, and transmission planning bodies. As such, grid integration is not performed by any one entity in the power system, but instead involves the actions of a variety of entities, some highly coordinated and others discrete. The burgeoning development of smart grids adds still more tools, options, and players to the mix. Many of these actors engage with various technology standards, practices, procedures, and policies for the operation of individual generators, RE clusters, substations, and the broader electrical energy system.

RE is implicated in all of these elements and is critical to transforming energy grids to meet the environmental, economic, and social challenges of the future. Globally, RE's share of electricity generation will increase substantially over the next two decades and beyond. Indeed, this is already occurring: governmental action at the international, national, and subnational levels has created a wide variety of laws and policies to promote RE development. These include:

- Carbon taxes: taxation of greenhouse gas emissions, so as to internalize the climate disruption costs of fossil fuel use
- Cap-and-trade systems: provision of tradable annual emission allowances to greenhouse gas emitters coupled with reduction in the quantities of allowances issued each year
- Renewable energy goals: mandates requiring load-serving entities to source a specified proportion of energy sold from renewable sources
- Feed-in tariffs (FiTs): guaranteed wholesale prices for RE coupled with a requirement that load-serving entities take renewable power whenever it is available

- Tax credits: credits against taxable income for generation or installation of RE
- The development of smart grids: advances in the architecture, functionality, and regulation of electricity grids so as to enable higher penetrations of RE
- Removal of long-standing fossil fuel subsidies

Expanding Energy Access

Energy demand in developing countries is growing rapidly (see Fig. B.2). IEA's New Policies Scenario projects electricity demand in non-OECD countries to increase at a compound average annual growth rate (CAAGR) of 3.5% to 2035. Total non-OECD electricity demand nearly triples from 8000 TWh in 2009 to almost 20,000 TWh by 2035 (see Fig. B.3). Asian electricity demand grows the most rapidly, with a 4.2% CAAGR in the same period. In addition to the needs outlined in the previous subsections for cleaner energy and more secure energy, the world simply needs *more* energy as more people in the developing world gain access to it.

As global energy demand increases, RE provides one means among many of the adding energy assets to the system alongside growth of other resources. IEA's New Policies Scenario projects a near tripling of global use of RE, from 3900 TWh in 2009 to 11,100 TWh in 2035, and growth in renewables accounts for nearly half of the total increase in generation by 2035. Indeed, under this scenario, a full third of global electricity generation will be supplied by RE (including hydroelectricity) by 2035. Figure B.4 provides a breakdown of incremental renewables growth by technology. Note the large increase in wind power.

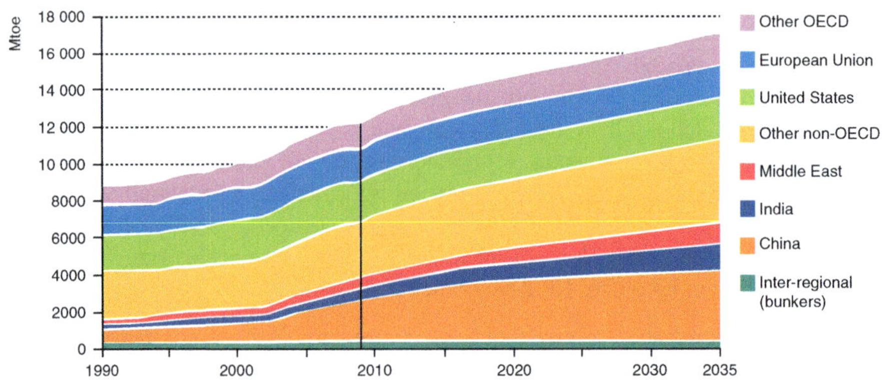

Fig. B.2 World primary energy demand by region in IEA's New Policies Scenario (Courtesy of International Energy Agency)

	1990	2009	New Policies Scenario 2035	New Policies Scenario 2009-2035**	Current Policies Scenario 2035	Current Policies Scenario 2009-2035**	450 Scenario 2035	450 Scenario 2009-2035**
OECD	**6 593**	**9 193**	**12 005**	**1.0%**	**12 554**	**1.2%**	**11 343**	**0.8%**
Americas	3 255	4 477	5 940	1.1%	6 119	1.2%	5 612	0.9%
United States	*2 713*	*3 725*	*4 787*	*1.0%*	*4 898*	*1.1%*	*4 505*	*0.7%*
Europe	2 321	3 088	4 028	1.0%	4 244	1.2%	3 802	0.8%
Asia Oceania	1 017	1 628	2 037	0.9%	2 191	1.1%	1 930	0.7%
Japan	*759*	*950*	*1 158*	*0.8%*	*1 225*	*1.0%*	*1 075*	*0.5%*
Non-OECD	**3 492**	**8 024**	**19 717**	**3.5%**	**21 798**	**3.9%**	**16 978**	**2.9%**
E. Europe/ Eurasia	1 585	1 280	1 934	1.6%	2 238	2.2%	1 742	1.2%
Russia	*909*	*791*	*1 198*	*1.6%*	*1 401*	*2.2%*	*1 057*	*1.1%*
Asia	1 049	4 796	13 876	4.2%	15 334	4.6%	11 666	3.5%
China	*559*	*3 263*	*9 070*	*4.0%*	*10 201*	*4.5%*	*7 447*	*3.2%*
India	*212*	*632*	*2 465*	*5.4%*	*2 590*	*5.6%*	*2 117*	*4.8%*
Middle East	190	600	1 393	3.3%	1 525	3.7%	1 264	2.9%
Africa	263	532	1 084	2.8%	1 152	3.0%	1 000	2.5%
Latin America	404	816	1 430	2.2%	1 550	2.5%	1 306	1.8%
Brazil	*211*	*408*	*750*	*2.4%*	*792*	*2.6%*	*675*	*2.0%*
World	**10 084**	**17 217**	**31 722**	**2.4%**	**34 352**	**2.7%**	**28 321**	**1.9%**
European Union	2 227	2 793	3 530	0.9%	3 716	1.1%	3 351	0.7%

*Electricity demand is calculated as the total gross electricity generated less own use in the production of electricity and transmission and distribution losses. **Compound average annual growth rate.

Fig. B.3 Electricity demand by region in IEA's WEO 2011 Scenario (Courtesy of International Energy Agency)

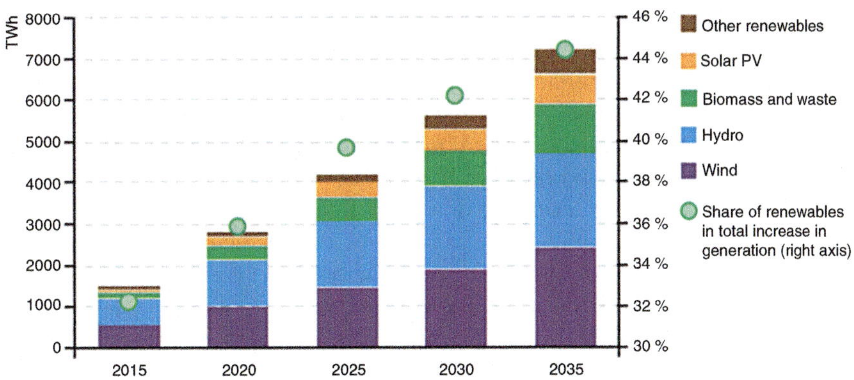

Fig. B.4 Incremental global renewables-based electricity generation relative to 2009 by technology in IEA's New Policies Scenario (Courtesy of International Energy Agency)

Decarbonization

As part of global demand growth for electricity and expansion energy access, there is a need for decarbonization. The need to address global climate change, a worldwide environmental phenomenon that will affect everyone on the planet, is the most public driving force for RE deployment. The Intergovernmental Panel on Climate Change (IPCC), the world's leading authority on climate change science, states in its Synthesis Report to the Fourth Assessment Report that "warming of the climate system is unequivocal, as is now evident from observations of increases in global average air and ocean temperatures, widespread melting of snow and ice and rising global average sea level" and that "most of the global average warming over the past 50 years is very likely due to anthropogenic Green House Gas (GHG) increases and it is likely that there is a discernible human induced warming averaged over each continent (except Antarctica)."

The MSB EEE Report notes that CO_2 emissions related to energy use account for 70% of total GHG emissions and that emissions related to electricity generation approach half of that (msb10). Consequently, governments have enacted policies to curb GHG emissions from the power sector. Because electricity generated from RE produces no GHG emissions, increasing penetrations of RE onto the electrical grid contribute to a *decarbonization* of the electricity system: a reduction in GHGs emitted per unit of energy produced. Energy system decarbonization in turn slows the increase in concentrations of GHGs in the atmosphere and thereby mitigates the resultant radiative forcing of the climate system.

In recent years, the ostensible progress of climate change policies has stalled at the international level, with a lack of hard commitments to emission reductions from some large emitters. Nevertheless, many countries have developed incremental policies to promote RE development in the absence of full international agreement. For example, nearly 30 states in the United States have enacted their own RE goals in the absence of federal action; Germany has long used aggressive feed in tariff (i.e., a government tax on import) requirements that oblige power companies to purchase renewably generated energy at fixed rates; and China has set a capacity goal of 150–180 GW of wind power and 20 GW of solar photovoltaic (PV) power for 2020. These goals and policies will result in significant growth in RE that will affect the operation of the power grid.

In summary, renewable energies, driven by climate change, fuel security, and other motives, will be providing more and more of our electricity in the future. They represent an opportunity and a risk. The opportunity is not the subject of the present paper; it is assumed simply that excellent reasons exist for the share of renewables in the energy mix to grow considerably and that they will therefore do so. The risk stems from characteristics of certain renewables which make them difficult to incorporate into our current electricity system. It is only the renewables (and their large-scale use) presenting that risk which are dealt with here, for together with many others, it is the IEC's responsibility to help the world community cope with the risk. The renewable energies in question are wind and solar—both photovoltaic and thermal—and the risk is that if they are present on a large scale, their variability and unpredictability will prevent the correct functioning of the whole electricity supply grid.

We have seen that the more renewables we feed into the grid, the more difficult the grid and its electrical properties will be to control and to operate efficiently. The risks include frequency and voltage fluctuations and outages, as well as major inefficiencies and waste. Much is already known and done to stay in control, but it is not enough for the 15%, 25%, or even 35% of variable renewables some grids will contain over the next decades. The "grid-friendly" renewable generation and "renewable-friendly" grids are both needed and suggests some methods for achieving them. These include improved forecasting of the likely energy available, flexibility, and reserves to guarantee supply and the grid's electrical characteristics, information and fast reactions to enable constant control, and enhanced transmission capability to adjust the grid without wasting energy. A constant in many of the methods is that the availability of large-scale EES will make them easier to apply, so the lessons from the IEC's preceding white paper on that subject have been very useful in the current one.

Two fundamental conclusions may be drawn. First, we understand, to a certain extent, what will be needed to cope with large-scale renewables in the grid—*but we do not yet have what we need.* Very considerable efforts will be needed to obtain it, whether it is knowledge, practical experience, tools, guidance, or investment. Second, neither theoretical knowledge nor practical experience is enough if it is applied by just those who know, or just those who have the experience, separately in their own domains. That is happening today, and it will obviously not be able to cope with the increase in renewables. Instead, it will be required to attack the problem together, across borders and areas of responsibility, basing the solutions on common research, tools, and infrastructure and in particular on common rules and international standards. The problem is too complex for any other approach to work.

For further information, read more details about this subject and refer to Ref. [1] by the International Electrotechnical Commission.

Reference

1. International Electrotechnical Commission white paper on Global Energy Interconnection

Appendix C: Energy Storage for Power Grids and Electric Transportation

Energy Storage for Power Grids and Electric Transportation: A Technology Assessment by Paul W. Parfomak and specialist in energy and infrastructure policy is partially republished here. Energy storage technology has great potential to improve electric power grids, to enable growth in renewable electricity generation, and to provide alternatives to oil-derived fuels in the nation's transportation sector. In the electric power system, the promise of this technology lies in its potential to increase grid efficiency and reliability—optimizing power flows and supporting variable power supplies from wind and solar generation. In transportation, vehicles powered by batteries or other electric technologies have the potential to displace vehicles burning gasoline and diesel fuel, reducing associated emissions and demand for oil [1].

Introduction

Federal policymakers have become increasingly interested in promoting energy storage technology as a key enabler of broad electric power and transportation sector objectives. The Storage Technology for Renewable and Green Energy Act of 2011 (S. 1845), introduced on November 10, 2011, and the Federal Energy Regulatory Commission's Order 755, *Frequency Regulation Compensation in the Organized Wholesale Power Markets*, are just two recent initiatives intended to promote energy storage deployment in the United States. Numerous private companies and national laboratories, many with federal support, are engaged in storage research and development efforts across a very wide range of technologies and applications.

The report attempts to summarize the current state of knowledge regarding energy storage technologies for both electric power grid and electric vehicle applications. It is intended to serve as a reference for policymakers interested in understanding the range of technologies and applications associated with energy storage,

B. Zohuri, *Hybrid Energy Systems*,
https://doi.org/10.1007/978-3-319-70721-1

comparing them, when possible, in a structured way to highlight key characteristics relevant to widespread use. While the emphasis is on technology (including key performance metrics such as cost and efficiency), this report also addresses the significant policy, market, and other nontechnical factors that may impede storage adoption. It considers eight major categories of storage technology: pumped hydro, compressed air, batteries, capacitors, superconducting magnetic energy storage, flywheels, thermal storage, and hydrogen.

Energy storage technologies for electric applications have achieved various levels of technical and economic maturity in the marketplace. For grid storage, challenges include round-trip efficiencies that range from under 30% to over 90%. Efficiency losses represent a trade-off between the increased cost of electricity cycled through storage and the increased value of greater dispatchability and other services to the grid. The capital cost of many grid storage technologies is also very high relative to conventional alternatives, such as gas-fired power plants, which can be constructed quickly and are perceived as a low-risk investment by both regulated utilities and independent power producers. The existing market structures in the electric sector also may undervalue the many services that electricity storage can provide. For transportation storage, the current primary challenges are the limited availability and high costs of both battery electric and hydrogen-fueled vehicles. Additional challenges are new infrastructure requirements, particularly for hydrogen, which requires new distribution and fueling infrastructure, while battery electric vehicles are limited by range and charging times, especially when compared to conventional gasoline vehicles.

Substantial research and development activities are underway in the United States and elsewhere to improve the economic and technical performance of electricity storage options. Changes to market structures and policies may also be critical components of achieving competitiveness for electricity storage devices. Removing nontechnical barriers may be as important as technology improvements in increasing adoption of energy storage to improve grid and vehicle performance.

Energy Stage Technology

Energy storage technology has a great potential to improve electric power grids, to enable growth in renewable electricity generation, and to provide alternatives to oil-derived fuels in the nation's transportation sector. In the electric power system, the promise of this technology lies in its potential to increase grid efficiency and reliability—optimizing power flows and supporting variable power supplies from wind and solar generation. In transportation, vehicles powered by batteries or other electric technologies have the potential to displace vehicles burning gasoline and diesel fuel, reducing associated emissions and demand for oil.

In recent years, federal policymakers have become increasingly interested in promoting energy storage technology as a key enabler of broad electric power and transportation sector objectives. In remarks about the STORAGE Act of 2011 (S. 1845) [2], which would provide investment tax credits for storage systems connected to the electric grid, businesses, and homes, Senate Energy and Natural Resources Committee Chairman Jeff Bingaman remarked:

> Deployment of storage technologies will make our nation's electricity grid more reliable while also enabling more efficient use of existing energy sources as well as new ones, such as wind and solar.... These technologies have the potential to cut electricity bills, reduce peak power demand and lower greenhouse gas emissions. [3]

Likewise, in a statement regarding new energy storage-related rules for wholesale electricity markets, Federal Energy Regulatory Commissioner John Norris stated:

> I believe today's final rule is a positive first step by the Commission in recognizing the unique characteristics and the value that storage resources offer.... As we move forward, I strongly believe that storage will become ever more critical as we look to integrate increasing amounts of variable energy resources. [4]

Referring to advanced batteries for electric transportation applications, Secretary of Energy Steven Chu reportedly stated:

> It's now within grasp, that you can get a battery where the business plans are one-third of the cost of today's batteries, where you can get ranges now that would allow cars instead of 100 miles on a single charge, go 300 or more miles on the same charge.... It's not a pipe dream 30 years from today or 20 years from today. It's in the next decade. [5]

Statements such as those above highlight not only the technical opportunities for energy storage in the grid and in electric transportation but also the attention being paid to energy storage technologies at the highest levels in the federal government. Nonetheless, many new energy storage technologies continue to face significant technological and economic challenges to their commercialization and widespread deployment. The recent bankruptcy of Beacon Power, one of the leading developers of flywheel energy storage technologies for the grid, is a prominent illustration of commercial barriers to grid storage technology. Public concerns about elevated fire risks from Chevrolet Volt electric car batteries, although shown to be exaggerated, are another [6].

By contrast, increasing investments by AES Corporation in utility-scale battery storage for power grids show continuing successful efforts to overcome technical challenges and market barriers to bring new storage technologies into the market [7].

Understanding the potential of energy storage in electric applications is complicated by a number of factors. The first is the wide range of storage technologies either commercially available, in development, or being researched. Because they are technologically diverse, it is difficult to gain a balanced understanding of the fundamental capabilities, costs, and comparative advantages of these different energy storage options. Second, there are multiple applications of energy storage, each with distinct operational requirements. Certain storage technologies may suit certain applications better than others. Finally, there are many aspects of market structure and economic regulation that affect energy storage deployment. Taken together, these factors make the development of an energy storage research and development portfolio challenging. While there is general consensus that storage technology improvements are needed, there are multiple potential pathways to such improvements that cut across different disciplines.

Energy Storage for Electric Grid Applications

It is possible to divide grid storage applications into two broad categories based on the length of time a storage device needs to provide service: (1) high-power applications where the device must respond rapidly and be able to discharge for only short-term periods (up to about 1 h) and (2) energy management-related applications where the device may respond more slowly but must be able to discharge for several hours or more. Ideally, all storage devices would be able to provide all services, but some technologies are technically restricted to provide only short-term services. However, many of these services have very high value in the grid, so short-term storage can still provide considerable benefits.

High-Power/Rapid Discharge Applications

The rapid response category can be further divided into short-term discharge—less than 1 min—used to provide grid stability and power quality and longer-term discharge, up to about an hour. Though important, short-term discharge services can often be provided by non-storage options such as power electronics. Furthermore, this class of grid services does not address the primary challenge of renewables integration, which requires minutes to hours of discharge time. Currently, capacitors and superconducting magnetic energy storage (SMES) are rapid response technologies capable only of providing short-term discharge. Research efforts for both technologies are focused on increasing energy density and decreasing cost, with capacitor efforts being directed in part toward vehicle applications. While SMES research has been active historically, current efforts are modest, and there is no clearly defined pathway for SMES to be competitive for applications requiring extended discharge.

Other grid applications require devices with up to about 1 h of discharge to provide services such as frequency regulation service (responding to random, rapid variations in demand) and contingency reserves (rapidly responding to a generator or transmission failure). Longer-term storage can also support renewables integration by providing the sub-hourly ramping requirements which will increase as greater amounts of variable generation sources are added to the grid. Flywheels have been deployed in significant demonstration projects providing frequency regulation. Several battery types have been demonstrated for both frequency regulation and operating reserves, including lithium-ion and various aqueous batteries (such as lead-acid, nickel-cadmium, and nickel-metal hydride). Most aqueous chemistries are considered mature technologies, but additional improvements are possible, even for 100 + −year-old lead-acid batteries. Research and development efforts on lithium-ion batteries are focused on reducing cost and weight for transportation applications, but these efforts should have spillover benefits to grid applications. In addition, there are certain lithium-ion configurations that are probably unsuitable for transportation applications but potentially suitable for the grid. A major effort by commercial vendors of rapid response technologies such as flywheels and lithium-ion batteries has been gaining access to markets for frequency regulation and full valuation of the response capabilities of the technology.

Energy Management Applications

Grid storage devices for energy management applications can provide continuous discharge for several hours or more. These devices would be potentially useful for shifting energy during periods of low demand (or high renewable supply) to periods of high demand (or low renewable supply). Many of them can also provide the same services as high-power/rapid discharge devices.

Pumped hydro storage (PHS) is the dominant technology in this category with about 22 gigawatts (GW), equivalent to about 22 large power plants, operating in the United States for decades. PHS has high reliability, high efficiency, and long lifetime but is dependent on the availability of suitable geologic conditions and requires long development times (~10 years including permitting). Based on sitting challenges and environmental opposition, PHS suffers from the perception that these issues will prevent large-scale deployment in the future. However, the actual technical potential is large, and the number of proposed plants exceeds the current installed capacity, with many of these proposed plants using *closed-cycle* designs that will not interact with existing water bodies and have the potential to reduce both opposition and licensing times. They may also use variable speed equipment improving their ability to provide rapid discharge services.

Compressed air energy storage (CAES) is technically mature and often considered the lowest-cost option for "bulk" electricity storage, although only one such facility is deployed in the United States. CAES is a hybrid technology which uses natural gas and typically requires a large underground formation. Major development efforts for CAES currently underway include demonstrating the technology in bedded salt and porous rock. The use of such geologic formations would open up much more of the country to CAES development. Other research and development activities include work on CAES cycles that do not require natural gas fuel.

Hydrogen- and other electricity-derived fuels are possible storage options with the advantage of long-term (even seasonal) storage. They currently are among the least efficient (well under 50%) and more expensive storage technologies available and have yet to be deployed beyond small demonstration projects. Fundamental research efforts are required to decrease the cost and increase the durability of electrolyzers and fuel cells. Most of the historic research on hydrogen has been as an alternative fuel for transportation.

Two classes of batteries are currently the primary candidates for electric grid applications—liquid electrolyte flow batteries and high-temperature batteries. High-temperature sodium-sulfur batteries are the most mature and commercially available, with over 270 MW deployed worldwide, including installations in the United States. They also have the advantage of relying on low-cost and abundant materials, although manufacturing costs have limited larger-scale use. Sodium-sulfur is the only high-temperature battery deployed at large scale, currently manufactured by a single company in Japan. There are several alternative high-temperature chemistries under various stages of research, development, and commercialization. Flow batteries are in the early stages of development and commercialization, with a few US demonstration projects of vanadium and zinc-bromine technologies, with several other technologies under development.

Thermal energy storage (TES) is often overlooked as an electricity storage technology option because it does not store and discharge electricity directly. However, in some applications, thermal storage can be functionally equivalent to electricity storage with efficiencies exceeding 90%, which is higher than most other storage technologies. There are two primary applications of TES for electricity. The first is storing thermal energy from the sun which is later converted into electricity. The currently deployed storage medium is a relatively low-cost molten salt.

The primary limitation is that TES is tied to a specific application, in this case concentrating solar power (CSP), which has the challenges of high cost and limited deployment locations, mostly in the desert southwest in the United States. The key research efforts include developing storage materials with higher working temperature, which, when combined with higher temperature CSP plants, will increase efficiency and decrease costs. CSP with thermal energy storage has been deployed in Spain. Construction of a 250 MW CSP/TES facility in the United States is expected to begin in 2012. The second application of TES is cold and hot storage in buildings. Cold storage, used to reduce peak demand from air conditioning, has been deployed on a relatively large scale. This is a commercially mature technology that provides firm system capacity at very high round-trip efficiency, with the capability of providing multiple grid services. The primary barrier to deployment is capturing the benefits of this distributed technology in the current regulatory and market environment.

Energy Storage for Transportation Applications

As with grid storage, energy storage for transportation applications can be loosely divided into two primary categories: high-power/rapid discharge and high energy/extended discharge. High-power devices provide short, rapid discharges for vehicle starting and acceleration. While they cannot provide continuous discharge for electrified transport, they can dramatically improve fuel efficiency, as demonstrated by the current generation of hybrid electric vehicles. Currently deployed technologies for these applications include lithium-ion- and nickel-based aqueous batteries. Technologies being explored are including capacitors, flywheels, and other battery types. Some of these technologies, such as capacitors, may also be used as a fast-responding "buffer" between the electric drive system and the battery or fuel cell in an electric vehicle (EV).

For high-energy applications, where stored electricity is actually used to provide a significant fraction of the driving energy, research and development efforts are currently focused on two technologies—hydrogen and batteries. Conceptually, hydrogen is a simple storage technology, produced by splitting water using electricity (among other options), storing hydrogen on board the vehicle, and then converting it to electricity to drive an electric motor via a fuel cell. (Internal combustion engines could also be used, but the low efficiency of that process is less attractive.)

The challenges of a hydrogen-based transportation system include the development of an entirely new fueling infrastructure including hydrogen delivery systems and filling stations, with needed safety standards and protocols. The low volumetric energy density of hydrogen makes storage challenging without extremely high-pressure tanks or advanced chemical storage still in the early research phase. Finally, fuel cells for vehicles remain expensive, with limited lifetimes. There have been demonstration fuel cell vehicle programs by several major auto manufacturers, with announced plans for commercial deployment as soon as 2015. However, substantial research efforts will be needed to reduce costs and improve performance for many of the technologies needed for large-scale hydrogen-based transportation. There are other electricity-to-fuel pathways under consideration but with limited research and development efforts in the United States. They face similar challenges of requiring new fuel infrastructure and currently face much higher costs than fossil fuel alternatives.

The primary alternative to electricity-based fuel production is battery electric storage in plug-in hybrid electric vehicles (PHEVs) and EVs. Most commercially available and proposed EVs and PHEVs (such as the Chevrolet Volt and Nissan Leaf) use lithium-ion batteries. Research and development efforts are focused primarily on reducing cost and increasing energy density as well as safety of lithium-ion technology. Earlier deployed technologies, such as lead-acid used in older EVs and nickel-metal hydride used in current hybrid electric vehicles (HEVs), are not considered likely candidates in future EVs due to fundamental limits of energy density. Concerns have been expressed about the large-scale availability of several metals used in lithium-ion batteries, as well as its concentration in a few geographic regions. In the longer term, lithium-metal and metal-air batteries are in the research and development phase, with the potential of much higher energy density than currently available battery types.

If there are further interests by the readers on all these subjects, we suggest to read Ref. [1] and others in the reference section of this appendix.

References

1. P.K. Parfomak, Energy storage for power grids and electric transportation: A technology assessment, Congressional Research Service, 27 Mar 2012
2. Storage Technology for Renewable and Green Energy Act of 2011 (S. 1845) introduced on November 10, 2011, by Senator Ron Wyden and co-sponsored by Senators Jeff Bingaman, Susan Collins, and Robert Menendez
3. Office of Senator Ron Wyden, Wyden, Collins, Bingaman Legislation Will Increase Investments in the Storage of Renewable Energy, Press release, 10 Nov 2011
4. Commissioner John R. Norris, Frequency Regulation Compensation in the Organized Wholesale Power Markets, Docket Nos. RM11-7-000 & AD10-11-000, Item No. E-28, Federal Energy Regulatory Commission, 20 Oct 2011
5. Michael Warren, Energy secretary Steven Chu on electric cars. *The Weekly Standard Blog*, 3 Apr 2011. http://www.weeklystandard.com/blogs/chu-electric-cars_556135.html
6. Jim Henry, Chevy Volt Battery fires threaten all electric vehicle makers, Not Just GM. *Forbes*, 12 Dec 2011; National Highway Traffic Safety Administration, NHTSA statement on conclusion of Chevy Volt investigation, Press release, 20 Jan 2012
7. AES peaker-sized battery proposals show company's vision of storage potential for the grid. *Electric Utility Week*, Platts, 2 Jan 2012.

Appendix D: Coping with the Energy Challenge

Over the next decades, the world will face increasing challenges to supply energy in sufficient quantities while reducing carbon emission levels. Saving energy and using energy more efficiently are keys to addressing these challenges. Ours is a connected world, and energy efficiency solutions will need to work together, safely, everywhere to make a real impact, not just in the developed world, but in developing countries as well [1].

Introduction

Without metrics all efforts to reduce and optimize energy consumption are doomed to remain small and insignificant. As the first IEC President, Lord Kelvin, always said: "If you cannot measure it, you cannot improve it!". This statement is especially true here: without measurement you cannot credibly demonstrate energy efficiency improvements. The IEC provides and will continue to provide many of the measuring standards that are the basis for benchmarking, energy audits, and compliance assessments.

But the IEC also holds an important piece of the solution for overall energy efficiency—smart electrification.

Electricity is the most easily controllable form of energy. The International Electrotechnical Commission (IEC) believes that electricity will be the most important contributor to climate change mitigation. It is easily controlled and weightless. It is easier to transport and distribute and cleaner at the point of use than most other energy sources, and it can be produced cleanly at the point of generation. It represents the most efficient way of generating and consuming power and the most intelligent approach for future global efforts to economize energy.

In the white paper in Ref. [1], the IEC is laying the foundation for the electrical energy efficiency discussion. To define where IEC's work needs to be focused, the IEC has studied the wide array of energy efficiency opportunities and technologies

B. Zohuri, *Hybrid Energy Systems*,
https://doi.org/10.1007/978-3-319-70721-1

that are available. Based on this, the IEC has developed a model projecting what it believes is likely to happen in the next 20 years.

> The global economy is set to grow fourfold between now and 2050, and national growth could approach tenfold in countries such as China and India.

This promises economic benefits and huge improvements in people's standards of living but also involves much more use of energy. Unsustainable pressure on natural resources and on the environment is inevitable if economic growth is not decoupled from energy demand and energy demand from fossil fuel consumption.

World population is expected to grow from an estimated 6.5 B in 2006 to 8.2 B in 2030, at an annual average rate of 1%. This rate will probably slow progressively over the projection period in line with the past trends: population expanded by 1.4% per year from 1990 to 2006. The population of non-OECD countries as a group continues to grow most rapidly.

Growing populations and industrializing countries create huge needs for electrical energy. In the reference scenario of the International Energy Agency (IEA), which assumes that there are no new governmental policies other than those of the mid-2008 (the so-called business-as-usual (BAU) scenario), projected world primary energy demand increases by 45% between 2006 and 2030—an average annual rate of growth of 1.6%—and doubles (i.e., a 100% increase) by 2050. Electricity demand will triple by 2050.

Today 1.6 billion people have no access to electrical energy; however, they will require electricity in the coming decades. Furthermore, most of the new inhabitants of the planet will live in today's developing countries. Therefore any measure envisaged affecting energy efficiency or consumption should take into account the fact that the new energy demand will be situated in those countries where energy distribution infrastructures are not yet at the right level to satisfy increasing demand.

In 2006, cities accounted for 67% of the world's energy consumption and 71% of global energy-related CO_2 emissions, at a higher rate per capita than the countryside. The combined power generation and heat sector absorb a growing share of global primary energy demand over the projection period. Its share reaches over 42% in 2030 compared with 38% in 2006.

Fossil fuels remain the leading sources of energy—roughly 80% in 2030. Coal remains the leading input for power generation and heat, its share of total inputs holding steady at about 47% over the outlook period. Oil remains the dominant fuel in the primary energy mix, but its share drops to 30% in 2030, from 34% in 2006, while the share of gas rises from 21% to 23%. Nuclear power's contribution falls from 16% in 2006 to 13% in 2030. Hydropower's share remains steady at 6%. Inputs from non-hydro renewables—photovoltaic (PV), wind, biomass, and waste—will grow worldwide at an average rate of 6.2% per year between 2006 and 2030, the fastest rate of all energy sources, with their share rising to 10% but still remaining limited as a source of energy in 2030.

According to the IEA World Energy Outlook 2008, *industry, transport, and buildings/services 2 are almost equal primary energy consumers* (*1/3 each*). If we examine not primary but electrical energy, it is essential to note that almost half is consumed by industry, with all other uses making up the remainder. In industry and

Table D.1 CO_2 emissions in the BAU scenario

	CO_2 emissions related to energy use (Gt)	Including: CO_2 emissions from electricity generation (Gt)
Today	28	10.8
2030	42	17.8
2050	62	29

buildings/services, electricity is dominant, with the fastest rate of growth. In transport, on the other hand, electricity is almost absent, but its developing use could be one important part of the solution. Growth is most rapid in industry and slowest in buildings and services.

Today CO_2 emissions related to energy use are at a level of 28 Gt (gigatons of CO_2 per annum), which represents 70% of total greenhouse gas (GHG) emissions. Electricity generation represents something approaching a half of this, at about 11 Gt.

If no specific action is taken (in the so-called reference or business-as-usual (BAU) scenario), the IEA projects in its *Energy Technology Perspectives 2008* that 42 Gt will be emitted in 2030 and 62 Gt in 2050; such a scenario could lead to a rise in global temperatures of up to 6 °C (see Table D.1).

We are faced with a double challenge: a purely energy challenge and in addition a climate challenge. A new strategy is needed, which cannot be local but must be global. It must decouple energy consumption from economic development and growth.

In short, the challenge is ensuring energy availability and preserving the environment. The key elements are the following:

1. *Stabilizing climate impact from fossil fuel use*
2. *Meeting the energy demand of a growing global population*
3. *Bringing electricity to the 1.6 billion people without access*
4. *Ensuring stable and secure energy access for all nations*
5. *Transporting electricity long distances from where it is generated to where it is used*

As part of this challenge, we need to look into framework for solution of controlling the CO_2 emission. Climate change migration is politically supported by most developed nations. Political commitments for CO_2 emission reduction will frame action for the next 30 years:

- Kyoto originally mandated a reduction of 8% of emissions with respect to the 1990 level over the period to 2012.
- The EU Spring Council in March 2007 fixed a reduction of at least 20% of the 1990 level as the basis, by 2020.
- Less than 50% of the 1990 level by 2050 is the intention, according to some countries
- Copenhagen follow-up, Bonn, Mexico, etc.

A key factor in the response must be electricity as part of parameters for the response to the challenge as well. In short it is considered to be:

- Thirty-one percent of global fossil fuel used each year goes to producing electricity.

- One-third of final energy use in industry comes from electricity, with a growth rate of 2.7%.
- Energy used in buildings/services also comes one-third from electricity, with a growth rate of 2.3%.
- Introducing electricity into transport will enable economies by allowing control.
- Electrification of various other uses of energy will also increase efficiency.

A useful statement of the problem might be the following: the more people there are, the more energy is used; the more energy is used, the more carbon dioxide is emitted; the more carbon dioxide is emitted, the more harm is done to the climate. A little more formally: at any point in time, total emissions of CO_2 are equal to the population, multiplied by the quantity of energy used per person, multiplied by the quantity of CO_2 emitted per unit of energy used:

CO_2	$= P \times [E/P] \times [CO_2/E]$
CO_2	= Quantity of CO_2 emitted
P	= Population
[E/P]	= Energy used per head of population
$[CO_2/E]$	= CO_2 emitted per unit of energy used

We will assume that P, population, is a given (see Sect. 1.2 of Ref. [1]). We must therefore act on the [E/P] and $[CO_2/E]$ quantities in order to reduce CO_2 emissions.

Note that acting on the [E/P] quantity is energy efficiency. It may be influenced in the short, medium, or long term. Short-term action may already give significant results. The two strategic elements are efficiency in electricity use and using electricity to replace a quantity of fossil fuel use.

Acting on the $[CO_2/E]$ quantity is decarbonization of energy, choosing energies, which emit less or no carbon (renewables, biofuels, carbon capture and storage (CCS), and nuclear energy). Results are medium- and long-term.

Some of the tactics which may be used are *investment*, investing to achieve reduction of energy use per person and CO_2 emission per unit of energy used; *technologies*, identifying those technologies and strategies which are most cost-effective in achieving CO_2 reduction (note that these technologies and strategies will be different in different countries); and *individual action*—investment by individuals as well as by governments (e.g., buying energy-efficient appliances or paying a premium on electricity prices to be used for investment) and changing of behavior to choose actions which use less energy.

The following actions are available to reduce CO_2 emissions related to electricity generation and use. In most cases, they concern mature technologies.

- Reduce energy used at end-use level by increased energy efficiency:
  - Today available and proven technologies can bring savings of up to 30%.
  - The issue is a massive implementation and not only with newly built but also with existing installations.
  - End-use behavior may be changed to reduce activities requiring much energy.

- Reduce transmission and distribution losses (9% today):
  - The benefit will be in line with the existing proportion.
- Improve generation efficiency (only one-third of primary energy used is available as electrical energy):
  - Existing power generation units will require time and resources to be converted.
  - Coal is still available and cheap in many countries
- Increase renewable and specifically decentralized generation, almost CO_2-free:
  - There are economic limitations (need to subsidize cost) and physical constraints (availability of land, wind, etc.).
- Change the fossil fuel mix toward less CO_2-emitting fuels (less coal, cogeneration, nuclear, combined gas cycle turbines, etc.):
  - As for generation efficiency, the existence of power plants which cannot be converted will delay real results.
- Make transportation, today 99% fossil fuel (oil)-dependent, more energy-efficient with electricity

Electricity is a key factor in energy efficiency, on condition that its use is evaluated and controlled. Measurement and evaluation depend crucially on a few basic concepts. Calculations should be in terms of electrical energy as far as possible, and verified, so as to realize the benefits of control. For the whole electrical energy cycle from generation to consumption, i.e., for each of generation, transmission, and distribution and in each application sector, electrical energy efficiency (EEE) indicators should be defined, and efficiency should be measured at each stage within each sector. For every value measured, the improvement which may be achieved by applying *best available technology (BAT)* should be recorded. The reduction in CO_2 emissions should be based on full explanatory information on generation resources and any additional resources used; performance information, such as efficiency of generation, storage, and transmission; and CO_2 emissions calculated by life cycle analysis (LCA) of the infrastructure processes.

In summary, two aspects are vital:

1. A systematic approach must be used, which takes the whole cycle into consideration.
2. Measurement and evaluation are necessary at each stage.

For more details and information, read Ref. [1].

Reference

1. International Electrotechnical Commission white paper on Coping with the Energy Challenges, The IEC's Role from 2010 to 2030

Index

A
Adiabatic compressed air energy storage for electricity (ADELE), 10, 15
Advanced adiabatic compressed air energy storage (AA-CAES), 16
Alabama's electric cooperative (AEC), 13
Alkaline membrane (AMFC), 140
Alternating current (AC), 142, 143, 220, 233
Ames Research Center's (ARC), 85
Arrhenius relationship, 81
Automatic generation control (AGC), 218

B
Boiling points (bp), 48
Brooklyn microgrid (BMG), 215
Business as usual (BAU), 207

C
California Energy Commission (CEC), 24
California Public Utilities Commission (CPUC), 23
Cardiovascular disease (CVD), 73
Carnot cycle, 65
Coiled tube air heat exchanger (CTAH), 21, 188
Combined cycle, 26
Combustion turbine (CT), 18
Compressed-air energy storage (CAES), 10, 11, 16, 22
Compressed natural gas (CNG), 118
Cosmic background explorer (COBE), 78
Cosmic microwave background (CMB), 79
Critical point, 40
Cryogenic energy storage (CES), 43, 93, 98–100
Cryogenic energy system (CES), 89, 90, 93, 95
Cryogenic fluid management (CFM), 85
Cryogens, 46

D
Department of Defense (DOD), 85
Department of Energy (DOE), 85, 161
Deutsches Elektronen-Synchrotron (DESY), 235
Direct current (DC), 142, 233
Direct-methanol fuel cells (DMFC), 140, 148

E
Electrical energy storage (EES), 90, 165–168
Electric Power Research Institute (EPRI), 17
Electric vehicles (EVs), 91, 244
Emergency room (ER), 80
Energy Efficiency and Renewable Energy (EERE), 141
Energy storage system (ESS), 1
Engineering and Physical Sciences Research Council (EPSRC), 93
European Space Agency (ESA), 79
European Transmission System Operation (ETSO), 218
European Union (EU), 34, 107

F
Feed-in tariff (FIT), 208

B. Zohuri, *Hybrid Energy Systems*,
https://doi.org/10.1007/978-3-319-70721-1

Firebrick resistance-heated energy storage (FIRES), 20, 187
Fluoride-salt-cooled high-temperature reactors (FHRs), 20, 184, 187
Forschungszentrum Karlsruhe (FZK), 235
Forward looking infrared (FLIR), 76
Fuel cell electric vehicles (FCEVs), 156
Fuel Cell Technologies Office (FCTO), 140, 152, 161
Fuel Cell Technologies Program (FCT Program), 149

G
General Electric (GE), 15
GEN-III reactors, 26
Geothermal Energy Association (GEA), 116
German Aerospace Center (DLR), 16
GNU Free Documentation License (GFDL), 117
Grashof number, 60
Greenhouse gas (GHG), 130

H
Heat recovery steam generator (HRSG), 21, 188
High-temperature gas-cooled reactor (HTGR), 20, 184
High-voltage direct current (HVDC), 221
Hubble space telescope (HST), 79
Hybrid electrical energy storage (HEES), 91
Hybrid energy storage system (HESS), 1
Hybrid energy system (HES), 6
Hybrid renewable energy systems (HRES), 1
Hydrogen energy storage (HES), 158
Hydrogen Materials—Advanced Research Consortium (HyMARC), 155
Hydrogen Storage Engineering Center of Excellence (HSECoE), 162

I
Idaho National Laboratory (INL), 182, 196
Inertial confinement fusion (ICF), 48
Information Administration (EIA), 207
Information and communications technology (ICT), 3, 173
Infrared Astronomical Satellite (IRAS), 78
Infrared Space Observatory (ISO), 79
In Situ Resource Utilization (ISRU), 85, 86
International Electrotechnical Commission (IEC), 176
International Renewable Energy Agency (IRENA), 209, 210
Internet of things (IoT), 175
Internet protocol (IP), 175
Inversion temperature, 68, 70

J
Japan Atomic Energy Research Institute (JAERI), 233–234

K
Kinetic energy recovery system (KERS), 232
Kinetic energy storage (KES), 230

L
Lambda line, 50
Large Hadron Collider (LHC), 87
Lawrence Livermore National Labs (LLNL), 234
Light-water reactor (LWR), 20, 182, 196
Liquid natural gas (LNG), 41, 48, 73, 74
Liquid nitrogen (LN), 46, 74
Liquid oxygen (LOX), 74, 84
Lithium-ion (Li-ion), 243
Low Earth orbit (LEO), 86
Low pressure (LP), 72

M
Magnetic confinement fusion (MCF), 48
Magnetic resonance imaging (MRI), 41, 74
Massachusetts Institute of Technology (MIT), 182, 196
Microbial fuel cells (MFCs), 253
Molten salt energy storage (MSES), 10
Molten salt reactor (MSR), 20, 187
Multi-layer insulation (MLI), 61, 62

N
National Renewable Energy Laboratory (NREL), 121, 150, 202
Near Infrared Camera and Multi-Object Spectrometer (NICMOS), 79
Normal boiling point (NBP), 40
North American Electric Reliability Corporation (NERC), 218
Nuclear Air-Brayton Combined Cycle (NACC), 183, 187
Nuclear magnetic resonance (NMR), 74

Nuclear power plants (NPP), 26
Nuclear Power Plants (NPPs), 98–100
Nüsselt number, 59

O
Off-pick, 1
On-pick, 1

P
Parts per million (PPM), 243
Personal digital assistants (PDAs), 240
Personal protective equipments (PPEs), 54
Photovoltaics (PVs), 6, 89, 110, 112, 125, 168, 240
Platinum group metal (PGM), 141
Plug-in hybrid electric vehicle (PHEV), 240, 243
Polymer electrolyte membrane (PEMFC), 140, 158
Power conditioning system (PCS), 233
Prandtl number, 60
Pressurized water reactor (PWR), 100
Program on Energy and Sustainable Development (PESD), 31
Public relation (PR), 243
Pumped hydro storage (PHS), 169
Pumped storage hydroelectricity (PSH), 227, 240

Q
Quality of life (QoL), 172

R
Renewable electricity standards (RES), 129
Renewable energy certificates (RECs), 129
Renewable energy power system (REPS), 1
Renewable Energy Standard Offer Program (RESOP), 126
Research and development (R&D), 140
Resistance temperature detectors (RTDs), 76
Return-on-investment (ROI), 6
Reynolds number, 59
Rheinisch-Westfälisches Elektrizitätswerk (RWE), 15

S
Small modular reactor (SMR), 26, 183, 197
Sodium-cooled Fast Reactors (SFRs), 184, 187
Solar PV (SPV), 8
Solar thermal energy (STE), 110
Standards Developing Organizations (SDOs), 174, 176
Starting-lighting-ignition (SLI), 239
State of the Art (SOA), 86
Superconducting magnetic energy storage (SMES), 232, 233
Superfluid helium on orbit transfer (SHOOT), 85
System Average Interruption Duration Index (SAIDI), 171
System Average Interruption Frequency Index (SAIFI), 171

T
Tokyo Electric Power Co. Ltd. (TEPCO), 249
Total cost of Ownership (TCO), 6
Triple point, 40

U
Uninterrupted power sources (UPS), 237
Uninterruptible power supply (UPS), 230
United Nation (UN), 107
United States Air Force (USAF), 235
United States of America (USA), 34

V
Very-high-temperature reactors (VHTR), 19
VYCON Direct Connect (VDC), 230

Z
Zero boil-off (ZBO), 86

Printed by Printforce, the Netherlands